Routing in the Third Dimension

Routing in the Third Dimension

From VLSI Chips to MCMs

Naveed A. Sherwani
Siddharth Bhingarde
Anand Panyam

Microprocessor Division
Intel Corporation

IEEE
PRESS

The Institute of Electrical and Electronics Engineers, Inc., New York

This book may be purchased at a discount from the publisher
when ordered in bulk quantities. For more information contact:

IEEE PRESS Marketing
Attn: Special Sales
P.O. Box 1331
445 Hoes Lane
Piscataway, NJ 08855-1331
Fax: (908) 981-8062

Printed in the United States of America

10   9   8   7   6   5   4   3   2   1

ISBN 0-7803-1089-6

IEEE Order Number: PC 4473

Library of Congress Cataloging-in-Publication Data
Routing in the third dimension : from VLSI chips to MCMs / Naveeed A.
   Sherwani, Siddharth Bhingarde, Anand Panyam.
     p. cm.
   Includes bibliographical references and index.
    1. Integrated circuits—Very large scale integration—Design and
construction—Data processing.   2. Multichip modules
(Microelectronics)—Design and construction—Data processing.
3. Computer-aided design.   I. Bhingarde, Siddharth.   II. Panyam,
Anand.
TK7874.75.R68   1995   94-44113
621.39′5—dc20   CIP

Naveed dedicates this book to his sister
Farkhunda Sherwani
While God (and life) has not been fair to her,
all of us are better humans because of her.

Siddharth dedicates this book to his parents
Shobhana and Late Bhargav Bhingarde

Anand dedicates this book to his parents
Varalakshmi and Krishnanjaneya Panyam
and his uncle
Prasad Akshintala

Contents

3. Channel Routing and Terminal Assignment   65

4. Routing Models   103

5. Basic Problems in Routing   139

6. Routing Algorithms for the Two-Layer Process   162

Preface

From a chip integrating a dozen transistors four decades ago to multichip modules (MCM) with several hundred chips, each integrating over a million transistors, very large scale integration (VLSI) technology has changed the world in general and the world of information processing in particular. Today's chips are capable of executing an amazing 400 million instructions per second. It is estimated that within a few years we will have chips capable of executing several billion instructions per second. The demand for higher speed and functionality is likely to increase with the development of interactive high-definition TV, personal global communications system, worldwide teleconferences, real-life graphics animation, virtual reality applications, accurate weather prediction systems, and a host of scientific and industrial applications. The demand for increasing the performance and functionality of the systems will lead to an increase in the number of chips or MCMs and transistors (chips) that need to be integrated on a chip (MCM), leading obviously to an increase in the chip die (MCM substrate) size. On the other hand, due to improvement in the fabrication process, the feature size will decrease, leading to a decrease in the die size of the chips and MCMs. However, the overall trend will be toward increasing die size with ever increasing density. This increasing density will require an increasing number of interconnections within a chip as well as at an MCM level.

Currently, on a chip with a couple million transistors, a router may have to make several million connections. This is not only a very time-consuming task, but the interconnections also occupy a large amount of real estate on the chip. As the number of transistors reach 100 million, the number of connections to be completed and real estate occupied by the interconnections will become the main focus of the design. The gains made by the fabrication technology by reducing the feature size may be lost to the increased area required for interconnection. At the MCM level, the crisis is even more acute. With over 100 chips, each supporting 400 I/Os, the three-dimensional routing problem is not only computationally complex, but it is also not very well studied. In the future, the performance of systems, as well as the number of components that can be integrated on a chip or an MCM, depends on the interconnections and packaging technologies. Interconnections and packaging will gain further importance as the feature sizes of the transistors decrease and die sizes keep increasing.

One of the main design goals in the layout of chips is to reduce the die size of the chip, since delay, yield, and cost are directly affected by the die size. To improve performance, yield, and cost, it would be necessary to reduce the die size without reducing the functionality of the chip. Given a specific fabrication process, which determines the minimum feature size, the only way to reduce the die size is to reduce or completely eliminate the real estate used by the interconnect or the routing footprint. The elimination of the routing footprint will reduce the area, thus reducing the worst case interconnect length and increasing the performance. At the MCM level, the design goal is to minimize area and maximize performance. The routing is accomplished in the third dimension.

This book focuses on routing problems as we move from a two-dimensional to a three-dimensional routing environment. At the chip level, we focus on the elimination of the routing footprint from a chip by routing the nets on top of the active regions, which are also called *over-the-cell* areas. Thus, this routing style is called over-the-cell (OTC) routing. In the OTC routing style, all the nets are laid on top of the cell and blocks, and routing does not use any real estate. If routing does not occupy any real estate in part of the chip, then the chip is referred to as a *zero-routing footprint chip*. In fact, from the layout point of view, the zero-routing footprint is the most challenging objective to achieve. In order to achieve the zero-routing footprint or reduce the routing footprint, research and development efforts must be spent on two key issues:

1. Development of new cell libraries
2. Development of new routers

In this book we discuss the fabrication and electrical issues involved in the development of new cell libraries, which will enable efficient use of the OTC routing area. We propose several new cell layout styles. We demonstrate the ideas by developing a moderate-sized cell library in these design styles. Cells designed in different design styles or cell models permit different usage of OTC areas. We discuss the tradeoff between the reduction in cell area and reduction in overall layout area. All the cells are characterized and their performance and areas are compared. Each cell model forces different kinds of restrictions on the router, and, as a result, a major portion of this book is devoted to the presentation of several different OTC routing algorithms. We present efficient algorithms for every design style proposed and take into account the restrictions imposed by the fabrication process. As the number of layers increases, we show how routing may be completely accomplished over the active regions leading to zero-routing footprint chips. In a multilayer environment, routing is a three-dimensional problem.

Finally, we concentrate on problems of interconnecting chips in an MCM. This problem is complicated not only by size, but also by the complex performance constraints that must be met. The MCM routing problem is a challenge that physical design algorithms have yet to answer.

We present not just our work in the last four to five years, but include all major contributions to routing within the scope of the book. Our algorithms are very soundly based on graph theoretic algorithms, and a complete chapter has been devoted to graph algorithms. All the algorithms are presented in a lucid and concise manner. All the algorithms have been implemented and tested on industrial benchmarks. The experimental data are then compared to determine the most efficient algorithm.

OTC routing has been an intense area of research in VLSI physical designs in the last decade, while MCM routing is just beginning to be addressed. This book provides the state-of-the-art cell design techniques and algorithms for OTC routing and an introduction to MCM routing techniques. While a significant part of this book is orginal research by the authors and their coworkers, an attempt has been made to include all the algorithms developed in this field. All important algorithms have been included with great detail, while others have been either included in a summary format or as a reference and a bibliographic note at the end of the chapter.

Overview of the Book

This book project evolved from a chapter in the first author's text book entitled *Algorithms for VLSI Physical Design Automation*, and the authors' and their coauthors' work on OTC and MCM routing since 1988. The book is organized to provide state-of-the-art information regarding routing algorithms and techniques. The first chapter outlines the need for the zero-routing footprint for chips and efficient algorithms for chips as well as MCMs. Chapters 2 and 3 are used to build the necessary background in graph algorithms and channel routing and pin assignment, which will be required for understanding the concept of the zero-routing footprint, OTC, and MCM routing. Chapter 4 gives an indepth discussion of parameters that must be considered for effective ultilization of routing resources. The remaining chapters discuss the various routing algorithms as we increase the number of layers from two in VLSI to over sixty layers in general MCMs.

Chapter 1 discusses the need for the elimination of the routing footprint from chip designs. It gives the user the perspective of what routing is and what role it plays in the physical design cycle. Concepts of channel routing are also discussed before the concept of OTC routing is introduced. For MCM designs, it builds a case for performance-driven routing.

Chapter 2 is used to build up the reader's background in graph theory and presents various graph algorithms that will be used in the various routing algorithms. An attempt has been made to provide the reader with a concise but sufficient summary on the background material in order to help the reader understand the material presented in the rest of the chapters. Chapter 3 deals with basic channel routing algorithms. These algorithms form the basis of many algorithms on the VLSI, advanced VLSI, and MCM levels. Chapter 4 introduces various routing models and discusses the various parameters that affect the utilization routing areas. At the chip level, we study the effect of cell library designs on routing and performance. At the MCM level, we study the effects of the fabrication process and die attachment techniques. Chapter 5 introduces several basic routing problems, including OTC routing problems in all the different cell models and MCM routing problems.

In Chapter 6, various approaches to solving OTC channel routing problems have been presented for the two-layer process. We consider routing algorithms for various cell models. Heuristic, approximation, and integer linear programming–based algorithms are presented. Detailed algorithms are presented with appropriate examples and experimental results. Finally, all models are compared. Efficient OTC routing algorithms for various cell models are discussed in Chapter 7 for the three-metal-layer process, when vias are not allowed over active areas. Approximation algorithms are presented for the boundary terminal model (BTM), while optimal terminal alignment algorithm for the center terminal model (CTM) and middle terminal model (MTM) forms the core of the routers for these models.

Chapter 8 deals with the development of OTC routing algorithms for the advanced three-layer metal process, which allows vias in active areas. Optimal algorithms for the CTM and MTM form the basis of very efficient routers in these models.

In Chapter 9 we present chip routing algorithms for a multilayer process i.e., process with four or five layers. This process is suitable for advanced VLSI chips as well as thin-film MCMs. We clearly show that, with the four- and five-layer processes, the zero-routing footprint is guaranteed for medium to dense circuits.

Finally, Chapter 10 gives the reader state-of-the-art MCM routing algorithms. Several techniques are presented and performance constraints characterizing this problem are discussed.

In each chapter we have attempted to present key algorithms in detail, while other algorithms are presented in a summary format. Each concept and algorithm is explained with the help of examples. All related concepts, which a reader may need for reference, are explained in the bibliographic notes at the end of the chapter. The book has a complete bibliography covering papers on OTC routing, MCM routing, graph theory, graph algorithms, and electronics. The book also contains an author and subject index for ease of reference and cross reference.

To the CAD Engineer

First and foremost, this book is written for CAD engineers who are interested in reducing the die size of chips by using OTC routing and improving system performance with efficient MCM routing. CAD engineers will find a comprehensive set of algorithms for a variety of cell models and fabrication processes. Even if cell models used by a certain company differ from the ones given in this book, there is enough coverage in terms of algorithms that we believe one will be able to acquire a sound background in this field. We have written this book with the hope of aiding the CAD engineer in implementing existing algorithms and developing new ones.

To the CAD Researcher

This is the first attempt to present all the material in OTC and MCM routing under one cover. CAD researchers already working in the field will benefit by using this book as a reference. On the other hand, researchers intending to start research in this field will find the background material, the complete and extensive list of references, and research problems very helpful for initiating research. Due to its nature, it can also be viewed as a book for individuals interested in combinatorics and mathematics. The book includes a comprehensive chapter on algorithms for interval, permutation, and circle graphs. These graphs are of great interest to theoreticians working in graph theory or graph algorithms.

To the Graduate Student in VLSI Design Automation

This book can also be used as a reference text for an advanced topics class or a research course in VLSI design automation. The book is well suited for such a class, since it is self-contained and all the required material for graph algorithms, channel routing, and pin assignment is presented. Students will find research problems stimulating.

Errors and Omissions

It can safely be said that no book is free of errors and omissions. We have made our best attempt to make this book error-free; however, it may still contain some errors. We would appreciate receiving reports on errors, comments, and suggestions. The comments can be mailed to:

> Intel Corporation, Microprocessor Division
> 5200 NE Elam Young Parkway
> Hillsboro, OR 97124-6097

or e-mail them to sid@ichips.intel.com., or sherwani@ichips.intel.com, or anand@ichips.intel.com.

We have made a sincere effort to include all the papers and book references that we could find. Readers are encouraged to report omissions in the book.

This book was typeset in Latex. Figures were made using 'xfig' and inserted directly into the text as .ps files using 'psfig'. The bibliography was generated using Bibtex and the index was generated by a program written by Siddharth Bhingarde.

Naveed A. Sherwani
Siddharth Bhingarde
Anand Panyam

Acknowledgments

Like all other research works, this book is a product of academic and intellectual contributions from multifarious sources. *Inter alia*, it is based on work presented in over 50 papers, with over 30 coauthors. Among individual contributors, acknowledgments are due to Professor Majid Sarrafzadeh of Northwestern University, who was associated with us during the inceptionary stages of our work on OTC routing. Acknowledgments are also due to Nancy Holmes, Bo Wu, and Shiv Kumar—all graduate students of the first author—whose work contributed to our knowledge of OTC and three-dimensional routing. A major portion of the initial work on two- and three-layer OTC routing was undertaken by Nancy. Bo investigated a new model for OTC routing, and Shiv studied the effect of performance on OTC routing. Our knowledge of OTC routing would have been restricted without their input.

We must acknowledge the help of our VLSI research group at Western Michigan University, nicknamed *The Nitegroup*. The members of the nitegroup gave generously in terms of their time and effort to help us during various phases of this project. We thank Aditya and Pramod for developing the cell library and providing insights into cell area/model tradeoffs. Nitegroup members Arun Shanbhag, Praveen Jayamohan, Sandeep Badida, John Qoing Yu, Sreekrishna Madhwapathy, Srinivasa Danda, and Aman Sureka helped with the editing of the final phase of the manuscript.

Several organizations played a key role in this project. Western Michigan University (WMU) provided the time for the authors. WMU's system and computer support are also gratefully acknowledged. The National Science Foundation provided the equipment for the VLSI Research Laboratory, which has played a key role in the development of the VLSI research program at WMU. We also wish to thank the Intel Corporation for supporting this project, and the ACM Special Interest Group on Design Automation (SIGDA) for providing the research assistantships.

We wish to thank Don Nelson, in particular, for his support of this project, and in general the VLSI research group at WMU. Thanks are also due to our department secretaries, Sue Moorian and Phyllis Wolfe, who cheerfully accepted the extra work that this project created for them.

IEEE Press also has played a key role in the development of this book. In this regard, we would like to thank Russ Hall, acquisitions editor at IEEE Press, for his encouragement and getting very critical reviews, which really improved the manuscript. We also thank Denise Gannon for helping us with the book production.

Finally, we wish to thank our spouses, mothers, fathers, and siblings for their support and patience. While it is our hands that write, they are the ink, the pen, and the paper.

Naveed Sherwani
Siddharth Bhingarde
Anand Panyam

1

Introduction

This book is devoted to the development of routing techniques to achieve minimum area layouts for very large scale integration (VLSI) systems. Minimum area layouts not only deliver the optimum performance, they also provide the most economical solutions. Optimizing performance is the most critical concept in the current VLSI systems design trend. Virtual reality, the information super highway, global communication networks, global weather prediction, and other grand challenges require high-performance VLSI systems. The performance of a processing system is measured in terms of millions of instructions executed per second (MIPS). A VLSI system may consist of a single chip or several chips mounted on one or several multichip modules (MCM). In the last two decades, there has been a remarkable increase in performance. In 1972, Intel's 8008 microprocessor was capable of only 0.03 MIPS. The performance was raised to 0.9 MIPS in the 80286 processor by 1982. In 1992, Digital released its Alpha microprocessor, which is capable of over 400 MIPS. It is clear that before the turn of the century, MIPS rating will give way to BIPS rating (that is, billions of instructions per second).

The minimum area layout of a VLSI system depends on many factors, two important factors among them are system architecture and the physical design of the system. Providing maximum functionality with a minimum number of transistors is the main objective of the system architecture phase. Logic minimization attempts to reduce the transistor count, since the layout area depends on the real estate occupied by the active part of the circuit. It is the passive part of the circuit (that is, interconnection) that consumes the remaining real estate. Given a functionality, the size of this passive real estate can be reduced by better layout schemes during the physical design phase of the system. The physical design cycle is the process of converting a circuit diagram of a system into a circuit layout. Due to its complexity, this process has been broken down into a number of steps. One important step in physical design is the interconnection of different components of a system. This step is called *routing* and plays a critical role not only in making the system functional, but also determining the system performance.

In this book, we will concentrate on routing methodologies and algorithms. However, there are many factors, such as design styles, partitioning, and placement algorithms, that also affect the area of the layout. These factors are related to more global considerations or to the system physical design cycle. As a result, it is important to have a good perspective of the entire physical design cycle.

In this chapter, we start with a brief discussion of the various steps involved in the physical design cycle and how they affect the minimum area layout of a system. Next, we discuss the factors involved in minimum area layouts at the chip and system levels. We will then survey the existing routing approaches for VLSI, thin-film MCMs, and general MCMs. We also discuss the new routing methodologies required to achieve the minimum area layouts.

1.1 System Physical Design Cycle

The various steps in the physical design cycle are shown in Figure 1.1 and are discussed below:

1. MCM or system partitioning.
2. For each chip in the system:
   [2.1] Partitioning
   [2.2] Floorplanning
   [2.3] Placement
   [2.4] Routing
   [2.5] Compaction
   [2.6] Fabrication.
3. MCM placement.
4. MCM routing.
5. MCM fabrication.

The focus of this book is on chip and MCM routing; however, to gain a more global perspective, we study all the steps briefly before describing the routing steps in detail. It is important to note that all the steps in the physical design cycle are vital in optimizing performance and minimizing the total routing space.

1.1.1 MCM or System Partitioning

An entire system may be too big to be fabricated as a single chip. It is therefore partitioned into several different chips, which are fabricated and then placed and

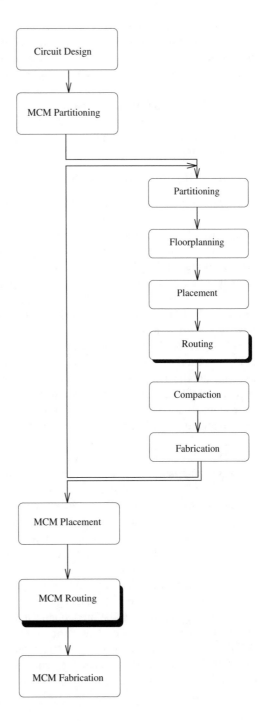

Fig. 1.1   The physical design cycle.

routed on an MCM. An MCM may contain as many as 100 chips. Depending on the fabrication process, each chip can contain a certain maximum number of transistors. The first task is to partition the given circuit into subcircuits, such that each subcircuit can be fabricated on a single chip and the number of subcircuits is less than or equal to the number of chips that the MCM can carry. It should be noted that the MCM designs require a performance-driven approach. This requirement necessitates that power consumption and timing constraints should be considered in the partitioning step, in addition to the traditional input/output (I/O) and area constraints for chip sites.

1.1.2 Design Style Selection

After a system has been partitioned into a set of subcircuits, each subcircuit is fabricated as a chip. So partitioning, floorplanning, placement, routing, and compaction steps are repeated for each chip in the system. Before we begin partitioning, an important decision at the chip-level physical design is to choose a suitable design style. Currently, two design styles are used for high-performance layouts: *full-custom* and *standard-cell* design styles.

In a full-custom design style, a circuit is partitioned into a collection of subcircuits called *functional blocks* (in short, blocks). The functional blocks do not have any restriction on their size or shape and can be placed at any location on the chip surface. The I/O pads located on the periphery of the chip are used to complete interconnections between different chips or interconnections between the chip and the board. The routing area is provided between the blocks. Usually several metal layers are used for routing. Currently, two metal layers are common for routing and the three-metal-layer process is gaining acceptance with feasible fabrication costs. In a hierarchical design of a circuit, each block in a full-custom design may be very complex and may consist of several sub-blocks.

The full-custom design style allows for very compact designs. However, the process of automating a full-custom design style has a much higher complexity than other restricted models. The automation process for a full-custom layout is still a topic of intensive research. Since blocks can be randomly positioned on the chip, the problem of optimizing area and interconnection of wires becomes difficult. Full-custom design is very time-consuming; thus, the method is inappropriate for very large circuits unless performance is of utmost importance. Full-custom is usually used for microprocessor layouts.

In the standard-cell design style, the layouts consist of rectangular cells of identical heights. Initially, a circuit is partitioned into several smaller blocks, each of which is equivalent to some predefined subcircuit (cell). The functionality and electrical characteristics of each predefined cell are analyzed, tested, and specified. A collection of these cells is called a *cell library*. Usually a cell library consists of 200 to 400 cells. Large libraries may contain as many as 1200 cells. Terminals on cells may be located either on the boundary or in the center of the cells. Cells are placed in rows and the space between two rows is called a *channel*. These channels are used to perform interconnections between cells.

Although both design styles have advantages, they suffer from some drawbacks as well. Full-custom allows reusability of blocks. However, these blocks may not be general-purpose; i.e., they are designed for specific functions. Thus, even though full-custom allows reusability, the extent to which it can be explored is limited. Moreover, this technique trades high performance with increased production cost. Standard cells are 100% reusable. They provide moderate performance at comparatively lower design costs. This is achieved by cell reusability and less complex steps of design cycle. From the above discussion it is apparent that no existing design style is completely adaptive to reusability, high performance, and low cost. Therefore, a "mix and match" of full-custom blocks and standard cells is used, in which the high-performance blocks are designed by using the full-custom design style, and the remaining layout is designed using the standard cells in order to reduce the design cost. The advent of this new design style has created new problems for physical design automation tools, since they need changes to suit the new "mix and match" design style. This integrated design style is called the *mixed block and cell* (MBC) design style.

The MBC design style is a combination of full-custom and standard-cell design styles, as shown in Figure 1.2(a). It primarily emphasizes reuse of existing layouts. Since existing chips form the basis for newer chips, large blocks from existing chips can be directly incorporated into new chips. The functionality between these blocks can be developed using standard cells. In this way, the design time required for development of high-performance chips is minimized. Almost 50% of the blocks are either taken from libraries directly without any modification, or some additional functionality is developed around these blocks. If a block has been redesigned, its area, shape, and port locations are well defined. These blocks are called *fixed* blocks. The remaining functionality is provided by standard cells. Typically, on a chip, one may have about ten such fixed blocks and about 10,000 standard cells. The MBC-based design style is also referred to as *structured custom*.

1.1.3 Chip-Level Partitioning, Floorplanning, and Placement

After selection of a design style, a circuit is partitioned. In partitioning, a circuit is decomposed into several smaller subcircuits. The basic idea is to reduce the complexity of each subcircuit for simplified design and minimized interconnections between subcircuits. If interconnection between subcircuits is small, it leads to smaller routing areas between the subcircuits. This is a very important phase in the physical design cycle, since reduced interconnections between the subcircuits is indeed a good starting point for zero-routing footprint chips. Several efficient algorithms have been proposed for partitioning [FF62, Dji82, LT79, Mil84, She93].

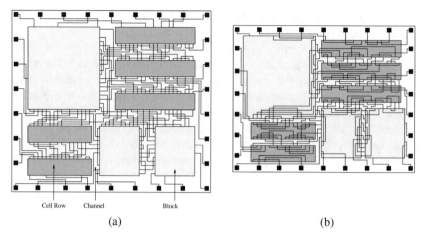

Cell Row   Channel   Block

(a)   (b)

Fig. 1.2   Mixed block and cell design style.

After partitioning the circuit, the next step in the physical design cycle is to determine the shapes and sizes of these subcircuits. This process of determining block shapes with an objective to reducing overall chip area is termed *floorplanning*. The key objective of this phase is to effect global chip-level area minimization. Floorplanning plays a major role in determining the final chip area. Several algorithms have been presented for floorplanning [MS86b, PMSK90, PD86, She93].

At the end of the floorplanning phase, the original circuit consists of a set of subcircuits or blocks. The process of determining the actual location of each block and cell is called *placement*. The placement phase traditionally estimates the routing area required and allocates the required areas between blocks and cells of rows for routing. Placement plays a critical role in determining the final die size of the chip. Several efficient placement algorithms have been presented, with emphasis on chip area minimization for high-performance circuits [Ake81, Pat81, DNA90, JJ83, GKP90, Leb83, She93].

1.1.4 Chip-Level Routing and Compaction

The last two steps in the physical design cycle are routing and compaction. The objective of this book is the development of efficient routing techniques. Therefore, the routing phase of the physical design cycle is discussed in detail in the subsequent sections.

The final step in the physical design cycle is compaction. After the completion of routing, the chip is functionally complete. However, due to nonoptimality of placement and routing algorithms, some vacant areas are present in the final layout. In order to minimize the die size, these vacant areas are eliminated without altering the functionality of the layout. This operation of layout area minimization is called *compaction*. For a detailed study on compaction, refer to (Chapter 10 in [She93]).

1.2 Minimum Layout Area at the Chip Level

In the last three decades, as shown in Figure 1.3, there has been a significant improvement in the chip area and the minimum on-chip feature sizes. The maximum chip area has increased by more than 50-fold, while the minimum on-chip feature size has decreased by the same order. Consequently, there has been a phenomenal increase in the scale of integration. As shown in Figure 1.4, from a mere 1,000 components per chip in the early 1970s, there are now more than a million components per chip.

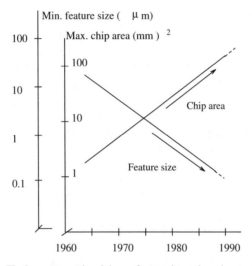

Fig. 1.3   The improvement in minimum feature size and maximum chip areas.

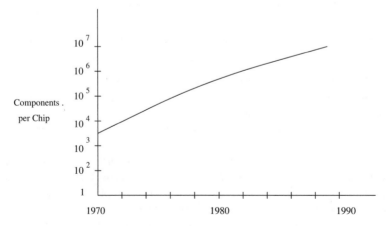

Fig. 1.4   The history of silicon integrated-circuit scale of integration.

While the increase in the scale of integration has resulted in complex chips with high computational powers, the incorporation of a large number of functions in a chip has led to a corresponding increase in the number of interconnections inside a chip, and hence large interconnection areas. These large interconnection areas result in increased chip areas and, correspondingly, deteriorated chip performance. Therefore, for a given chip functionality, the objective of a chip designer is to minimize interconnection areas and consequently the die size.

The minimization of interconnection areas is necessary to optimize the following factors:

- **Delay:** The performance of a chip depends on many factors, including its architecture, the clock distribution, the circuit family used, among others. For a given architecture, clock distribution scheme, and circuit family, the maximum wire length is the main factor that determines chip performance. For better performance, it is necessary to minimize the maximum wire length. As in the worst case, maximum wire length is directly proportional to the die size, so it is necessary to minimize the die size. With decreasing minimum feature size, there is a sharp rise in the delay per unit length of interconnection. For example, in a 7-μm n-type-channel metal oxide semiconductor (NMOS) technology, the per-unit resistance is 21 Ω/cm, but it rises to 2,440 Ω/cm for 0.35-μm complementary metal oxide semiconductor (CMOS) technology. Though per-unit capacitance is reduced, its effect is quite marginal. For example, in 7-μm NMOS technology, the per-unit capacitance is 5 pF/cm, and for a 0.35-μm CMOS technology, it decreases to 2 pF/cm.

- **Yield:** Smaller die sizes lead to a larger number of dies per wafer. Also, it increases the percentage utilization of a wafer by reducing the waste area along the circumference of the wafer and the waste area due to deformities in the wafer.

- **Cost:** The cost of a chip is dependent on the yield, which in turn depends on the die size. In other words, the smaller the die size, the lower the cost.

Thus, it is clear that improving performance, yield, and cost at the chip level necessitates reduction in the chip's die size without reducing the functionality of a chip. As a result, layout engineers spend an enormous amount of time and effort to obtain a die of absolute minimum size for a given circuit design or transistor count.

The total area (A_T) of a chip consists of the active area A_C used by cells and blocks, routing area (A_R), the unused area (A_U), and the pin area (A_P). In modern chips, unused area A_U is minimal due to very advanced placement tools, and pin area A_P is dictated by the connection requirements to the outside world. These regions have slight variations with the layout size. Therefore, reduction is only possible in the areas A_C and A_R. At a given hierarchical level, the active area can simply be considered as the summation of all the areas of blocks and cells; that is, $A_C = \sum_{c_i \in C} A_{c_i}$. The area (A_{c_i}) of a standard cell (or a block) c_i can be further divided into transistor area (A_{t_i}), internal routing area (A_{r_i}), and unused area (A_{u_i}). The routing area A_R may also be divided into two further subareas, the channel (or switch-box) routing area (A_{cr}) and the side routing area (A_{sr}). Therefore, $A_R = A_{cr} + A_{sr}$. In two-metal-layer designs, A_{cr} consumes about 10% to 15% of the total chip area, while the cells and blocks account for as much as 65% of the total chip area. Given a set of blocks and a standard-cell library, the absolute lower bound on the area of a chip is given by

$$A_L \geq \sum_{c_i \in C} A_{c_i} + A_P$$

where C denotes the multiset of chips in the design.

On the other hand, if we are allowed design flexibility in the cells, then the absolute lower bound on the area of a chip is given by

$$A_{L'} \geq \sum_{c_i \in C} A_{t_i} + A_P$$

The above discussion implies that for a given set of blocks and a given standard-cell library, the minimization of layout areas is possible only by reducing or eliminating the routing area, particularly the channel routing area (A_{cr}). In other words, physical design should ensure that the total area used by a chip does not exceed its theoretical lower bound based on the summation of the active areas. We refer to a chip with area equal to the theoretical lower bound as a *zero-routing footprint chip*. As discussed earlier, reduction in die size leads to chips with better performance, higher yield, and lower costs. The minimum die size for a given circuit is obtained by minimizing or eliminating the routing footprint.

For a given circuit, the routing area depends on the following factors:

- **Block and Cell Design Styles:** The shapes and sizes of the individual components are specified by the design styles. (A set of interconnected transistors realizing a logic function is called a block or a cell.)
- **Relative Placement:** The placement of blocks and cells is a very critical phase in chip design. The area required for communication between the blocks depends on the placement of the blocks and cells and the number of interconnections between them. Hence, the goal of an efficient placement algorithm is to place the blocks relative to each other depending on their connectivity, and to minimize the need for large real estate on the chip.
- **Fabrication Process:** The fabrication process plays a vital role in determining the area required for routing. It determines the number of metal layers available for interconnections.
- **Routing Techniques:** Efficient routing algorithms use minimum area for completing the routing.

Since performance optimization at a chip level is also closely related to die size minimization, an attempt should be made to minimize the die size by eliminating the routing areas. After each chip has been fabricated with a minimum or zero-routing footprint, we need to place and route the chips together on an MCM to form a complete system.

1.3 Minimum Layout Area at the System Level

In order to improve the layout area of an MCM, we must reduce the number of chips. However, this must be accomplished without increasing the area of each. Clearly there is tradeoff between the functionality in each chip and the number of chips. In addition, the performance also depends on the size of the MCM.

The performance of the chip is bound by the time required for computation by the logic devices and the time required for the data communication. However, as shown in Figure 1.5, in the past three decades there has been a 1,000-fold improvement in the average speed of computation time per gate. Hence, the interconnection delays, once insignificant compared to computational delays, are now on the order of gate delays and therefore cannot be neglected. As a result, the system performance depends on the performance of individual chips and communication between chips (that is, on the effectiveness of the MCM). To improve performance, one must reduce the amount of interchip communication and devise effective methods for interchip communication. One method of minimizing interchip communication is to make individual chips more capable; that is, more and more devices have to be integrated into a single chip. However, that leads to an increase in chip areas.

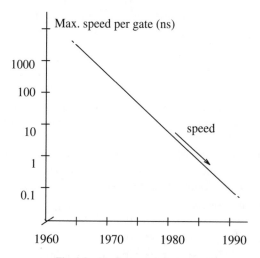

Fig. 1.5   Maximum speed per gate.

MCM technology has been advancing rapidly in recent years, since it is the only feasible packaging technology that provides the performance capability that can meet the increasing demand for high-performance systems. As MCM technology develops, the number of chips mounted on an MCM substrate and the number of layers available for routing increase dramatically. For example, an MCM with more than 100 chips and 63 layers has been recently reported [RRT92]. Therefore, both the capacity of and demand on routing resources has increased rapidly and will continue to increase. Though MCM technology is developing rapidly, the CAD tools for the MCM lag behind. In particular, routing tools are incapable of handling the complexity of MCM routing problems, thereby creating a bottleneck for the further development of MCM technology.

A full-custom design of an MCM takes significant engineering efforts, and due to the lack of a mature infrastructure, high-density and high-performance MCMs are still expensive to fabricate, and the cost increases with the number of mask layers. Irrespective of the types of MCM technology used, bare chips are to be attached to the substrates. Bare chips are attached to the MCM substrates in three ways: wire bonding, tape automated bonding (TAB), and flip-chip bonding. In wire bonding, the back side of a chip (nondevice side) is attached to the substrate and the electrical connections are made by attaching very small wires from the I/O pads on the device side of the chip to the appropriate points on the substrate. The wires are attached to the chip by thermal compression. In an MCM, all the chips are placed on a substrate and routing is done in layers on top of the chips. In essence, the footprint of an MCM is already minimized and no further minimization is possible. Therefore, the performance must be improved by careful placement of chips and with performance-driven routing. The MCM is a true three-dimensional routing environment, with over 60 layers available for routing. The confines of VLSI routing with a restricted number of layers, which characterized the problem at chip level, no longer applies. New algorithms need to be developed, which account for the enormous size, complex performance constraints, and three-dimensional nature of MCM routing. Until we have good MCM routing tools, the gains made at the chip level will be lost at the MCM or the system level.

1.4 Routing in the Third Dimension

While at first glance it may appear that VLSI and MCM routing problems are two essentially different problems, there is indeed a continuum based on the number of layers. In the early days of VLSI, only a single metal layer was available for routing, and at that stage polysilicon (poly) and M1 layers were used to complete the interconnections. The approach in that routing era was essentially planar and/or two-dimensional. As processes improved, a second metal layer was added; however, the two-dimensional approach developed earlier continued. With more advances of process technology, a third metal layer was made available for routing.

Recently, a four-metal process has been added and true three-dimensional chip-level routing with a zero-routing footprint is now possible. In the meantime, thin-film MCMs were developed, which use a four or more layer process similar to VLSI. The routing approach is similar for multilayer VLSI chips and thin-film MCMs. For general (or thick-film) MCMs, the routing environment consists of several dozen layers and is quite different from the VLSI or thin-film MCM routing environment. This is due to the fact that partial or restricted use of the third dimension becomes almost complete or unlimited use.

In summary, with two and three layers, the routing at chip level was essentially between the blocks, although some routing is done on top of active areas. With four layers, routing is three-dimensional; that is, it is entirely on top of the active areas. However, the use of the third dimension is restricted, since the number of layers allowed is fixed and small. The same style of routing is used in thin-film MCMs, while thick-film MCMs are characterized by true three-dimensional routing. Thus, one sees a gradual shift from two dimensions to three dimensions as we go from two-layer chip routing to MCM routing. This can be clearly seen from Figure 1.6.

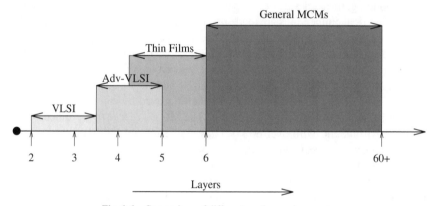

Fig. 1.6   Comparison of different routing environments.

This book primarily focuses on the study of algorithmic techniques for routing at several different levels. We are interested in studying the transition in algorithmic techniques as we graduate from a two-dimensional VLSI environment to a three-dimensional MCM environment. At the chip level, we study improvement of performance by achieving a minimal or zero-routing footprint, and at the MCM level, we study performance-constrained routing, since the footprint is already at minimum. We aim to study routing problems at all three levels; chip, multilayer chip/thin-film MCM, and thick-film MCMs. At the chip level, we propose to reduce area by use of better cell libraries and efficient over-the-cell (OTC) routing techniques. We primarily concentrate on cell-based or MCB-based designs. At the multilayer chip or thin-film level, one is freed from the confines of cells and blocks to a great extent, but routing must be completed in the given number of layers.

At the general MCM level, we assume that minimum area has been achieved (by efficient placement algorithms) and we concentrate on optimization of performance as the basic objective of routing.

We start with a review of existing routing approaches at the VLSI chip level. Then we study the new routing approaches that are needed to minimize the routing footprint on a chip. Finally, we discuss conventional and new approaches to MCM routing.

1.5 VLSI Chip Routing Techniques

On completion of the placement phase, to make the layout functional, all the interconnections need to be established. This process of establishing the interconnections is called *routing*. The objective of routing depends on the type of chip under consideration. For general-purpose chips, it is sufficient to minimize the total wire length. For high-performance chips, total wire length may not be a major concern. Instead, we may want to minimize the longest wire to minimize the delay in the wire and therefore maximize its performance.

The major shift in routing approaches is due to changes in fabrication processes. As the number of layers increases, the routing graduates from a planar problem to a two-dimensional problem, and finally to a three-dimensional problem when four and five layers are available. This section describes the chip-level routing problems in general.

1.5.1 Conventional Routing Approaches

In a typical two-phase routing approach, the routing area is first partitioned into smaller regions. This is the task of a global router. The global-routing problem is to assign wires to different regions and the detailed-routing problem is to find the actual geometric path for each wire in a region. The complexity of the routing problems varies due to many factors, including the shape of the routing region, the number of layers available, and the number of nets. Let us take a look first at basic terminologies of routing problems. A *channel* is a routing region bounded by two parallel rows of terminals. Without loss of generality, it is assumed that the two rows are horizontal. The top and the bottom rows are also called *top boundary* and *bottom boundary*, respectively, as shown in Figure 1.2(a). However, the shape of the region is perhaps the most important factor. Routing algorithms need to consider parameters such as number of terminals per net, width of nets, different types of nets, restrictions on vias, shape of the routing boundary, and number of routing layers.

The channel routing problem is to find the interconnections of all the nets in the channel including the connection sets so that the channel uses the minimum possible area. A solution to a channel routing problem is a set of horizontal and vertical segments for each net. This set of segments must make all terminals of the net electrically equivalent. In the grid-based model, the solution specifies the channel height in terms of the total number of horizontal segments required for routing. The channel routing problems and related algorithms are discussed in detail in Chapter 3.

The main objective of channel routing is to minimize the channel height. Additional objective functions, such as minimizing the total number of net intersections used in a multilayer routing solution and minimizing the length of any particular net, are also used.

Conventional channel routers are developed to minimize the channel areas. Since, advances in fabrication technologies have now allowed the use of additional areas for routing, the primary routing objective is not just to minimize the channel areas, but to efficiently utilize areas over the active elements of the circuits for routing, as discussed in the next section.

1.5.2 New Routing Approaches

The existing routing approaches minimize chip sizes by minimizing channel areas. With the advent of new fabrication technologies, the new routing approaches aim at eliminating the channel areas by using the metal layers over the active areas of the blocks and cells for routing. Intracell and intrablock routing is normally accomplished in the polysilicon and the first metal layer. However, the remaining metal layers are unused. Hence, these unused metal layers can be used for intercell and interblock connections. The new approaches presented in the following subsections aim at achieving a zero-routing footprint by routing over the active areas.

1.5.2.1 Area routing approach

In this approach, we place the blocks adjacent to each other without considering the area required for interconnections. The area routing approach considers that all routing layers are available for interconnections over the entire chip, and there are no restrictions on the usage of *vias*. An example of the area routing approach is shown in Figure 1.7(a). This approach is used in routing MCMs.

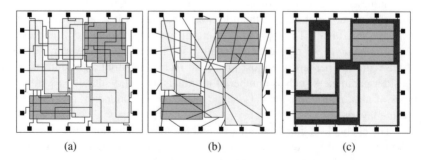

| (a) | (b) | (c) |

Fig. 1.7   Three approaches to OTC routing.

In the area routing approach, the routing algorithms consider the entire routing area to be a stack of superimposed grid planes. The algorithms use greedy and maze routing techniques to complete the routing. The number of layers in the stack depends on the fabrication process. For a two-layer fabrication process, the area routing algorithms use planar routing approaches in OTC areas, since the cells normally use the first metal layer for intracell routing. However, for processes with more than three metal layers, the OTC areas can be routed in a nonplanar fashion when vias are allowed in OTC areas.

A modified area router presented in [KK90] assumes that four layers are available for routing. Routing is completed in two steps. A selected group of nets is routed in the between-cell areas using existing channel routing algorithms and the first two routing layers. Then the remaining nets are routed over the entire layout area, between-cell and OTC areas, using a two-dimensional router and the next two routing layers. The router used for OTC routing recognizes arbitrarily sized obstacles due, for example, to power and ground routing or sensitive circuits in the underlying cells.

The global router generates the interconnection paths based on the net densities in different regions. However, for dense circuits, since the blocks are placed adjacent to each other, this approach may encounter *routability problems*. The routability problem arises when the detailed router fails to assign tracks to some interconnecting wires. This problem does not arise in MCM routing, since MCMs typically have a large number of routing layers and additional layers are added to overcome the routability problems.

The area routing approach may fail to generate a valid routing solution for dense circuit layouts since, unlike the MCM, a large number of metal layers are not available to complete the routing when routability becomes the criterion. Furthermore, the area of the MBC design cannot be increased because the blocks are placed close to each other without prior allocation of channel areas.

1.5.2.2 Topological approach

In MBC layouts, some blocks are predesigned. These blocks restrict the routing of vias in the metal layers directly above them, since it may deteriorate the cell performance in the block.

A circuit consists of a set of modules and a set of nets. Each net specifies a subset of points on the boundary of modules, called *terminals*. The region that is not occupied by modules can be used for routing. The solution of a routing problem consists of connections of terminals of nets by a set of curves, and two end points of each curve must belong to the same net. In a k-layer routing problem, a curve or a piece of curve may be assigned to any of the k layers. Vias are assigned at points where a curve changes layer and no two distinct curves intersect in the same layer. The physical dimensions of the wires, pins, and vias are not considered in topological routing.

Topological routing has been studied primarily for minimization of vias, and it is generally known as the *topological via minimization* (TVM) problem (also known as the *unconstrained via minimization* problem). The TVM problem is to find a topological routing and layer assignment of wires so that the total number of vias is minimized [AS91, CL90, Hsu83, LSL90, MS92, RKN89, SL89, SL90, SHL90, Xio88]. The general TVM problem in k layers (k-TVM) may be stated as follows: *Given a set of nets, number of layers* k, *and terminal locations, find a* k-*layer topological routing solution that completes the interconnections of all nets using the minimum number of vias.* The TVM problem is known to be NP-hard (discussed in Chapter 2) [SL89]; however, polynomial algorithms are known for special cases.

In the general TVM problem, we assume that each layer is free of obstacles, and the objective is to complete the routing with minimum vias. From the OTC point of view, we will have to define the blockages in each layer and complete the routing avoiding the blockages. Layers M1 and M2 are likely to have significant blockages, while M3 and higher layers are likely to be relatively free of them.

Once the topological solution is constructed, we need to convert it into a geometric solution. This requires assigning actual geometric locations to pins, wires, and vias. However, we have observed that topological routing often uses very long wires for many nets and causes high congestion in the routing region. Since the geometric routing problem has a fixed area routing region, it may not be possible to transform a high-congestion topological routing solution to a geometric routing solution. Hence, the topological approach will not guarantee a valid geometric routing solution for a given interconnection list. If the routing region is restricted, then polynomial algorithms are possible. In [Mal90], the problem of transforming single-layer topological routing into geometric routing was considered. In [HWF92], efficient algorithms for transforming two-layer topological channel routing into geometric routing were presented.

For general regions, we need a topological routing solution that is guaranteed to be transformable into an actual geometric routing solution. This can be achieved by allowing some extra vias to keep the topology as close to the actual geometric solution as possible so that the final topological routing solution can be easily transformed into an actual geometric routing solution. This problem for some restricted routing regions has been considered in [HS91]. Currently, no efficient algorithm (even heuristic) is known for the general regions. The main concern with the application of topological routing techniques to general chip regions is the issue of routability. The routability problem is inherent in processes with less than five metal layers for routing and hence can be overcome by the use of a large number of routing layers.

In the topological approach, we can generate the interconnections for the entire layout using a global router, as shown in Figure 1.7(b), without considering the restrictions imposed by the predesigned blocks. After completing the global routing for the entire chip topologically, we can now examine the routing topology over the predesigned blocks for net intersections. These intersections can be "combed" out of the block areas, so that the routing over the predesigned blocks is planar. However, for dense circuits, the task of relocating the intersections to outside the predesigned block areas may result in routability problems. Hence, the topological approach will not always guarantee a routing solution. However, for a large number of layers, the routability may not be a concern, as is the case in multilayer VLSI and thin-film MCMs processes.

1.5.2.3 Channel-based approach

The routability problems inherent in the area and topological routing approaches are surmounted in the channel-based routing approach. In this approach, the placement algorithms consider the interconnection areas while placing the blocks. The interconnection densities between two blocks determine the separation areas between them. The global routing is accomplished only in the channel areas. Global routing through channels is a well-understood problem and several efficient global-routing algorithms have been presented [SK87, VK83, LS88, YCHK87, She93]. These algorithms generate interconnection paths between various blocks in the separation areas. An example of the channel-based routing approach is shown in Figure 1.7(c). The routing solution of the global-routing problem is the channel assignment of all the interconnections. The net assignments to the channels are along the edges of the adjacent blocks. The objective of the detailed router is to transform the nets from the channel to the nets over the block areas. The predesigned blocks do not permit the intersection of nets over the active areas. However, the predesigned blocks have terminals at the boundaries, and the periphery of the block is used for routing the I/O pins to the terminals on the boundaries. These areas, called *moat* areas, do not have any active areas and hence permit the intersection of nets. Therefore, the detailed routers may restrict the vias to these moat areas on the block and complete planar routing over the rest of the blocks.

The main objective of the detailed router is to minimize the separation areas between blocks. This is accomplished by transforming the nets from channel areas to areas over the blocks (moat areas in the case of predesigned blocks) and cells. The transformation is based on the net intersections and the restrictions imposed on routing over the blocks. The channel-based approach never fails to generate a routing solution, since the areas required for interconnections are considered during the placement phase.

The key advantage of the channel-based approach over the area and topological routing approaches is that the channel routing problems and their associated concepts are well understood and well studied while the others are new problems. The computational time required for the well-defined channel-based approaches is significantly less than the computational time required for the less studied area-based approaches, due to the restricted number of paths available for nets in channel-based approaches. In this book we follow the channel-based approach so that the existing global routers can be used, without incurring additional expenses in the development of new area-based and topological routers.

The routing footprint can be eliminated only by efficient utilization of metal layers for routing in the areas above the blocks and cells. In the next section we introduce the technique for routing over the active areas and discuss various issues to be considered for efficient utilization of the areas over the blocks and cells for routing.

1.5.3 Minimum Chip Layout Area Using OTC Routing

In single- or two-metal-layer technology, it is not possible to achieve a zero-routing footprint. This is due to the interconnection requirements between active chip elements. However, with the advent of the three-layer metal process, it is now possible to route all nets above the active areas of cells and blocks and potentially avoid using any real estate for routing. *OTC routing* refers to the routing of a subset of nets in the unused metal layers over the cell. In MBC designs, *over-the-block* (OTB) routing refers to routing in metal layers over the block regions. In this book, we use over-the-cell areas and over-the-block areas interchangeably and refer to them as OTC areas. Currently, this has been an area of intense research. (See Chapter 5.) The basic OTC routing algorithm consists of three steps:

1. Routing over the cell.
2. Choosing the net segments in the channel.
3. Routing in the channel.

When channels are eliminated, the second and third steps are redundant. In standard-cell designs, the problem of OTC channel routing refers to routing a subset of nets over the cell regions to minimize channel height, thereby reducing the overall layout height. Our objective is to minimize or eliminate channel areas by maximizing the routing in the OTC areas.

So far we have understood that OTC routing can eliminate the routing footprint. We have seen the routing process as a part of the physical design cycle. We have also briefly understood the general routing problems. In this section, we look at OTC routing and the problems associated with this style of routing.

The internal routing of the blocks and cells is usually accomplished in the first metal layer. Hence, depending on the technology, either one or two metal layers over the cells are not used. These areas can be used for routing interblock nets, which were previously routed in the channel, thereby reducing the channel areas.

The OTC routing concept will gain more prominence with the increase in the number of routing layers. For a two-metal-layer process, the goal of the OTC routing algorithm is to route as many nets as possible in OTC areas. This results in a reduced number of nets to be routed in the channel, thus leading to a significant reduction in chip areas. In OTB routing, we assume block terminals to be located at its periphery and use OTC techniques to solve the OTB routing problem. The reduction in chip areas between the conventional routing approach and the channel-based approach is illustrated in the example shown in Figure 1.8.

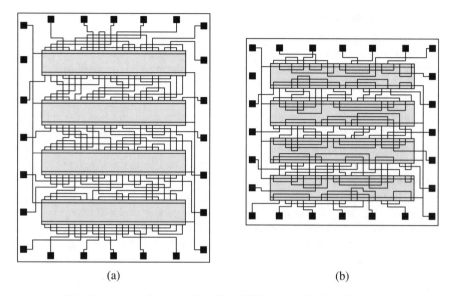

(a)   (b)

Fig. 1.8   An example layout: **(a)** without OTC routing; **(b)** with OTC routing.

For high-performance circuits, critical nets could be routed in the OTC areas with the objective of reducing the critical net length. However, in high-performance circuits, routing high-frequency clock signals in an OTC area may lead to performance degradation in the active elements directly beneath the clock lines due to coupling effects. To overcome this problem, the circuit blocks are characterized by the presence of stray metal in OTC areas.

The input to the OTC channel routing problem is the same as in case of the channel router. But in the case of OTC routing, if N represents the set of nets, the objective is to route a set N' of nets over the cell, where $N' \subseteq N$, such that the channel height is minimized.

The channel-based OTC routing approach uses the conventional global-routing techniques and their associated routers.

From these preliminary discussions, we understand that the primary objective of a chip designer in developing a high-performance, complex-functionality, and cost-effective chip is by minimizing the die size. The die size is minimized by achieving a zero-routing footprint, which in turn is achieved by OTC routing. Then what is the hindrance in achieving a zero-routing footprint? Well, this is the focus of the book. First, we must study some routing issues that limit the utilization of OTC areas for routing. These issues are discussed briefly in the next section and are addressed in detail in the rest of the book.

1.5.4 Issues in Over-the-Cell Routing

There are three key issues that affect the utilization of OTC and OTB areas, as discussed below.

1.5.4.1 Fabrication issues

The fabrication technology used in chip processing plays a vital role in achieving the zero-routing footprint. The fabrication technologies dictate two key routing parameters: number of metal layers available for routing, and permissibility of vias over the active areas.

The increase in the number of metal layers decreases the routing area, thus leading to reduction in the overall chip area. But the addition of each metal layer requires the generation of three additional masks, and consequently increased fabrication cost. But when the chip area and performance are the main design criteria, additional layers are used. Currently, two- and three-metal-layer processes are more popular in the industry.

The availability of more than one metal layer for routing leads to the concept of stacked vias and vias over active regions. Use of vias over active regions may cause oxide undercut and step-coverage problems. These problems are discussed in Chapter 4. Hence, some processes do not allow the use of vias over active areas or the use of stacked vias. When the use of vias over active areas is not allowed, then the routing over the active areas should be planar; that is, the nets are not allowed to switch layers over the active areas. But some processes overcome the oxide undercut and step-coverage problems by using planarization techniques and permit the use of vias over the active areas. The unrestricted use of vias leads to compact designs, since the nets can switch layers at any location. The fabrication issues are discussed in detail in Chapter 4.

1.5.4.2 Circuit/layout design and electrical issues

The effectiveness of OTC routing is also dependent on the layout and electrical issues such as circuit layout style, design rules, and permissible coupling and noise levels. The design rules specify the separation gaps between various elements of a circuit layout. These rules ensure proper connection and isolation between various circuit elements. These rules specify the minimum separation gaps between any two elements. These are necessary for electrical isolations. The separation/overlap rules are usually specified in terms of λ. The actual value of λ varies depending on the fabrication process. The circuit layout style determines the parameters such as location of terminals, and the location of power and ground lines. These parameters determine the routing flexibility and routing strategies for OTC routing. These parameters and their effect on utilization of OTC areas for routing are discussed in detail in Chapter 4. The efficiency of the utilization of OTC areas for routing is also based on circuit specifications such as the permissible noise and coupling levels. For some high-performance circuits, both the width and separation gaps between metal lines are varied. The system clock nets require the minimization of skew and clock delays. These conditions necessitate the clock routing to be completed with added priority over other signal nets. Hence, the utilization of OTC areas for routing is determined by these electrical and layout parameters.

1.5.4.3 Routing issues

The routing environment is dictated by the *routing model*, which specifies the shape and size of the routing region in terms of the number of metal layers available for routing, usage of vias in OTC areas, location of power and ground lines, terminal locations, and the routing boundaries. The routing algorithms consider these specifications as input and generate a routing solution based on the restrictions imposed by the routing model.

The fabrication-related issues such as the number of metal layers available for routing and the restrictions on usage of vias in OTC areas determine the routing *style*. The cell and block designs require at least one metal layer for internal routing. Therefore, in a two-metal-layer process, the usage of vias in OTC areas is not a significant factor, since only one metal layer is available for routing in OTC areas. However, in a three-layer process, the restrictions on the usage of vias in OTC areas play a very significant role in minimizing the routing areas. The routing approaches for the two cases are entirely different and there is a significant change in the routing areas between the two cases. When vias are allowed in OTC areas, the routing areas are either eliminated or considerably reduced, as opposed to areas obtained when vias are not allowed. These changes in routing approaches and the difference in the routing areas when vias are allowed or restricted in OTC areas are extensively discussed in Chapters 5, 6, 7, and 8.

The terminal geometries and terminal alignment also affect the utilization of OTC areas for routing. This factor is discussed in detail in Chapters 5, 6, 7, and 8. For a given technology and given fabrication process, the key to achieving minimal or zero-routing footprint is physical design. Another important factor is the design style. The design style dictates the distribution of routing resources and therefore determines the overall die size.

In the MBC design style, the routing regions can be classified into one of the three types discussed below.

- **Cell-Cell:** This is the routing region defined by two rows of standard cells. Here, the OTC area available may have one to two metal layers. In addition, vias may or may not be allowed in these regions, depending on the fabrication technology.

- **Cell-Block:** This routing region is defined when a row of standard cell is adjacent to a block. The OTC area available on the cell rows is similar as that available on any of the cell rows as described in the previous case. However, the situation may be different over the block, since it may have different regions with a different number of layers, some of which may allow vias and some not.

- **Block-Block:** This routing region is defined when two blocks are adjacent. The OTB areas on both blocks have similar characteristics as the OTB area described in the cell-block case. This is a conventional channel routing problem if OTB routing is not allowed. This is the most general case of OTB routing when we decide to do OTB routing.

The cell-block and block-block routing regions can be transformed into instances in cell-cell routing regions by applying some restrictions to the routing in OTC areas. In this book, we present algorithms for OTC routing for the cell-cell routing regions. Since the block-block and cell-block types of routing can be considered as the subsets of cell-cell routing problems, we use the terms OTB and OTC interchangeably.

1.6 Multilayer VLSI/Thin-Film MCM Routing Approaches

The multilayer VLSI process consists of four to five metal layers. In the thin-film MCM process, we have a similar (or slightly larger) number of layers for routing. Both of these environments allow more freedom in the third dimension, but a number of layers are still restricted, and therefore the approaches used in these two environments are similar.

1.6.1 Conventional Routing Approaches

The multilayer VLSI and thin-film MCM routing approaches are based on generalizations of VLSI routing approaches.

The *maze routing approach* is the most commonly used routing method in three-dimensional maze routing. Although this method is conceptually simple to implement, it suffers from several problems. First, the quality of the maze routing solution is very sensitive to the ordering of the nets being routed, and there is no effective algorithm for determining good net ordering in general.

1.6.2 New Routing Approaches

Two new approaches have been reported for multilayer VLSI and thin-film MCMs.

1. **Topological Routing Approach:** A multilayer router based on rubber-band sketch routing is used. In this approach, hierarchical top-down partitioning is done to perform global routing for all nets simultaneously. The router refines this global routing with successive refinement to help correct mistakes made before more detailed information is discovered. Layer assignment is performed during the partitioning process to generate routing that has fewer vias and is not restricted to one-layer one-direction. The detailed router uses a region connectivity graph to generate shortest path rubber-band routing. The routing approach is suitable primarily for rout-

ing MCM substrates which consist of multiple layers of free (channel-less) wiring space.

2. **Virtual-Channel-Based Approach:** In this approach, the routing surface is partitioned into several routing channels. One layer is used to "bring" terminals to the channel boundaries in such a way that all channels have uniform density. This reduces the overall routing problem to that of several channels. The channels cannot be expanded and depend on a good early estimate to ensure completion.

1.6.3 Issues in Multilayer VLSI/Thin-Film Routing

Following are the key issues of routing in multilayer VLSI and thin-film MCMs.

1. **Problem Complexity Issues:** Advanced VLSI, chip-level, and thin-film routing problems tend to be very large problems and tax the computing as well as memory resources to the maximum. Conventional approaches, which are either too computer-intensive or too memory-intensive are unsuitable for this environment.

2. **Routability Issues:** While the number of routing layers is larger than in the ordinary VLSI routing environment, the routability is still a concern. So sequential approaches, which are based on routing one net at a time, are unsuitable for this environment, since they cannot guarantee completion.

1.7 General MCM Routing Techniques

The general MCM routing environment is distinctly different from printed circuit board (PCB) and VLSI routing environments, and it can be characterized as a truly three-dimensional routing medium. Another aspect of the MCM routing problem is the very large number of nets and the large grid size. An MCM with over 7,000 nets and grid size of over $2,000 \times 2,000$ has been reported in [KC92]. It is expected that the number of layers, number of chips mounted on a substrate, number of nets, and grid size will continue to increase with fabrication technology advances. Performance is the main criterion for MCMs and the cost of an MCM is directly proportional to the number of layers used in the design. Thus, minimizing the total number of layers used is also an objective of MCM routing. Crosstalk is a parasitic coupling between neighboring lines due to the mutual capacitances and inductances.

Most of the existing routing tools for general MCMs are based on PCB routing tools. However, PCB routing tools cannot handle the density of MCM routing problems, while the VLSI routing tools are suitable for routing problems with a limited number of layers, typically two or three.

1.7.1 Conventional MCM Routing Approaches

As stated earlier, most existing approaches for general MCM routing are based on PCB routing techniques. This is due to the fact that most vendors simply converted their PCB tools to thin-film MCMs, and no specialized tools were originally developed. Apart from maze routing, which is mentioned above, only one approach has been conventionally used for general MCMs, which is in fact a carryover from PCBs.

In the *multiple-stage routing approach*, the MCM routing problem is decomposed into several subproblems. The close positioning of chips and high pin congestion around the chips require the separation of pins before routing can be attempted. Pins on the chip layer are first redistributed evenly with sufficient spacing between them so that the connections between the pins of the nets can be made without violating the design rules.

Other PCB routing techniques, such as decomposition into several single-row routing problems, are also applicable.

1.7.2 MCM Routing Approaches

There are two new approaches for general (thick-film/cofired/laminate) MCM routing problems:

1. **Integrated Pin-Distribution Routing Approach:** Instead of distributing the pins before routing, this approach redistributes pins along with routing in each layer. The basic idea is to perform planar routing on a layer-by-layer basis. After the routing on one layer, the terminals of the unrouted nets are propagated to the next layer. The routing process is continued until all the nets are routed.

   The important feature of this approach is the computation of the planar set of nets for each layer. The algorithm tries to connect as many nets as possible in each layer. For the nets that cannot be completely routed in a layer, the algorithm attempts to partially route those nets so that they can be completed in the next layer with shorter wires.

2. **Three-Dimensional Routing Approach:** The three-dimensional routing approach differs from all the other approaches primarily in problem decomposition. The three-dimensional approach maintains the characteristics of the three-dimensional routing problem, while the others convert the three-dimensional problem into a two-dimensional or two-layer routing problem.

1.7.3 Issues in MCM Routing

MCMs are developed for high-performance circuits. Therefore, the primary criterion in MCM routing is performance. However, due to the presence of a large number of nets, the routing task is quite complex. The fabrication-related issues such as chip bonding, use of staggered vias, and minimum separation constraints between nets are also to be considered during MCM routing. The various issues are discussed below.

1.7.3.1 Performance constraints

The interconnections among terminals are made with the help of metal wires laid out on the routing layers beneath the chip layer. The minimization of the number of routing layers is the driving force for most VLSI and PCB routers. However, since more and more layers are readily available and a host of performance constraints are introduced by the MCM routing problems, the number of layers is not as critical as in VLSI routing problems. Instead, the objective of the MCM routing approach should be to satisfy the performance constraints while minimizing the number of layers required for routing. The performance constraints include the manufacturability constraint, net-length constraint, net-separation constraint, and via constraint.

1.7.3.2 Problem complexity

The large size of the input, which plagues the multilayer VLSI/thin-film routing environment, is of concern in general MCM routing environment as well. Not only are the die size and the number of layers (and therefore the routing grid size) large, but the number of nets to be routed is also very large.

1.7.3.3 Fabrication issues

An MCM with dense routing regions is difficult to fabricate, resulting in the low fabrication yield. Therefore, a good MCM routing tool should achieve uniform wiring over the entire substrate. The major fabrication-related issues are the net-separation constraint and via constraint. The via constraint is introduced to allow the specification of the maximum number of stacked vias allowed for a net so as to satisfy the fabrication requirement.

1.8 Purpose of This Book

In this book, we study routing techniques for VLSI, multilayer VLSI/thin-film MCMs, and general MCMs. We aim to answer the following questions:

- What is the effect of number of layers on routing techniques?
- How can we achieve a zero-routing footprint for a given chip?
- What are the constraints posed by the fabrication process and the design styles?
- What are the new design styles developed to maximize the utilization of OTC areas?
- What are the routing techniques for various design styles?
- How can we achieve minimum routing areas for a given design and fabrication process?

- How can we satisfy complex performance and fabrication constraints in MCMs?
- What factors do the futuristic routing styles depend on?

Because many of the routing algorithms depend on graph algorithms, it is imperative to understand the concepts of channel routing, the graphs and related algorithms used for routing. Summaries of the necessary and sufficient graph theoretic fundamentals and channel routing algorithms are discussed in Chapters 2 and 3, respectively. The routing models are discussed in Chapter 4. Basic problems in OTC routing are discussed in Chapter 5. The routing algorithms for different cell models for the two-layer process (resp. three-layer process) are discussed in Chapter 6 (resp. Chapter 7). The algorithms for the multilayer VLSI and thin-film MCM processes are discussed in Chapter 8. Finally, the routing algorithms for general MCMs are discussed in Chapter 10.

1.9 Summary

Routing problems for VLSI, multilayer VLSI, and general MCMs differ, due to differences in the number of layers. There is a gradual shift from the two-dimensional routing approach to the three-dimensional routing approach as we move from VLSI to MCMs. High-performance at the chip level requires minimization of die size for a given functionality. A smaller die size not only improves performance by reducing the interconnection distances and on-chip delay, it improves yield as well. However, reduction of die size of chip area is a complicated problem, which involves all phases of physical design. In addition, utilization of chip area depends on the design style. At the MCM level, the objective is to optimize performance by satisfying timing and manufacturing constraints.

The purpose of this book is to develop algorithms for the VLSI process, multilayer VLSI process, and general MCMs. At the chip level, we aim to eliminate routing areas in MBC designs. We will examine factors that affect the use of OTC and OTB areas. Finally, we attempt to develop routing algorithms that, used with proper placement tools, produce zero-routing footprint chips. At the MCM level, we aim to develop routing algorithms that satisfy the complex performance constraints to optimize an MCM-based system.

2

Graphs and Basic Algorithms

Many VLSI physical design problems, including OTC routing problems, can be modeled using graphs. One significant advantage of using graphs to formulate problems is that the graph problems are well studied. Certain special classes of graphs, such as interval graphs, permutation graphs, and circle graphs, are frequently used to model OTC routing problems. The purpose of this chapter is to present algorithms related to these special classes of graphs. These algorithms are used to develop efficient OTC routers. First we explain our terminology and give a brief overview of NP-hardness and different algorithmic techniques used when the problem is computationally hard. The remaining sections consist of algorithms of three main classes of graphs used in OTC routing. We begin with basic graph terminology.

2.1 Graph Terminology

We have adopted the standard graph terminology used in various books published in the field of graph theory [Gol80, She93]. For a detailed study of graph problems, we refer the reader to an excellent book by Golumbic on algorithmic graph theory [Gol80]. However, for the sake of completeness, the following material from [She93] is presented in this section.

A *graph* is a pair of sets $G = (V, E)$, where V is a set of vertices and E is a set of pairs of distinct vertices called *edges*. If G is a graph, V and E are the vertex and edge sets of G, respectively. A vertex u is adjacent to a vertex v if (u, v) is an edge (i.e., $(u, v) \in E$). The set of vertices adjacent to v is $\mathrm{Adj}(v)$. An edge

$e = (u, v)$ is *incident* to the vertices u and v, which are called the *ends* of e. The degree of a vertex u is the number of edges incident with the vertex u.

A complete graph of n vertices is a graph in which each vertex is adjacent to every other vertex. We use K_n to denote such a graph. A graph H is called the *complement* of a graph $G = (V, E)$ if $H = (V, F - E)$, where F is a set of edges in a complete graph.

Given a graph $G = (V, E)$, an *independent set* is a subset $V' \subseteq V$ such that no two vertices in V' are joined by an edge in E.

A graph $G' = (V', E')$ is a *subgraph* of a graph G if and only if $V' \subseteq V$ and $E' \subseteq E$. If $E' = \{(u, v) \mid (u, v) \in E$ and $u, v \in V'\}$, then G' is a *vertex-induced subgraph* of G. Unless otherwise stated, by subgraph we mean vertex-induced subgraph.

A *walk* of a graph G is an alternating sequence $P = v_0, e_1, \ldots, v_{k-1}, e_k, v_k$ of vertices v_i and edges e_i such that $e_i = (v_{i-1}, v_i)$, $1 \le i \le k$. A *tour* is a walk in which all edges are distinct. A *path* is a tour in which all vertices are distinct. The *length* of a path (walk, tour) P given above is k. A path is a (u, v)-path if $v_0 = u$ and $v_k = v$. A *cycle* is a path of length k, where $k > 2$ and $v_0 = v_k$. A cycle is called an *odd cycle* if its length k is odd, and it is called an *even cycle* if its length is even. Two vertices u and v in G are *connected* if G has a (u, v)-path. A graph is connected if each vertex is connected to every other vertex in the graph. Each *component* of the graph G is called a maximally connected subgraph of G. An edge $e \in E$ is called a *cut edge* in G if its removal from G increases the number of components of G. A *tree* is a connected graph having no cycles. A complete subgraph of a graph is called a *clique*. That is, given a graph $G = (V, E)$, a clique is a subset $V' \subseteq V$ such that every pair of vertices in it are adjacent.

A *directed graph* is a pair of sets $(V, \vec{E})$, where V is a set of vertices and $\vec{E}$ is a set of ordered pairs of distinct vertices, called *directed edges*. We use the notation $\vec{G}$ for a directed graph, unless it is clear from the context. A directed edge $\vec{e}$ is defined by an ordered pair (u, v) of vertices, where u and v are called the *head* and the *tail* of the edge $\vec{e}$. The *in-degree* of u denoted by $d^-(u)$ is equal to the number of in-edges of u. Similarly, the *out-degree* of u denoted by $d^+(u)$ is equal to the number of out-edges of u. An *orientation* for a graph $G = (V, E)$ is an assignment of a direction to each edge. An orientation is called *transitive* if, for each pair of directed edges (u, v) and (v, w), there exists a directed edge (u, w). If such a transitive orientation exists for a graph G, then G is called *transitively orientable*. The definitions of subgraph, path, and walk can also be extended to directed graphs. A *directed acyclic graph* is a directed graph having no directed cycles. A vertex v is called a *descendant* of a vertex u if there is a directed (u, v)-path in $\vec{G}$. A rooted tree (or *directed tree*) is a directed acyclic graph in which all vertices except the root have an in-degree equal to one and the *root* has in-degree zero. The root of a rooted tree T is denoted by *root(T)*. The *subtree* of tree T rooted at v, is the subtree of T induced by the descendants of v. A *leaf* is a vertex in a directed acyclic graph having no descendants.

A *bipartite graph* is a graph G whose vertex set can be partitioned into two subsets X and Y, such that each edge has one end in X and the other end in Y; such a partition (X, Y) is called *bipartition* of the graph. A *complete bipartite graph* is a bipartite graph with bipartition (X, Y) in which each vertex of X is adjacent to every vertex of Y. A graph in which $|X| = m$ and $|Y| = n$ is denoted by $K_{m,n}$. A graph is bipartite if and only if it does not contain an odd cycle.

We define the *coloring* of a graph G to be an assignment of colors (elements of some set) to the vertices of G, one color to each vertex, such that adjacent vertices are assigned distinct colors. The *chromatic number* of a graph G is defined as the minimum number of colors required to color the graph. A graph or a subgraph is called *k-colorable* if it can be colored by k colors.

A graph is called *planar* if it can be drawn in the plane so that no two edges cross. For example, K_4 is a planar graph, while K_5 is nonplanar. Notice that there are many different ways of "drawing" a planar graph. A drawing may be obtained by mapping a vertex to a point in the plane and mapping edges to paths in the plane. Each such drawing is called an *embedding* of G.

An interesting class of graphs based on the notion of cycle lengths are the *triangulated graphs*. If $C = v_0, e_0, v_1, e_1, \ldots, v_k, e_k, v_0$ is a cycle in G, a *chord* of C is an edge e in $E(G)$ connecting vertices v_i and v_j such that $e \neq e_i$ for any $i = 1, \ldots, k$. A graph is *chordal* if every cycle containing at least four vertices has a chord. Chordal graphs are also known as *triangulated graphs*. A graph $G = (V, E)$ is a *comparability graph* if it is *transitively orientable*. A graph is called a *cocomparability graph* if the complement of G is transitively orientable.

Triangulated comparability and cocomparability graphs actually form different subclasses of perfect graphs. The graph shown in Figure 2.1(a) is triangulated. The six-cycle graph shown in Figure 2.1(b) is a comparability graph. Figure 2.1(c) shows a cocomparability graph, and a transitively oriented graph is shown in Figure 2.1(d). A graph $G = (V, E)$ is called perfect if the size of the maximum clique in G is equal to the chromatic number of G, and this is true for all subgraphs H of G.

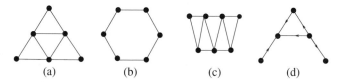

(a)   (b)   (c)   (d)

Fig. 2.1 (a) Triangulated graph, (b) comparability graph, (c) cocomparability graph, (d) transitively oriented graph.

2.2 General Algorithmic Approaches

Some of the techniques that are used to develop routing algorithms are the greedy technique, divide and conquer technique, dynamic programming, and integer programming. These techniques and algorithms may be found in a good computer science or graph algorithms text, as a result, we omit the discussion of these techniques and refer the reader to the excellent text by Cormen, Leiserson, and Rivest [CLR90].

All the techniques mentioned above and their hybrids have been used for various OTC routing problems. Since the number of nets in a routing may be quite large, the algorithm must have low time and space complexities to be practical. As a result, for large channels, even $O(n^3)$ algorithms may be intolerable. One must also be concerned about low-complexity algorithms, which have rather high constants, since these algorithms lead to excessive run times in practical applications.

We also list general algorithms, which are used in routing problems.

1. Graph search algorithms:
   Depth-first search
   Breadth-first search.

2. Minimum cost spanning tree.

3. Shortest path algorithm.

4. All pairs shortest path algorithm.

5. Matching algorithms.

6. Network flow algorithms.

These algorithms, along with illustrative examples, may be found in [CLR90].

2.3 NP-Hard Problems

A large number of problems in computer science do not have known polynomial time algorithms. This is also true for physical design problems in general, and routing problems in particular. In the last two decades, significant research has led many researchers to believe that no polynomial time algorithm may exist for many of these problems. Researchers classify solvable problems into two classes: P and NP. The problems that form class P can be solved by a deterministic Turing machine in polynomial time. A deterministic Turing machine is essentially equivalent to a conventional computer. Many well-known problems in computer science, such as minimum cost spanning tree, single-source shortest path, and graph matching, all belong to class P. On the other hand, the NP class consists of problems that can be solved in polynomial time by a nondeterministic Turing machine. A nondeterministic Turing machine can be thought of as a parallel computer, which starts up as many new processors as needed whenever a decision has several different outcomes. Each new processor pursues the solution for one of these outcomes. From a practical viewpoint, a nondeterministic Turing machine is not a realistic model for a computer. It is quite clear that class NP contains class P. In other words, problems in class NP are harder than problems in class P. The obvious question is: Can there ever exist a polynomial time complexity algorithm for a problem in class NP? This question has been a topic of intense research for the last 20 years and has led to the concept of NP-completeness. We will not present the concept of NP-completeness in detail. Instead, we give an informal definition and refer the reader to the most comprehensive text on this subject by Garey and Johnson [GJ79]. If every problem in class NP can be reduced to a problem $\mathcal{P}$, then problem $\mathcal{P}$ is in class NP-complete. Several thousand problems in computer science, graph theory, combinatorics, operations research, computational geometry, and VLSI design have been proven to be NP-complete. (For a list of problems, refer to [GJ79].) If a problem is known to be NP-complete, then it is unlikely that a polynomial time algorithm exists for that problem. A problem may be stated in two different versions. For example, we may ask if there exists a subgraph H of a graph G, which has a specific property and has size k or larger. Or we may simply ask for the largest subgraph of G having a specific property. The former type is the *decision* version, while the latter type is called the *optimization* version of the problem. The optimization version of problem $\mathcal{P}$ is called NP-hard if the decision version of problem $\mathcal{P}$ is NP-complete.

Routing problems encountered in VLSI are computationally difficult; in fact, most of the optimization problems related to the routing of VLSI chips are NP-hard. At the same time, routing problems must be solved efficiently in order to complete VLSI layouts. Hence, the routing problems are solved by compromising on the quality of the solution. In other words, we need to develop algorithms that, although they may not produce optimal solutions, produce a solution that may be close to optimal.

2.4 Algorithmic Approaches to NP-Hard Problems

If a problem is known to be NP-complete or NP-hard, algorithm designers are left with the following four choices.

1. Develop an exponential algorithm based on an exhaustive search of the complete search space to find the solution.

2. Develop heuristic algorithms.

3. Develop optimal algorithms for special cases of the problem.

4. Use approximation techniques.

In this book, examples of all these types of algorithms will be presented. Here we give a brief introduction of these methods.

In practice, it may not be necessary to solve the problem in its full generality. In other words, it may be possible to simplify the problem by applying some restrictions. Let us consider three examples.

- **Development of Different Design Styles in the VLSI Physical Design:**
  The basic idea is to restrict the full custom to standard cell or gate array design in hopes of simplified placement and routing problems. Consider the

placement problem in standard-cell and full-custom design styles. Conceptually, it is much easier to place cells of equal heights in rows, rather than placing arbitrarily sized rectangles in a plane.

- **An Optimal Solution of the Clock Routing Problem:** The clock routing problem is rather hard for full-custom designs, but can be solved in $O(n)$ time for symmetric structures such as gate arrays.

- **Determination of the Maximum Set of Independent Vertices in a Graph:** This problem is known to be NP-complete for general graphs. However, it is solvable in polynomial time for many special classes of graphs, such as circle graphs and permutation graphs, which are important in OTC routing. Thus, by restricting the problem of finding the maximum set of independent vertices to special classes of graphs, the problem can be solved.

We can afford to solve a problem using an exponential time complexity algorithm only if the input size is small. In many routing problems, some connections may be critical to chip performance and therefore it is necessary to spend extra resources to solve that problem optimally. Another reason for using exponential algorithms is the application of hierarchical, dynamic, or divide and conquer approaches to problem solving. All these approaches reduce the problem size and allow us to use exponential algorithms in some cases. Integer programming is a frequently used method, which in the worst case uses exponential time complexity. In practice, however, it has been observed that routing algorithms based on integer programs work very efficiently on moderate-sized problems. With the advent of very high performance workstations, it is now feasible to solve many restricted problems using exponential algorithms.

As pointed out above, in many cases the sheer size of the problem may exclude the possibility of using exponential algorithms and the problem may be computationally hard, even after applying restrictions. This situation is a real challenge to algorithm designers, and two options typically used in this situation are discussed below.

2.4.1 Algorithms Based on Heuristics

A heuristic algorithm produces a correct solution but does not guarantee the optimality of the solution. Suppose the optimal number of wires that can be routed in a certain problem is ten. A given heuristic algorithm may route one wire at a time. It will create a solution in which the wires will be routed without errors. However, we cannot say exactly how many wires may be routed, since that depends on the order in which the wires are selected for routing. It is unlikely that an optimal number of wires will be routed by such an algorithm.

The majority of problems in VLSI physical design tend to be NP-hard by nature, and hence significant research is directed toward the development of heuristic algorithms. A heuristic algorithm is considered to be good if it generates near-optimal solutions for a majority of practical instances of the problem. The average run times and memory requirements of these algorithms must be low. In an effort to minimize space and time complexities, in some cases fast heuristic algorithms that generate close to optimal results are developed for optimally solvable problems with high time complexities.

In the absence of a guarantee of optimality, one must test a heuristic algorithm on various benchmark examples to verify its effectiveness. A set of benchmarks has been established for most major open problems, so that algorithm designers can compare their results. For example, the *Deutsch Difficult* example is a benchmark for channel routing algorithms. The benchmark-driven development of heuristic algorithms has its drawbacks. First of all, it is difficult to capture all aspects of a problem with a limited set of benchmarks. Usually, as soon as benchmarks are widely available, researchers knowingly (or unknowingly) start developing algorithms tailored for the benchmarks. This drawback can be avoided by using a large set of benchmarks. That, however, makes comparison between algorithms more difficult. Of course, the problem of who will develop the benchmarks still remains to be tackled for many problems, and hence the availability and wide acceptance of benchmarks.

Once a heuristic algorithm is developed that performs well on all benchmarks, an attempt is made to determine the worst case performance of the algorithm.

2.4.2 Algorithms with Performance Bounds

An algorithm is called an *approximation algorithm* if it produces a correct solution to the given problem and guarantees that the solution is within a certain percentage of the optimal solution. That is, while an approximation algorithm may not produce an optimal result, it guarantees that the results will never be worse than a lower bound determined by the *performance ratio* of the algorithm. The performance ratio γ of an approximation algorithm is defined as

$$\gamma = \frac{\text{The solution produced by the algorithm}}{\text{The optimal solution for the problem}}$$

For example, let us assume that the size of the clique in a certain graph is 20. If we use a 0.75 approximation algorithm to find a maximum clique in this graph, we are guaranteed to find a clique that has at least 15 vertices.

In practice, the approximation algorithm may perform much better than the lower bound, since the worst case on which the lower bound is based may not occur too often.

As the size of VLSI circuits grows, the approximation algorithm becomes critical to the circuit performance and algorithm efficiency. In this book, we will stress the development of approximation algorithms wherever possible. Later in this chapter, we will present several approximation algorithms, including an approximation algorithm for finding the maximum k-partite ($k \geq 2$) subgraph and a generalized maximum independent set in a circle graph.

2.5 Routing-Related Graph Problems

Graphs are widely used to model various routing problems. The routing problems involve complex interaction between wire segments, regions in which the segments are routed and layers to which regions are assigned. Once the problem has been properly modeled, one needs to find a maximum set of wires that may be routed in a layer or that are intersecting. In this section, we define various classes of graphs and outline the relationship between these classes of graphs. We also list various graphs that are important for routing problems in general and OTC in particular.

2.5.1 Intersection Graphs

The graphs related to VLSI routing correspond to intersections of polygons and line segments. Two terminal nets are very naturally represented as wires, line segments, or strings. Multiterminal nets may be represented by a polygon. The shape of the routing region and restrictions on the routing give rise to many classes of graphs. An intersection graph $G = (V, E)$ is defined by the intersection of objects, where each object is represented by a vertex and an edge is defined if the two objects intersect.

The simplest class is defined by the intersection of horizontal line segments (interval graphs), and the most complex are defined by the intersection of topological wires (strings) in plane (string graphs). The various classes of graphs are shown in Table 2.1 and their relationships are shown in Figure 2.2 [JK91].

In OTC routing, circle, interval, and permutation graphs find extensive applications. We will define all these graphs in detail in the subsequent sections.

Table 2.1: Graphs Related to VLSI

Graph Type	Intersections of
Interval	Segments of a line.
Permutation	Straight line segments with end points on two parallel lines.
Circular arc	Segments of a circle.
Cocomparability	Curves between two parallel lines lying entirely in between these lines.
Chordal	Subtrees in trees (also called triangulated graphs).
Circle	Chords of a circle (also called overlap graphs).
Polycircle	Polygons inscribed in a circle.
Segment	Straight line segments in a plane.
Plano convex	Convex sets in a plane.
String	Arc-connected sets in a plane.
2-direction	Straight line segments with segments being parallel to at most k directions.

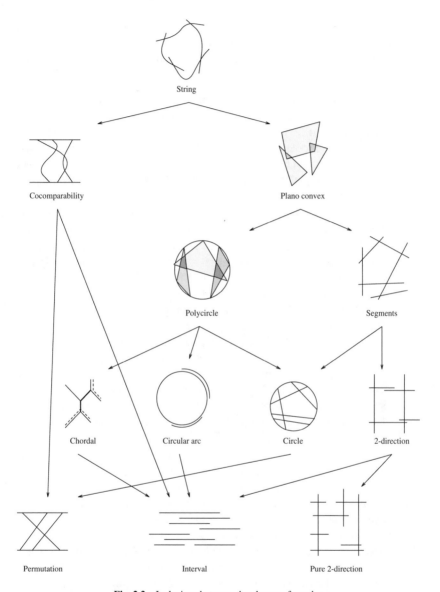

Fig. 2.2   Inclusions between the classes of graphs.

2.5.2 Graph Problems

Given a graph representing a routing problem, we are generally interested
in routing a set of wires in a plane (layer) or a set of layers, and we may ask to
reduce the total number of intersections of certain kinds. In this section, we will

briefly state the definitions of the basic OTC-related graph problems, and in the subsequent sections, we will present efficient algorithms to solve these problems for interval, circle, and permutation graphs. An extensive list of problems and related results may be found in [GJ79].

1. **Maximum Independent Set Problem.** Given a graph $G = (V, E)$, find the independent set in G that has the largest cardinality among all independent sets in G. This set is called *maximum independent set* (MIS) for G. The MIS problem is NP-hard for general graphs and remains NP-hard for many special classes of graphs [GJ79, GJS76, MS77, Pol74]. When weights are associated with each vertex, the problem is called *weighted maximum independent set* (WMIS) for G.

2. **Maximum k-Independent Set.** An interesting variant of the MIS problem, called *maximum k-independent set* (MKIS), arises in various routing problems. The objective of MKIS is to select a subset of vertices with the largest cardinality that can be partitioned into k-independent sets. That is, the selected subset is k-colorable. In VLSI applications of this problem, there are two cases to be considered:
   (a) Determines k-MIS for an unweighted graph.
   (b) Determines k-MIS for a weighted graph, called *weighted maximum k-independent set* (WMKIS), where a weight is assigned to each vertex. The objective of WMKIS is then to select a k-independent set with the largest total weight.

   This problem is NP-hard for general graphs and remains NP-hard for many special classes of graphs, such as planar graphs and circle graphs [SL89]. However, the problem is solvable in polynomial time for chordal graphs (and hence interval graphs [YG87]) and comparability and cocomparability graphs (and hence permutation graphs [SL93]).

3. **Maximum Clique Problem.** Given a graph $G = (V, E)$, find the clique in G that has the largest cardinality among all cliques in G. This clique is called the *maximum clique*. The problem is NP-hard for general graphs and for many special classes of graphs. However, this problem is solvable in polynomial time for circle graphs [Gav73], interval graphs, and permutation graphs [Gol80]. A variant of the maximum clique problem, called *maximum k-clique*, is to select a subset of vertices with the largest cardinality that can be partitioned into k vertex-disjoint cliques.

4. **Graph K-Colorability Problem.** Given a graph $G = (V, E)$, find the smallest integer K such that G is K-colorable. The minimum number of colors needed to color a graph is called the *chromatic number* of the graph. This problem is NP-complete for general graphs and remains so for all fixed $K \geq 3$. It is polynomial for $K = 2$, since that is equivalent to bipartite graph recognition. It also remains NP-complete for $K = 3$ if G is the intersection graph for straight line segments in the plane [EET89]. For arbitrary

K, the problem is NP-complete for circle graphs. The general problem can be solved in polynomial time for comparability graphs [EPL72], and for chordal graphs [Gav72].

As discussed earlier, many problems in OTC routing can be transformed into the problems discussed above. Most commonly, these problems serve as subproblems. Therefore, it is important to understand how these problems are solved. We will review the algorithms for solving these problems for three classes of graphs, namely, interval, permutation, and circle graphs. These graphs will be defined in subsequent sections. Table 2.2 states the status of various graph problems. (The table is reproduced from [Joh85].) In the table, we use the following abbreviations:

MEM   Member
MCQ   Maximum clique
CHR   Chromatic number
HAM   Hamiltonian circuit
MBS   Maximum bipartite set
MIS   Maximum independent set
DOM   Dominating set
P   Polynomial time solvable
N   NP-complete

In Table 2.2, T refers to a restriction that trivializes the problem. By trivialize, it is meant that it (a) forces only one possible answer, (b) renders the problem solvable by a simple greedy algorithm or a connectivity test, (c) renders the problem solvable by an exhaustive search over a polynomially bounded set of alternatives. O indicates open and may well be hard. O! indicates that it is a famous open problem. O? indicates that it is apparently open, but possibly easy to resolve.

Table 2.2: Complexity Results for Some Famous Graph Problems

GRAPH	MEM	MIS	MCQ	CHR	HAM	DOM
Trees/forests	P (T)	P [GJ79]	P (T)	P (T)	P (T)	P [GJ79]
Planar	P [GJ79]	N [GJ79]	P (T)	N [GJ79]	N [GJ79]	N [GJ79]
Perfect	O!	P [LS81]	P [LS81]	P [LS81]	N [NS80]	N [Boo80]
Chordal	P [TY84]	P [Gol80]	P [Gol80]	P [Gol80]	N [CS85]	N [Boo80]
Split	P [Gol80]	P [Gol80]	P [Gol80]	P [Gol80]	N [CS85]	N [CN85]
Comparability	P [Gol80]	P [Gol80]	P [Gol80]	P [Gol80]	N [NS80]	N [Dew83]
Bipartite	P (T)	P [GJ79]	P (T)	P (T)	N [NS80]	N [Dew83]
Permutation	P [Gol80]	P [Gol80]	P [Gol80]	P [Gol80]	O	P [FKar]
Interval	P [BL76]	P [GLL82]	P [GLL82]	P [GLL82]	P [Kei85]	O?
Circular arc	P [Tuc80]	P [GLL82]	P [Hsu85]	N [GJMP80]	P [Ber83]	O?
Circle	P [GHS86]	P [GJ79]	P [Hsu85]	N [GJMP80]	P [Ber83]	O?

Notice that most of the problems have polynomial time complexity algorithms for comparability, cocomparability, and chordal graphs. This is due to the fact that these graphs are *perfect graphs* [Gol80]. Perfect graphs admit polynomial time complexity algorithms for maximum clique, MIS, and several other problems. Note that chromatic number and maximum clique are equivalent problems in perfect graphs. (See bibliographic notes at the end of this chapter.)

Interval graphs and permutation graphs are defined by the intersection of different classes of perfect graphs, and therefore are themselves perfect graphs. As a result, many problems that are NP-hard for general graphs are solvable in polynomial time for these graphs. On the other hand, circle graphs are not perfect and generally speaking are much harder to deal with as compared to interval or permutation graphs.

2.6 Interval Graphs

Several special graphs are used in OTC routing, which are related to the representation of electrical connections as line segments. We can represent line segments as intervals of a real line. The relationship between lines leads to an interesting class of graphs, called *interval graphs*. Among all classes of graphs defined on a set of lines, interval graphs are perhaps the best known and have been studied extensively [Gol80].

An interval I_i is represented by its left and right end points, denoted by l_i and r_i, respectively. Given a set of intervals $\mathcal{I} = \{I_1, I_2, \ldots, I_n\}$, we define an interval graph $G_I = (V, E_I)$, where $V = \{v_i \mid v_i \text{ represents interval } I_i\}$ and two vertices are joined by an edge if and only if their corresponding intervals have a nonempty intersection. An example of an interval graph is shown in Figure 2.3.

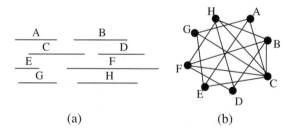

Fig. 2.3   **(a)** Set of intervals; **(b)** an equivalent interval graph.

Because interval graphs are tightly structured, many problems that are NP-hard for general graphs are polynomial for interval graphs [Gol77]. Interval graphs are perfect graphs and two famous classes of perfect graphs, namely, chordal and cocomparability graphs, can be used to characterize interval graphs.

Theorem 1   *A graph G is an interval graph if and only if G is a chordal graph and the complement of G is a comparability graph.*

Figure 2.4 shows the relationship of interval graphs to chordal and cocomparability graphs. The figure shows an example of a chordal graph, which is not cocomparability, and a cocomparability graph, which is not chordal.

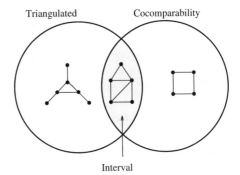

Fig. 2.4 Relationship of interval graphs to other classes of perfect graphs.

Several problems, including maximum clique, MIS, and graph recognition, can be solved using linear time complexity algorithms [Gol80].

The maximal cliques of an interval graph can be linearly ordered so that for every vertex $v \in V$, the cliques containing v occur consecutively [GH64]. Such an ordering of maximal cliques is called a *consecutive linear ordering* and it can be determined in $(|V| + |E|)$ for interval graphs [BL76].

Interval graphs arise in many routing problems, including channel routing, single-row routing and OTC routing. One is generally interested in how many intervals can be assigned to routing tracks, how many intervals can be routed in a given region, or finding the size of the region to accommodate all the intervals. These problems translate to find, MIS, MKIS, or maximum clique, and algorithms for these problems are present in the following.

2.6.1 Maximum Independent Set

Gupta, Lee, and Leung [GLL82] presented an optimal algorithm for determining the MIS in an interval graph. The algorithm first sorts the $2n$ end points in ascending order of their values. The list is then scanned from left to right until the first end point is encountered. The vertex corresponding to this interval is placed into the MIS being constructed, and the intervals containing this point are deleted. This process is repeated until no intervals remain. The time complexity of the algorithm is dominated by interval sorting. If n is the total number of intervals, the sorting can be accomplished in $O(n \log n)$.

Theorem 2   *Given an interval graph, the MIS can be found in $O(n \log n)$ time, where n is the total number of vertices in the graph.*

The weighted version of the MIS problem (WMIS) requires selection of the maximum weight independent set from a collection of intervals, where each interval is assigned a weight. In [HTC92], a linear time algorithm for WMIS is proposed.

2.6.2 Maximum k-Independent Set

In [YG87], an optimal algorithm for finding the MKIS in an interval graph was presented. The algorithm is greedy in nature and the basic idea is as follows. The input is a set of intervals $\mathcal{I} = \{I_1, I_2, \ldots, I_n\}$ sorted in increasing order of their right end points. For each interval, the vertex corresponding to the interval is added to the growing MKIS set if and only if the size of the maximum clique in the subgraph induced by the MKIS set and this vertex is less than or equal to k; otherwise the vertex is ignored. This algorithm is presented in Figure 2.5.

Algorithm MKIS_INTERVAL($\mathcal{I}, k$)

Input: A set of intervals $\mathcal{I}, k$
Output: MIS

---

    begin
        SORT-INTERVAL($\mathcal{I}$);
        $M = \phi$;
        for $i = 1$ to n **do**
            if CLIQUE($G(M \cup I_i)) \leq k$ **then** $M = M \cup I_i$.
        return M;
    end.

Fig. 2.5   Algorithm MKIS_INTERVAL.

The subroutine CLIQUE refers to a function that returns the size of the maximum clique in the induced subgraph $G(V')$, where V' is a set of vertices that form the subgraph. The example shown in Figure 2.6 illustrates the algorithm for $k = 2$. The first step is to sort the intervals in increasing order of their right end points, and the sorted list is $\{1, 3, 2, 4, 5, 6\}$. Initially the MKIS $M = \phi$. The first two numbers 1 and 3 in the sorted list are included in M. Since the third number 2 overlaps both 1 and 3, inclusion of 2 in M would result in a clique of size 3, which is greater than k. Hence, interval 2 is not added to the set M. Since the next interval 4 overlaps 1 and 3, it is not included in the set M. The process is continued for all the intervals, and an interval is included in the list if and only if it does not lead to an increase in the size of the maximum clique. The MKIS for the example is $\{1, 5, 3, 6\}$.

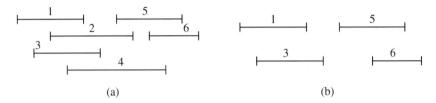

Fig. 2.6   **(a)** Set of intervals; **(b)** MKIS for interval graph in (a).

It is easy to show that this greedy algorithm finds the optimal MKIS in an interval graph. The details of the proof can be found in [YG87].

Theorem 3   *The MKIS of an interval graph $G = (V, E)$ can be found in $O(n^2)$ time.*

Weighted maximum k-independent set (WMKIS) can be found in polynomial time, since the problem can be written as an integer linear programming problem on the node-clique matrix, which is totally unimodular [YG87]. Using the best known polynomial time algorithm for solving a unimodular integer linear programming problem, this problem can be solved in $O(n^{3.5})$ [Kar84]. However, for $k = 2$ (resp. 3), the MW2IS (resp. MW3IS) can be solved in $O(n^2)$ (resp. $O(n^3)$) time by using the algorithm presented in [HTC92]. Since the complement of interval graphs are comparability graphs, $O(kn^2)$ in [SL93] can be used to find MWKIS in interval graphs.

2.6.3 Maximum Clique and Minimum Coloring

Since interval graphs are perfect, the minimum coloring is equivalent to the maximum clique problem in interval graphs. The algorithm shown in Figure 2.7 finds a maximum clique in a given interval graph. The input to the algorithm is a set of intervals $\mathcal{I} = \{I_1, I_2, \ldots, I_n\}$ representing an interval graph. Each interval I_i is represented by its left end point l_i and right end point r_i. Since there are n intervals, there are $2n$ end points. The algorithm shown above runs in $O(n \log n)$ time. This algorithm can easily be extended to find the maximum clique itself. A well-known channel routing algorithm called *left-edge algorithm* (discussed in Chapter 3) is based on this algorithm.

Algorithm MAX-CLIQUE_INTERVAL($\mathcal{I}$)

Input: A set of Intervals $\mathcal{I}$
Output: Size of maximum clique

```
begin
      Sort list of end points
      for i = 1 to n do
          A[l_i] = L;
          A[r_i] = R;
      cliq = 0;
      max_cliq = 0;
      for i = 1 to 2n do
          if A[i] = L then cliq = cliq + 1;
              if cliq > max_cliq then max_cliq = cliq;
          else cliq = cliq − 1;
      return max_cliq;
end.
```

Fig. 2.7   Algorithm MAX-CLIQUE_INTERVAL.

2.7 Permutation Graphs

Pnnueli, Lempel, and Even [PLE71] introduced the class of permutation graphs. Permutation graphs can be regarded as a class of intersection graphs by considering the *matching diagrams*. A matching diagram can be constructed by first labeling the terminals with numbers $1, 2, \ldots, n$ left to right and then labeling the bottom-row terminals with a given permutation $\pi_1, \pi_2, \ldots, \pi_n$ in sequence from left to right. Now connect the top-row terminal i to the bottom-row terminal with $\pi_j = i$. The segments i and j intersect if and only if i and j appear in reversed order in π (e.g., $\pi_i > \pi_j$ when $i < j$).

Permutation graphs are frequently used in routing and can be defined by a matching diagram. We define a *permutation graph* $G_P = (V, E_P)$, where the vertex set V is the same as defined above and E_P is a set of edges defined below:

$$E_P = \{(v_i, v_j) \mid \text{if line } i \text{ intersects line } j\}$$

An example of a permutation graph for the matching diagram in Figure 2.8(a) is shown in Figure 2.8(b). It is well known that the class of containment graphs is equivalent to the class of permutation graphs [Gol80].

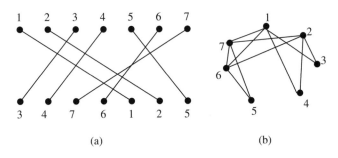

Fig. 2.8   (a) Matching diagram; (b) permutation graph.

Since the complement of a permutation graph can be derived from the sequence $\bar{\pi}$, which is the reverse of the sequence π, the complement of a permutation graph is also a permutation graph. The complement of the graph G is denoted by $\bar{G}$.

Lemma 1   *If G is a permutation graph, then $\bar{G}$ is a permutation graph.*

A clique in a permutation graph G is an independent set in $\bar{G}$, which is also a permutation graph, and vice versa. Therefore, the independent set problem and the clique problem are equivalent in permutation graphs. A permutation graph is also transitively orientable. In fact, comparability and cocomparability graphs can be used to characterize permutation graphs. The following theorem was derived in [PLE71].

Theorem 4   *A graph G is a permutation graph if and only if both G and $\bar{G}$ are comparability graphs.*

Figure 2.9 shows the relationship of permutation graphs to other classes of perfect graphs. The class of permutation graphs is also a structured class of graphs similar to interval graphs. Most problems that are polynomial for interval graphs are also polynomial for permutation graphs. Permutation graphs are widely used in planar channel routing algorithms. In the following sections, we present the outline of several important algorithms related to permutation graphs. Note that the MIS problem and maximum clique problem are equivalent in permutation graphs.

2.7.1 Maximum Clique and Minimum Coloring

The permutation graphs are perfect graphs. Thus, the minimum coloring in a permutation graph is equal to the size of the maximum clique in the graph. Let π be the permutation of integers $1, 2, \ldots, n$ corresponding to a permutation graph G. Then a decreasing subsequence in π corresponds to a clique in G. It can be easily seen that the following lemma holds.

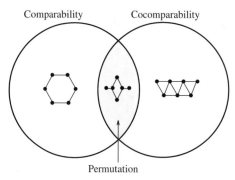

Comparability   Cocomparability

Fig. 2.9 Relationship of permutation graphs
to other classes of perfect graphs.

Permutation

Lemma 2   *The maximum decreasing subsequence in the permutation π corresponds to the maximum clique in the permutation graph G.*

Then the maximum clique problem is equivalent to finding the maximum decreasing subsequence in π. The maximum decreasing subsequence in π can be found by using a set of queues. Assume that the number of queues is unlimited and the queues are ordered. Initially the queues are empty. Each time a number i is picked up from π starting at the first number. The number i is compared to the numbers in the queues, and it will enter the first queue in which it will be less than all the numbers already in the queue. If there is no queue in which the number i is less than all the numbers in the queue, then the number i enters the first empty queue. The queue with the largest size contains the maximum decreasing subsequence in π, thus the maximum clique. The above algorithm appeared in [Gol80]. The time complexity of the algorithm is $O(n \log n)$.

Let us determine the maximum clique for a matching diagram shown in Figure 2.10(a). The permutation for the matching diagram is $\pi = \{3, 4, 7, 6, 1, 2, 5\}$. The first element, 3, is picked up and placed in the first queue. (See Figure 2.10(b).) The second number, 4, is picked up and placed in the second queue, since it is greater than a number in the first queue. Then 7 is placed in the third queue. After that, 6 is placed in the third queue, since it is less than 7. This process is repeated until all the numbers in the permutation have been placed in a queue. The maximum clique is then $\{7, 6, 5\}$, as shown in Figure 2.10(b).

2.7.2 Maximum *k*-Independent Set

The complement of a permutation graph is a permutation graph. Hence, the MKIS problem in a permutation graph G is equivalent to the maximum k-clique problem in $\bar{G}$. In this section, we discuss an $O(kn^2)$ time algorithm for finding the maximum k-clique in a permutation graph presented by Gavril [Gav87]. In fact, this algorithm is very general and applicable to any comparability graph.

The basic idea of the algorithm is to convert the maximum k-clique problem in a comparability graph into a network flow problem. (See [Tar83] for an excellent

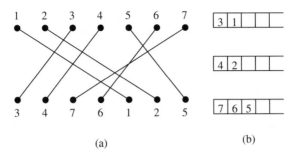

Fig. 2.10   **(a)** Matching diagram; **(b)** maximum clique.

survey of network flow algorithms.) First, a transitive orientation is constructed for a comparability graph $G = (V, E)$, resulting in a directed graph $\vec{G} = (V, \vec{E})$. A directed path in $\vec{G}$ is also called a *chain*. Note that each chain in $\vec{G}$ corresponds to a clique in G, since G is a comparability graph. Next, each vertex in V is split into two vertices. Assume that $V = \{v_1, v_2, \ldots, v_n\}$. Then each vertex v_i corresponds to two vertices x_i and y_i in a new directed graph $\vec{G}_1 = (V_1, \vec{E}_1)$. There is a directed edge between x_i and y_i for all $1 \leq i \leq n$. A cost of -1 and capacity of 1 are assigned to the edge (x_i, y_i) for all $1 \leq i \leq n$. In addition, there is a directed edge between y_i and x_j if there exists a directed edge (v_i, v_j) in $\vec{G}$. A cost of 0 and capacity of 1 are assigned to the edge (y_i, x_j). Four new vertices s (source), t (sink), s', and t' are introduced, and the directed edges (s', x_i) and (y_i, t') for all $i \in V$, (s, s'), and (t', t) are added. A cost of 0 and capacity of 1 are assigned to the edges (s', v_i) and (v_i, t'). A cost of 0 and capacity of k are assigned to the edges (s, s') and (t', t). The graph $\vec{G}_1 = (V_1, \vec{E}_1)$ so constructed is called a *network*, where $V_1 = \{x_i, y_i | 1 \leq i \leq n\} \cup \{s, s', t', t\}$ and $\vec{E}_1 = \{(x_i, y_i) | 1 \leq i \leq n\} \cup \{(y_i, x_j) | (v_i, v_j) \in \vec{E}\} \cup \{(s', x_i), (y_i, t') | 1 \leq i \leq n\} \cup \{(s, s'), (t', t)\}$.

Then the maximum k-clique problem in graph G is equivalent to the minimum-cost maximum-flow (min-cost max-flow) problem in network $\vec{G}_1$. The flow in a directed graph has to satisfy the following.

1. The flow $f(e)$ associated with each edge of the graph can be assigned a value no more than the capacity of the edge.

2. The net flow that enters a vertex is equal to the net flow that leaves the vertex.

The absolute value of flow that leaves the source (e.g., $|f(s, s')|$) is called the *flow* of $\vec{G}_1$. The min-cost max-flow problem in the directed graph $\vec{G}_1$ is to find the assignment of $f(e)$ for each edge $e \in \vec{E}_1$ such that the flow of $\vec{G}_1$ is maximum and the total cost on the edges that the flows pass is minimum. Notice that the capacity on the directed edge (s, s') is k. Thus, the maximum flow of $\vec{G}_1$ is k. In addition, flow that passes x_i or y_i has value 1, since the capacity on the directed edge (x_i, y_i) is 1 for each of $1 \leq i \leq n$. Note that a flow in $\vec{G}_1$ corresponds to a chain in $\vec{G}$. The maximum flow in $\vec{G}_1$ is k, thus the maximum number of the chains in $\vec{G}$ is k, and vice versa. The absolute value of the cost on each flow is equal to the number of vertices on the chain corresponding to the flow. Thus, the minimum cost on all flows results in a maximum number of vertices in the chains in $\vec{G}$, and hence a maximum number of vertices in the cliques in G.

An example of a permutation graph G is given in Figure 2.11(b). The transitive orientation of G is given in Figure 2.11(c), while the network $\vec{G}_1$ is shown in Figure 2.11(d). The min-cost max-flow while $k = 2$ is highlighted in Figure 2.11(d). The chains corresponding to the min-cost max-flow are highlighted in Figure 2.11(c). The maximum 2-clique is {5, 6, 7} and {2, 4}.

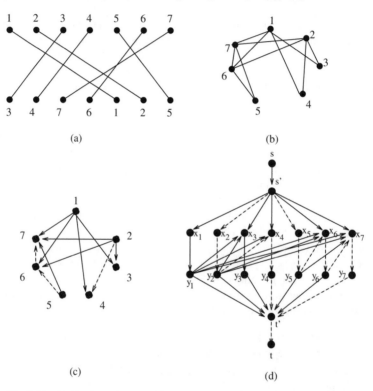

Fig. 2.11   (a) Matching diagram; (b) permutation graph; (c) transitive orientation of the graph; (d) network flow graph.

The time complexity of the algorithm is dominated by the time complexity of the algorithm for finding the min-cost max-flow in a network which is $O(kn^2)$, where n is the number of vertices in the graph [Law76]. Note that the complement of an interval graph is a comparability graph. Thus, the above algorithm can be used to determine the MKIS in an interval graph. The weighted version of the MKIS problem can be solved by the $O(kn^2)$ algorithm presented in [SL93].

Just like Gavril's algorithm presented above, the algorithm in [SL93] is based on network flow.

2.8 Circle Graphs

Circle graphs are defined as the intersection graph of chords of a circle.

Given a set of chords $C = \{c_1, c_2, \ldots, c_n\}$, we define a *circle graph* $G_C = (V, E_C)$ as

$$V = \{v_i \mid v_i \text{ represents chord } c_i\}$$
$$E_C = \{(v_i, v_j) \mid \text{ if } c_i \text{ and } c_j \text{ intersect}\}$$

Figure 2.12(b) shows a circle graph obtained from the set of intersecting chords shown in Figure 2.12(a). We refer to the circle with the set of chords as the circle diagram.

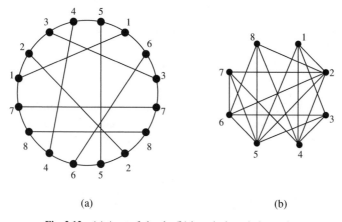

(a)   (b)

Fig. 2.12   **(a)** A set of chords; **(b)** its eqivalent circle graph.

Circle graphs are equivalent to overlap graphs. Let us also define an overlap graph.

Given a set of intervals $\mathcal{I} = \{I_1, I_2, \ldots, I_n\}$, we define an *overlap graph* $G_O = (V, E_O)$ as

$$V = \{v_i \mid v_i \text{ represents interval } I_i\}$$
$$E_O = \{(v_i, v_j) \mid l_i < l_j < r_i < r_j\}$$

In other words, each vertex in the graph corresponds to an interval and an edge is defined between v_i and v_j if and only if the interval I_i overlaps with I_j but does not completely contain or reside within I_j.

In order to see that circle graphs and overlap graphs are equivalent, note that one can create a circle representation from a set of intervals choosing a point p and draw lines from end points of each interval to p. When these drawn lines intersect the circle, they define the chord corresponding to an interval (see Figure 2.13).

An interval representation can be created from a circle by simply wrapping the intervals around the circle and then drawing the chord between two end points of each interval. As a consequence of this observation, we use the terms circle and overlap graphs interchangeably and note the difference when necessary.

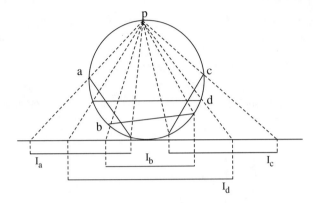

Fig. 2.13 Equivalence of circle and overlap graphs.

The class of circle graphs properly contains the class of permutation graphs. This follows directly from the concept of a matching diagram being a special case of a circle diagram. Figure 2.14 shows the relationship of circle graphs to other classes of perfect graphs, such as comparability graphs and cocomparability graphs.

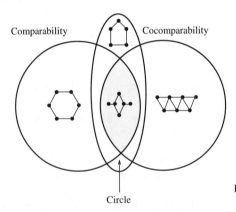

Fig. 2.14 Relationship of circle graphs to other classes of graphs.

Circle graphs are less structured than interval and permutation graphs. Several problems, maximum bipartite subgraph problems, which can be solved in polynomial time for interval and permutation graphs, are NP-complete for circle graphs. However, there are still many problems such as maximum clique and MIS that can be solved in polynomial time for circle graphs [Gav73], which are NP-complete for general graphs. But the chromatic number problem for circle graphs remains NP-complete [GJMP78]. However, like interval graphs and permutation graphs, circle graphs can also be recognized in polynomial time [GHS86]. In the following, we review the circle graph algorithms used in OTC routing.

2.8.1 Maximum Independent Set

The algorithms for MIS and maximum clique were first presented by F. Gavril, in 1973 [Gav73]. The algorithms had a time complexity of $O(n^3)$. An algorithm with $O(n^2)$ time complexity using dynamic programming techniques was presented in [Sup87].

Let C be a set of n chords of a circle. It is assumed that all the chords have unique end points. The $2n$ end points of n chords are numbered from 0 to $2n - 1$ clockwise around the circle. We could refer to a chord as a net. If $N_l \in C$, then $N_l = \{k, j\}$ and we refer to k and j as terminals of net N_l. Let G_{ij} denote the subgraph of the circle graph $G = (V, E)$, induced by the set of vertices

$$V = \{v_l \in V \mid N_l = \{a, b\} \ \& \ i \le a, b \le j\}$$

Let $M(i, j)$ denote an MIS of G_{ij}. If $i \ge j$, then G_{ij} is the empty graph, and hence $M(i, j) = \phi$. In particular, $M(i, j)$ is computed for each pair i, j; $M(i, j_1)$ is computed before $M(i, j_2)$ if $j_1 < j_2$. To compute $M(i, j)$, we have to consider two cases:

1. There exists a net $N_l = \{k, j\}$ and k is not in range $[i, j - 1]$, then $M(i, j) = M(i, j - 1)$.

2. There exists a net $N_l = \{k, j\}$ and k is in the range $[i, j - 1]$, then we have two cases:

   (a) If $N_l \in M(i, j)$, then by definition of an independent set, $M(i, j)$ contains no nets $N = \{a, b\}$ such that $a \in [i, k - 1]$ and $b \in [k + 1, j]$.

   Therefore, $M(i, j) = M(i, k - 1) \cup \{N_l\} \cup M(k + 1, j - 1)$

   (b) If $N_l \notin M(i, j)$, then $M(i, j) = M(i, j - 1)$.

Thus, $M(i, j)$ is set to the larger of the two sets $M(i, j - 1)$ and

$$M(i, k - 1) \cup \{N_l\} \cup M(k + 1, j - 1)$$

The algorithm is more formally stated in Figure 2.15.

Theorem 5   *The algorithm MIS_CIRCLE finds a maximum independent set in a circle graph in time $O(n^2)$.*

Proof: At each step of the algorithm, there are only two possible choices for $M(i, j)$. The MIS of G_{ij}, $M(i, j)$ may either consist of $M(i, k-1) \cup M(k+1, j-1) \cup N_l$, where $N_l = \{k, j\}$, as shown in Figure 2.16(a), or consist of $M(i, j - 1)$, as shown in Figure 2.16(b). The larger set of $M(i, k - 1) \cup M(k + 1, j - 1) \cup v_{kj}$

Algorithm MIS_CIRCLE($\mathcal{C}$)

Input: A set of chords $\mathcal{C}$
Output: MIS

---

begin
    for $j = 0$ to $2n - 1$ **do**
        for $i = 0$ to $j - 1$ **do**
            Let $N_l = \{k, j\}$ be a net.
            if $i \le k \le j\text{-}1$ and $|M(i, k - 1)| + 1 +$
                $|M(k + 1, j - 1)| > |M(i, j - 1)|$ **then**
                $M(i, j) = M(i, k - 1) \cup \{N_l\} \cup$
                    $M(k + 1, j - 1)$;
            else $M(i, j) = M(i, j - 1)$
end.

Fig. 2.15   Algorithm MIS_CIRCLE.

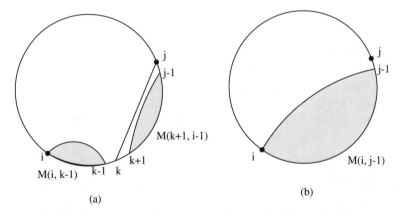

Fig. 2.16   (a) $M(i, j) = M(i, j - 1)$; (b) $M(i, j) = M(i, k - 1) + N_{kj} + M(k + 1, j)$.

and $M(i, j - 1)$ becomes $M(i, j)$. Since we need to find $M(i, j)$ between all pairs of end points on the circle, the complexity of the algorithm is $O(n^2)$.

The weighted version of the MIS problem can also be solved rather easily by considering a simple extension of the algorithm MIS_CIRCLE. Let $w(N_l)$ be the weight of the net N_l and let $W(i, j)$ denote a maximum weighted independent set of G_{ij}. In the MIS problem discussed above, replace $M(i, j)$ with $W(i, j)$, and in the three cases, replace N_l with $w(N_l)$ to solve the WMIS problem. The formal statement of the WMIS algorithm, WMIS_CIRCLE, can be stated in a similar fashion as that of MIS_CIRCLE.

Theorem 6   *The algorithm $WMIS\_CIRCLE$ finds a weighted maximum independent set in a circle graph in time $O(n^2)$.*

2.8.2 Maximum *k*-Independent Set

A set consisting of k disjoint sets is called a *k-independent set*, and a set that has the maximum number of vertices among all such k-independent sets is called a *maximum k-independent set* (MKIS). When $k = 2$, the MKIS problem is also called a *maximum bipartite subgraph* problem, and for general k, it is also called a *maximum k-colorable subgraph* problem. The problem of finding the MKIS is NP-complete even when $k = 2$ [SL89]. An approximation algorithm for finding the MKIS in the context of a planar routing problem in an arbitrary routing region, which is equivalent to finding the MKIS in circle graphs, was presented in [CHS93]. Holmes, Sherwani, and Sarrafzadeh presented approximation algorithms for $k = 2$ [HSS91b] and $k = 4$ [HSS91a]. We present an approximation algorithm for finding the MKIS in a circle graph.

The algorithm decomposes the MKIS problem into a series of MIS problems. First, determine the MIS S_1 for a given graph $G_1 = (V_1, E_1)$. Now delete S_1 from the set of vertices V_1 and find MIS S_2 from the remaining set of vertices. Repeat k times and the union of the vertices in all of the k maximum independent sets gives the MKIS. The algorithm is illustrated in Figure 2.17. Since the MIS in a circle graph can be determined in $O(n^2)$, the total time complexity to find the MKIS is given by $O(kn^2)$.

Algorithm MKIS_CIRCLE $(G(V, E), k)$

Input: A circle graph G, integer k
Output: MKIS of G

---

begin
    Let $V' = V$;
    for $i = 1$ to k **do**
        $S_i = \text{MIS}(V')$;
        $V' = V' - S_i$;
        return $S_1 \cup S_2 \cup \cdots \cup S_k$;
end.

Fig. 2.17   Algorithm MKIS_CIRCLE.

For any heuristic algorithm H_k for the MKIS, the *performance ratio* of H_k is defined to be Ψ_k / Ψ_k^*, where Ψ_k is the size of the k-independent set obtained by the algorithm H_k, and Ψ_k^* is the MKIS in the same graph. The lower bound on the performance ratio is established based on the following results:

Lemma 3   *Let Ψ_k^* be the size of the optimal solution of the maximum k-colorable subgraph. Let ψ_i be the size of the independent subset $S_i (1 \leq i \leq k)$ chosen by the algorithm MKIS_CIRCLE at the i-th iteration. Then we have*

$$\psi_1 \geq \frac{\Psi_k^*}{k}$$

$$\psi_2 \geq \frac{\Psi_k^* - \psi_1}{k}$$

$$\psi_3 \geq \frac{\Psi_k^* - (\psi_1 + \psi_2)}{k}$$

$$\vdots$$

$$\vdots$$

$$\psi_k \geq \frac{\Psi_k^* - (\psi_1 + \psi_2 + \cdots + \psi_{k-1})}{k}$$

Proof: Let $G = (V, E)$ be the given graph. Let $S^* = S_1^* \cup S_2^* \cup \ldots \cup S_k^*$ be the set of vertices of a maximum k-colorable subgraph of G, where each S_i^* is an independent set of vertices from V and $S_i^* \cap S_j^* = \phi$ for $1 \leq i, j \leq k$, and $i \neq j$. At the end of the $(i - 1)$-th iteration of the algorithm MKIS_CIRCLE, the set of remaining vertices in the graph is

$$V' = V - \bigcup_{j=1}^{i-1} S_j$$

Note that

$$|S_1^* \cap V'| + |S_2^* \cap V'| + \cdots + |S_k^* \cap V'| = |S^* \cap V'|$$

$$= |S^* \cap (V - (S_1 \cup S_2 \cup \cdots \cup S_{i-1}))|$$

$$\geq |S^* - (S^* \cap (S_1 \cup S_2 \cup \cdots \cup S_{i-1}))|$$

$$\geq |S^*| - |(S_1 \cup S_2 \cup \cdots \cup S_{i-1})|$$

$$= \Psi_k^* - (\psi_1 + \psi_2 + \cdots + \psi_{i-1})$$

By pigeonhole principle, there exist at least a $j (1 \leq j \leq k)$ such that

$$|S_j^* \cap V'| \geq \frac{\Psi_k^* - (\psi_1 + \psi_2 + \cdots + \psi_{i-1})}{k}$$

Since $S_j^* \cap V' \subseteq V'$ is an independent set of vertices in G, and $S_i \subseteq V'$ is the maximum independent subset of V' according to the algorithm MKIS_CIRCLE, we have

$$\psi_i = |S_i| \geq |S_j^* \cap V'| \geq \frac{\Psi_k^* - (\psi_1 + \psi_2 + \cdots + \psi_{i-1})}{k}$$

$\square$

Lemma 4   *Let $\Psi_k = \psi_1 + \psi_2 + \cdots + \psi_{i-1}$, where ψ_i is the size of the independent set S_i produced by the algorithm MKIS_CIRCLE. Then we have*

$$\Psi_k \geq [1 - (1 - \frac{1}{k})^k]\Psi_k^*$$

Proof: Let

$$x_1 = \frac{\Psi_k^*}{k}$$

$$x_2 = \frac{\Psi_k^* - x_1}{k}$$

$$x_3 = \frac{\Psi_k^* - (x_1 + x_2)}{k}$$

$$\vdots$$

$$\vdots$$

$$x_k = \frac{\Psi_k^* - (x_1 + x_2 + \cdots + x_{k-1})}{k}$$

It is then easy to show that

$$x_i = \frac{k-1}{k}x_{i-1} = \frac{(k-1)^2}{k^2}x_{i-2} = \cdots = \frac{(k-1)^{i-1}}{k^{i-1}}x_1 = \frac{(k-1)^{i-1}}{k^{i-1}}\frac{\Psi_k^*}{k}$$

Therefore, we have

$$\sum_{i=1}^{k} x_i = \sum_{i=1}^{k} \frac{(k-1)^{i-1}}{k^{i-1}}\frac{\Psi_k^*}{k} = [1 - (1 - \frac{1}{k})^k]\Psi_k^*$$

Now we show by induction that

$$\sum_{i=1}^{l} \psi_i \geq \sum_{i=1}^{l} x_i \qquad (2.1)$$

holds for every $1 \leq l \leq k$. When $l = 1$, the inequality (2.1) holds, since $\psi_1 = x_1$. Assume that inequality (2.1) holds for l. Then, for $l + 1$, we have

$$\sum_{i=1}^{l+1} \psi_i$$

$$\geq \sum_{i=1}^{l} \psi_i + \frac{\Psi_k^* - \sum_{i=1}^{l} \psi_i}{k}$$

$$= \frac{\Psi_k^* + (k-1)\sum_{i=1}^{l} \psi_i}{k}$$

$$\geq \frac{\Psi_k^* + (k-1)\sum_{i=1}^{l} x_i}{k}$$

$$= \sum_{i=1}^{l} x_i + \frac{\Psi_k^* - (k-1)\sum_{i=1}^{l} x_i}{k}$$

$$= \sum_{i=1}^{l+1} x_i$$

Therefore, we have

$$\Psi_k = \sum_{i=1}^{k} \psi_i \geq \sum_{i=1}^{k} x_i = [1 - (1 - \frac{1}{k})^k]\Psi_k^*$$

$\square$

From the above two lemmas, we can conclude the following:

Theorem 7   *Let ρ_k be the performance ratio of the algorithm MKIS_CIRCLE for a k-colorable subgraph problem. Then,*

$$\rho_k \geq 1 - (1 - \frac{1}{k})^k$$

It is easy to see that the function $f(x) = 1 - (1 - \frac{1}{x})^x$ is a decreasing function. Moreover,

$$\lim_{x \to \infty} [1 - (1 - \frac{1}{x})^x] = 1 - e^{-1}$$

When k is known, we use the formula in lemma 2 to obtain a more precise performance ratio for MKIS_CIRCLE. In particular, the performance ratio of MKIS_CIRCLE is at least 75% for the maximum 2-colorable subgraph problem and 70.4% for the maximum 3-colorable subgraph problem. Table 2.3 shows the performance ratio of the algorithm for the maximum k-colorable subgraph problem for some small values of k.

The approximation result presented above for circle graphs is equally applicable to any class of graphs where the problem of finding the MIS is polynomial

Table 2.3: Performance Ratio of MKIS_CIRCLE for Different Values of k

Value of k	Performance Ratio
2	75.0%
3	70.4%
4	68.4%
5	67.3%
∞	63.2%

time solvable. The weighted version of this problem can be solved by modifying S_i in the algorithm MKIS_CIRCLE. In the algorithm MKIS_CIRCLE, by setting the independent set $S_i = \text{WMIS}(V')$, where WMIS is the weighted maximum independent set, we obtain the WMKIS in a circle graph.

2.8.3 Maximum Clique and Minimum Coloring

In [Gav73], Gavril presented an algorithm for finding the maximum clique in circle graphs. The algorithm first finds induced subgraphs for each vertex. The induced subgraph $G_v = (V_v, E_v)$, where $V_v = \{v\} \cup \{\text{Adj}(v)\}$, and $E_v = \{(u, v) | u \in V_v\}$, of vertex $v \in V$ in a circle graph $G = (V, E)$ is known to be a permutation graph. For each induced subgraph G_v, the maximum clique C_v is determined using the algorithm discussed in Section 2.7.1. The maximum clique in the circle graph G, is given by the maximum among the cliques; that is, $\max\{C_v\}$ for all $v \in V$.

Figure 2.18 illustrates the induction of a permutation graph from a circle graph for the determination of maximum clique. Figure 2.18(a) shows a circle graph. The circle is bisected into two arcs at chord a. (See Figure 2.18(b).) The chords that are not intersecting with chord a are shown in dotted lines, and these do not contribute to the maximum clique in this subgraph. The two arcs are stretched until they form a matching diagram as shown in Figure 2.18(c). The chords shown in the dotted lines in Figure 2.18(b) are omitted in the matching diagram and the maximum clique for the permutation graph corresponding to the matching diagram is determined in $O(n \log n)$ time. This process is repeated for all the chords, and the maximum of all the cliques gives the maximum clique for the circle graph. Therefore, for n chords in a circle, the maximum clique can be determined in $O(n^2 \log n)$ time complexity.

The algorithm described above by Gavril originally required $O(n^3)$ time. Buckingham [Buc80] presented $O(n \log n + m \log \omega)$, where m is the number of edges in the graph and ω is the size of the maximum clique. Like Gavril's algorithm, Buckingham's algorithm also consists of solving n subproblems, each consisting of finding the maximum clique of a permutation graph. However, Buckingham solves each subproblem more efficiently by applying the algorithm for finding the

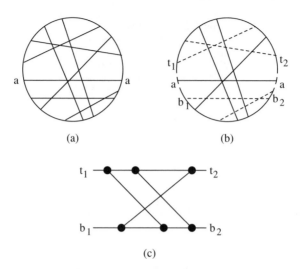

(a)   (b)

(c)

Fig. 2.18   (**a**) Circle graph; (**b**) circle is bisected into two arcs at chord a; (**c**) maximum
clique in the bisected circle graph.

longest increasing subsequence of a sequence. In [RU81], an $O(n^2)$ algorithm
is developed. In this paper, the problem is decomposed into $O(n)$ subproblems,
just as is the case with Gavril's and Buckingham's decomposition. However, the
decomposition scheme is different. When the number of edges is relatively small,
$O(n \log n + \min(m, n.\omega))$ given in [MNKF90] is the most efficient.

2.8.4 Maximum k-Density Independent Set in an Overlap Graph

Another variation of the MIS problem in overlap graphs is called k-*density*
MIS (k-DMIS). Given a set of intervals, the objective of k-DMIS is to find the
maximum weighted independent set of intervals with respect to overlap property
such that the interval graph corresponding to that set has a clique with a size of at
most k.

In [CPL90], an optimal algorithm is presented for finding k-DMIS in an
overlap graph. In this section, we present an overview of the algorithm presented
in [CPL90], which is essentially an extension of the algorithm MIS_CIRCLE
[Sup87].

From a routing perspective, k-DMIS is equivalent to assigning a maximum
number of intervals to k tracks such that if interval (i, j) is assigned to track s, then
no interval assigned to tracks $1, 2, \ldots, s-1$ should intersect columns i and j. Let
$M(i, j, s)$ denote the solution of the k-DMIS problem resulting from restricting
the intervals to the range of $[i, j]$ and allowing s tracks for routing. Suppose
that the width of the channel is W. Then the k-DMIS problem in a channel is to
find $M(1, W, k)$. In order to compute $M(i, j, s)$, we use dynamic programming.
Notice that to compute $M(i, j, s)$, we have the following three cases:

1. If i is vacant, then

$$M(i, j, s) = M(i + 1, j, s)$$

2. There exists a net N_l with terminals i and i', but $i' \notin [i, j]$. Then,

$$M(i, j, s) = M(i + 1, j, s) \qquad \text{if} \quad i' \notin [i, j]$$

3. There exists a net N_l with terminals i and i' such that $i' \in [i, j]$; then we have the following two subcases:

Including the net N_l in the solution leads to

$$M(i, j, s) = M(i + 1, j, s)$$

Excluding the net N_l from the solution results in

$$M(i, j, s) = M(i + 1, i' - 1, s - 1) + M(i' + 1, j, s) + N_l$$

As shown in Figure 2.19, if $i' \in [i, j]$, we need to check if including N_l will lead to a better solution. Therefore, we must compare the solutions with or without net N_a and take the maximum of the two. Therefore,

$$\begin{aligned} M(i, j, s) = \max\{ & (M(i + 1, j, s), M(i + 1, i' - 1, s - 1) \\ & + M(i' + 1, j, s) + N_l) \qquad \text{if} \quad i' \in [i, j] \} \end{aligned}$$

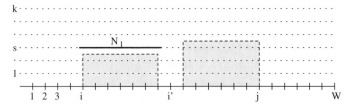

Fig. 2.19   An example of maximum k-density independent set in overlap graph.

For the example shown in Figure 2.20(a), the maximum k-density independent set is shown in Figure 2.20(b). Based on the above equations, we can obtain the following result:

Theorem 8   *The k-DMIS problem for overlap graphs can be solved in $O(kn^2)$ time, where n is the number of terminals and k is the number of available tracks.*

The weighted version of the problem can be solved by a simple extension of the k-DMIS problem. Let $W(i, j, s)$ denote the weight of $M(i, j, s)$ and $w(N_l)$ denote the weight of the net N_l. In the solution for the k-DMIS problem, replace $M(i, j, s)$ with $W(i, j, s)$ and N_l with $w(N_l)$. Therefore,

$$\begin{aligned} W(i, j, s) = \max\{ & (W(i + 1, j, s), W(i + 1, i' - 1, s - 1) \\ & + W(i' + 1, j, s) + w(N_l)) \qquad \text{if} \quad i' \in [i, j] \} \end{aligned}$$

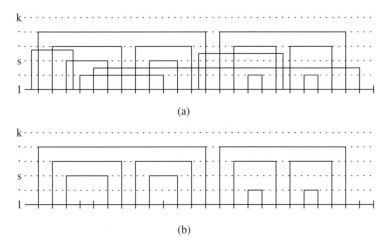

Fig. 2.20   **(a)** Overlap graph; **(b)** maximum k-density independent set in the overlap graph.

Both MIS_CIRCLE and the algorithm given above can be generalized to work with multiterminal nets, as described in the following section.

2.8.5 Maximum Independent Set for Multiterminal Nets

The algorithm for finding the MIS in circle graphs can be extended to poly-circle graphs [Con90]. In VLSI routing, poly-circle graphs essentially model a switch-box routing problem with multiterminal nets. In this section, we discuss the algorithm for finding the MIS in circle graphs in the presence of multiterminal nets. This algorithm is a direct extension of the MIS_CIRCLE algorithm.

We are given $2N$ end points on a circle, and we can partition this set into several subsets, such that each subset (corresponding to a net) must be made electrically equivalent by drawing a polygon between all the terminals of the net. Let m be the maximum number of terminals in a net. Figure 2.22 shows a three-terminal net. Figure 2.23 shows three possible ways net segments may be selected. Notice that the MIS in all three cases is identical.

Let $M(i, j)$ denote a maximum independent set of G_{ij}. If $i \geq j$, then G_{ij} is the empty graph, and hence $M(i, j) = \phi$. The algorithm is an application of dynamic programming. In particular, $M(i, j)$ is computed for each pair i, j; $M(i, j_1)$ is computed before $M(i, j_2)$ if $j_1 < j_2$. To compute $M(i, j)$, let N_l be the unique net with a terminal located at j. Let $k_1, k_2, \ldots, k_m$ be the terminals of net N_l and $k_1 < k_2 < \cdots < k_m$. If $i < k_l \leq j$ for all $1 \leq l \leq m$, then N_l is in the range of $[i, j]$. Otherwise, N_l is not in the range of $[i, j]$. We have to consider the following three cases to solve for $M(i, j)$.

1. If N_l is not in the range of $[i, j]$, then

$$M(i, j) = M(i, j - 1)$$

Algorithm MIS_MULT()

Input: A set of $2N$ endpoints
Output: MIS for multi-terminals nets

begin
    for $j = 0$ to $2N - 1$ **do**
        for $i = 0$ to $j - 1$ **do**
            find k such that $(i < k < j)$ and each terminal
            of N_l is located with range (k, j)
            if the net N_l is in the range of $[i, j]$ **then**
                if $|(\cup_{l=1}^{m} M(k_l + 1, k_l - 1)) \cup M(i, k_1 - 1) \cup N_l|$
                $> |M(i, j - 1)|$ **then**
                    $M(i, j) = (\cup_{l=1}^{m} M(k_l + 1, k_l - 1)) \cup M(i, k_1 - 1) \cup N_l$
            else $M(i, j) = M(i, j - 1)$
end.

Fig. 2.21   Algorithm MIS_MULT.

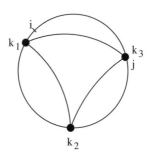

Fig. 2.22   MIS in the presence of multiter-
           minal nets.

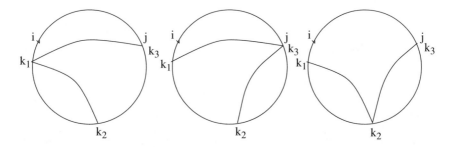

Fig. 2.23   Three possible ways of selecting segments of a three-terminal net.

2. If net N_l is in the range of $[i, j]$ and $N_l \in M(i, j)$, then

$$M(i, j) = M(i, k_1 - 1) \cup M(k_1 + 1, k_2 - 1) \cup \ldots \cup M(k_{m-1} + 1, k_m - 1) \cup N_l$$

3. If net N_l is in the range of $[i, j]$ and $N_l \notin M(i, j)$, then

$$M(i, j) = M(i, j - 1)$$

Thus, $M(i, j)$ is set to the larger of the two sets

$$M(i, j - 1) \quad \text{and}$$
$$M(i, k_1 - 1) \cup M(k_1 + 1, k_2 - 1) \cup \cdots \cup M(k_{m-1} + 1, k_m - 1) \cup N_l$$

At each step of the algorithm, there are only two possible choices for $M(i, j)$. The MIS of G_{ij}, $M(i, j)$ may either consist of

$$(\cup_{l=1}^{m} M(k_l + 1, k_l - 1)) \cup M(i, k_1 - 1) \cup N_l$$

as shown in Figure 2.24(a), or consist of

$$M(i, j - 1)$$

as shown in Figure 2.24(b), depending on which set is larger. The algorithm is formally stated in Figure 2.21.

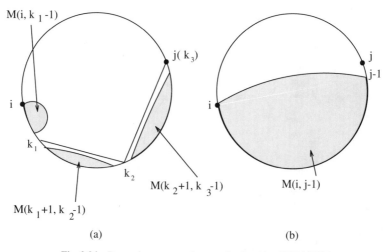

Fig. 2.24　Dynamic programming step in algorithm MIS_MULT.

Theorem 9　*The algorithm MIS_MULT finds a maximum independent set for multiterminal nets in a circle graph in* $O(mn^2)$, *where m is the maximum number of terminals in a net.*

The weighted version of the problem can be solved by a simple extension to the algorithm MIS_MULT. Let $W(i, j)$ denote the WMIS and $w(N_l)$ denote the weight of the net N_l. Now to solve for $W(i, j)$, replace $M(i, j)$ with $W(i, j)$ in the algorithm MIS_MULT, and in the three cases, replace N_l with $w(N_l)$. Thus, the WMIS for multiterminal nets in a a circle graph is given by

$$W(i, j) = \max\{M(i, j - 1),$$
$$M(i, k_1 - 1) \cup M(k_1 + 1, k_2 - 1) \cup \cdots \cup M(k_{m-1} + 1, k_m - 1) \cup N_l\}$$

2.9 Summary

In this chapter, we have presented efficient algorithms for circle, permutation, and interval graphs. These graphs are used to model OTC routing problems. The complexity of algorithms is summarized in Table 2.4. Efficient algorithms presented in this chapter form crucial subroutines in the OTC routing algorithms. Further developments in OTC routing depend on improvement in our understanding of permutation, circle, and interval graphs and the development of efficient algorithms for these classes of graphs.

It must be noted that algorithms presented in this chapter find applications not only in OTC routing, but also in general routing, scheduling, operations research, and computational geometry.

To get a clear perspective of OTC routing algorithms, it is imperative to understand channel routing. In the next chapter, we review the necessary channel routing algorithms essential for solving OTC routing problems.

Table 2.4:   Time Complexity Comparison

Graph Type	MIS	Clique	MKIS
Interval	$O(n \log n)$	$O(n \log n)$	$O(n^2)$
Permutation	$O(n \log n)$	$O(n \log n)$	$O(kn^2)$
Circle	$O(n^2)$	$O(n^2 \log n)$	NP-hard

PROBLEMS

2-1. The MWKIS problems for interval graphs can be solved in $O(n^{3.5})$ using integer linear techniques. In a special case, when $k = 2$, an $O(n^2)$ algorithm has been presented in [HTC92]. This algorithm can be extended to provide an $O(n^3)$ algorithm for the case of $k = 3$. Does there exist an algorithm for the MWKIS with complexity less than $O(n^{3.5})$ for $k \geq 4$?

2-2. The MKIS problem for permutation graphs can be solved in $(n^2 \log n)$ for the general case and in $\theta(n \log n)$ for $k = 2$. Does there exist a $\theta(n \log n)$ algorithm for the MKIS in permutation graphs when $k \geq 2$?

2-3. Does there exist an $O(n \log n)$ for the MIS problem in circle graphs? Or does there exist a lower bound of $O(n^2)$ for the MIS problem in circle graphs?

2-4. MKIS_CIRCLE produces a solution that is within 0.75 of the optimal $k = 2$. This is the best possible if we use the greedy approach used in MKIS_CIRCLE. However, if we construct the maximum 2-independent directly, better results may be possible. Develop an approximation algorithm for finding the MKIS in circle graphs, with performance bounds better than 0.75.

2-5. Consider the generalized version of the k-DMIS problem in an overlap graph. In this problem, we assign a thickness to each net, and the objective is to find the MIS such that density of the solution is no more than k.

2-6. Consider the following generalized version of the MKIS problem in interval graphs. For each interval we are provided a set of tracks to which it can be assigned. Select the MKIS such that each interval in it is properly assigned.

BIBLIOGRAPHIC NOTES

The *perfect graph* was discovered in the early 1960s by Claude Berge. Since then, many classes of graphs have been shown to be perfect. Many of these classes of graphs arise in many real-world applications, such as scheduling, optimization of disk storage, routing in VLSI circuits, analysis of genetic structures, and synchronization of parallel processes.

A graph satisfying the following properties is called a *perfect* graph.

$$(P_1) \qquad \omega(G_A) = \chi(G_A) \quad \text{for all } A \subseteq V$$

and

$$(P_2) \qquad \alpha(G_A) = k(G_A) \quad \text{for all } A \subseteq V$$

where $\omega(G_A)$ is the clique number, $\chi(G_A)$ is the chromatic number, $\alpha(G_A)$ is the stability number, and $k(G_A)$ is the clique cover number of the graph G_A.

By duality, it is clear that a graph G satisfies (P_1) if and only if its complement $\bar{G}$ satisfies (P_2). The third equivalent condition was conjectured by Berge in [Ber61] and was finally proved by Lovasz [Lov72].

$$(P_3) \qquad \omega(G_A)\alpha(G_A) \geq |A| \quad \text{for all } A \subseteq V$$

Perfect graphs admit polynomial time algorithms for many problems that are NP-complete for general graphs. This is due to the good structural properties of perfect graphs. In particular, one can develop an intersection model for many of these graphs. The intersection model allows insight into the structure of the graph and leads to polynomial algorithms.

Perhaps the most referenced book on the algorithmic nature of perfect graphs is *Algorithmic Graph Theory and Perfect Graphs* by Martin C. Golumbic. This book presents most basic algorithms related to interval, permutation, and circle graphs known at the time of printing in 1980. The book also presents a basic characterization theorem related to the perfect graph classes mentioned above.

The concept of characterization of interval graphs was independently developed by G. Hajös and the well-known molecular biologist Seymour Benzer. In 1957, G. Hajös posed the following problem. Given a finite number of intervals on a straight line, a graph associated with this set of intervals can be constructed in the following manner: each interval corresponds to a vertex of a graph, and two vertices are connected by an edge if and only if the corresponding intervals overlap at least partially. The question is whether the graph so constructed is isomorphic to one of the graphs just characterized [Haj57].

In 1959, the well-known molecular biologist Seymour Benzer raised the following issue related to gene structures: Are the subelements within the genes linked together in a linear order analogous to the higher level of integration of the genes in the chromosomes [SBe59]?

Interval graphs have a *linear structure* due to their linear ordering of maximal cliques. This admits very simple, greedy algorithms for interval graphs. Interval graphs can be recognized in $(|V| + |E|)$ time using the *PQ-tree* data structure [BL76]. Interval graphs have been extensively studied and almost all problems on interval graphs have been solved in linear time.

Permutation graphs are similar to interval graphs. An $O(n^2 \log n)$ time algorithm was presented in [RS93] for the generalized M2IS problem in permutation graphs. The maximum two-chain problem in permutation graphs was solved in $\theta(n \log n)$ time [LSL92].

In [SBS87] Jeremy Spinard, Andreas Brandstadt, and Lorna Stewart introduced the bipartite permutation graph and two characterizations of graphs and also presented some polynomial algorithms for a number of NP-complete problems for general graphs.

In [BK87], an algorithm with a time bound of $O(n^2)$ for the weighted independent domination problem on permutation graphs is presented. A polynomial time solution for the weighted feedback vertex set problem in permutation graphs is also developed.

Efficient algorithms for finding the minimum independent dominating set in a permutation graph and finding the shortest maximal increasing subsequence are discussed in [AK89].

In [CS90], Charles J. Colbourn and Lorna K. Stewart proved that the minimum cardinality-connected dominating set in a permutation graph can be solved in $O(n^2)$ time when the permutation module is not given, and the minimum cardinality Steiner tree can be found in $O(n^3)$ time.

S.K. Dhall, Y. Liang, C. Rhee, and S. Lakshmivaran [YLL91] presented a new approach for solving the domination problem on permutation graphs. Their algorithm takes $O(n(m + n))$ time, where m is the number of edges and n is the number of vertices when the permutation module is not given. The algorithm is particularly good for the sparse permutation graphs.

In [CW92], Maw-Shang Chang and Fu-Hsing Wang presented an $O(n \log \log n)$ algorithm presented for the above two problems on permutation graphs. Since a permutation graph is a subclass of an overlap graph, by using an interval model of an overlap graph, these two problems can be solved in $O(n \log \log n)$ time and $O(n)$.

Recently, Kuo-Hui Tsai and Wen-Lian Hsu [TH93] presented an $O(n \log \log n)$ algorithm to solve the minimum cardinality dominating set problem in permutation graphs. The previous algorithm was the $O(n^2)$ dynamic algorithm in 1985 by Faber and Keil.

Circle graphs have found a variety of applications, particularly, switch-box routing, via minimization, and topological routing [CHS93, HSS91b, SL89].

One of the first papers to consider a circle graph was presented by S. Even and A. Itai [EI71]. The problem of realizing a given permutation through a network of stacks in parallel is translated into a coloring problem of a suitable graph. It is shown that if we allow the unloading of stacks before loading is complete, the resulting graph (called the *union graph*) is equivalent to a circle graph. The number of stacks necessary to realize the permutation is equal to the chromatic number of the graph.

Sen, Deng, and Guha [SDG92] give an integer linear programming formulation for the general graph partition problem. They show that the partition problem can be solved in polynomial time if the graph is a tree. The problem on circle graphs remains NP-complete. They observe that they always obtain an integral solution to the linear programming problem when the graph is perfect. They conjecture that an optimal integral solution for the perfect graph always exists. If the conjecture is true, it is likely to lead to the development of a polynomial time algorithm for the problem.

A parallel algorithm for finding a near-maximum independent set in a circle graph is presented in [TCLH90]. The algorithm is modified for predicting the secondary structure in ribonucleic acids. The proposed system, composed of an n neural network array (where n is the number of edges in the circle graph of the number of possible base pairs), not only generates a near-maximum independent set but also predicts the secondary structure of ribonucleic acids within several hundred iteration steps.

3

Channel Routing and Terminal Assignment

Routing is typically done in two phases called *global routing* and *detailed routing*. In the first phase, the routing area is partitioned into smaller regions and a "coarse" route is determined for each net using a global router. Since the global router only establishes connections between different regions, the detailed router has to find the actual geometric path for each wire in a region. The complexity of the routing problems vary due to many factors, including shape of the routing region, the number of layers available, and the number of nets.

Routing regions are typically rectangular and are classified on the basis of number of sides that have terminals of nets to be routed. A two-sided routing region is called a *channel* and its associated routing problem is called a *channel routing problem.* As stated earlier, a key to successful three-dimensional routing in all design styles is the basic understanding of channel routing algorithms, since all design styles have well-defined channels and OTC routing is very closely related to channel routing.

The channel routing problem is very well studied and hundreds of papers on various aspects of this problem are available in the literature (See Chapter 7 [She93]). The main result of this extensive research has been the development of efficient routing algorithms. In fact, algorithms have been developed that, in practice, can complete the routing of a channel very close to its theoretical optimal. As to why optimal algorithms cannot be developed, one has to note that single- and multilayer routing problems are NP-complete [Szy85], even when the routing region has a simple shape. For this reason, routing algorithms are heuristic in nature, although recently there is an increased emphasis on the development of approximation algorithms.

Our approach to three-dimensional routing, particularly with a small number of layers, is conceptualized from the theory of channel routing. Therefore, it is important to understand the basic theory of channel routing and the main algorithms developed for this problem.

Another important aspect of detailed routing is pin assignment. The pin assignment process defines the exact pin (geometric location) for each terminal of each net. Chip-level and block-level pin assignment is done during the placement phase. However, pin assignment during placement must be "coarse" since exact paths of wires are not known. Therefore, it is typical to perform some fine-tuning to pin assignment, particularly to pins in the channels, before detailed routing. This fine-tuning can sometimes lead to significant area improvements. Several OTC routing algorithms also require reassignment of pins, and as a result one needs to be conversant with the channel pin assignment algorithms as well.

In this chapter, we review channel routing and pin assignment algorithms, which are used in the development of OTC routing algorithms in later chapters. Before formally presenting the channel routing problems and associated algorithms, we describe the basic terminology, important considerations, and models used in channel routing and channel pin assignment problems.

3.1 Basic Concepts and Terminology

The basic terminology is fairly standard for routing problems. An attempt has been made to ensure that all required terms are defined here. The reader is encouraged to consult [She93] for further details.

A straight piece of conductor material in a layer is called a *segment*. In most routing problems, segments are restricted to being horizontal or vertical. An electrical connection is made between terminals of a net by laying down a series of segments. Different detailed-routing strategies have been developed with a variety of objectives, but all the detailed-routing problems share some common characteristics. These characteristics deal with routing constraints. For example, wires must satisfy some geometric restrictions, which often concern wire thickness, separation, and path features. One obvious restriction present in all routing problems is intersection; that is, no two wires from different nets are allowed to cross each other on the same layer. A primary objective function of a router is to minimize the total routing area. Various secondary objective functions have also been considered. One such objective is to minimize the number of vias. Vias are difficult to fabricate due to the mask alignment problem. In addition, vias increase delay and are therefore undesirable in high-performance applications. Other objective functions include minimization of the average length of a net and minimization of the number of vias per net.

3.1.1 Routing Parameters

In general, the routing problem has many parameters. These parameters are usually dictated by the design rules and the routing strategy. Following is the list of parameters that are important from an OTC routing perspective.

1. Number of terminals per net.

2. Widths of nets.

3. Different types of nets.

4. Restrictions on vias.

5. Shape of the routing boundary.

6. Number of routing layers.

Let us briefly review these parameters. The number of terminals of a net has a significant impact on the routing methodology. This is due to the fact that the routing of two-terminal nets is very well studied, while the problem of routing multiterminal nets is relatively less understood. However, a majority of nets are two-terminal nets. There are a few nets with very large number of terminals, such as clock nets. Since two-terminal net routing problems are combinatorially easier and a large number of nets are two-terminal nets, routing algorithms, for the sake of simplification, assume all nets to be two-terminal nets. More recently, algorithms that can directly handle multiterminal nets have also been developed. The second parameter important for routing is the width of a net, which depends on the layer the net is assigned and its current carrying capacity. Usually, power and ground nets have different widths and routers must allow for such width variations. In OTC routing, power and ground routing is not a concern, due to the regular and fixed routing styles of ground and power.

Vias are undesirable for two reasons: mask alignment and oxide breakdown problems. The chip is fabricated one layer at a time, and the masks for the various layers must align perfectly to fabricate features such as vias which exist in two layers. Perfect alignment of masks is difficult, and thus vias were normally only allowed between adjacent layers. Even between two layers, minimization of vias reduces mask alignment problems. Improvements in the chip manufacturing technology have reduced mask alignment problems, and today stacked vias (vias passing through more than two layers) can be fabricated. However, vias still remain a concern in routing problems and must be minimized to improve yield and performance. As discussed in Chapter 4, some processes do not allow vias over active areas, since there is a possibility of oxide breakdown, which may lead to electrical malfunction.

A boundary is an edge of the routing region containing the terminals. Most detailed routers assume that the boundaries are regular (straight). Even simple routing problems that can be solved in polynomial time for regular boundaries become NP-hard for the irregular boundary routing problem. Some recent routers [Che86, CK86, VCW89] have the capability of routing within irregular boundaries.

3.1.2 Routing Models

For ease of discussion and implementation of net-routing problems, it is often necessary to work at a more abstract level than the actual layout. In many cases, it is sufficient to use a mathematical wiring model for the nets and the rules that they must obey. For instance, wires are usually represented as paths without any thickness, but the spacing between these wires is increased to allow for the actual wire thickness and spacing in the layout. In the most common model for routing, a rectilinear (or possibly octilinear) grid is superimposed on the routing region and the wires are restricted to following paths along the grid lines. A horizontal grid line is called a *track* and a vertical grid line is called a *column*.

Routing problems can also be modeled based on the layer assignments of horizontal and vertical segments of nets. This model is applicable only in multilayer routing problems. If any net segment is allowed to be placed in any layer, then the model is called an *unreserved-layer model*. When certain types of segments are restricted to particular layer(s), then the model is called a *reserved-layer model*. Most of the existing routers use reserved-layer models. In a two-layer routing problem, if layer 1 is reserved for vertical segments and layer 2 is reserved for horizontal segments, then the model is called a *VH model*. Similarly, a HV model allows horizontal segments in layer 1 and vertical segments in layer 2. Two-layer models can be extended to three-layer routing models: VHV (vertical-horizontal-vertical) or HVH (horizontal-vertical-horizontal). In the VHV (HVH) model, the first and third layers are reserved for routing the vertical (horizontal) segments of nets and the second layer is reserved for routing the horizontal (vertical) segments. The HVH model is preferred to the VHV model in channel routing because, in contrast with the VHV model, the HVH model offers a potential 50% reduction in channel height.

An unreserved-layer model has several advantages over the reserved-layer model. This model uses a smaller number of vias and, in fact, in most cases can lead to an optimal solution (i.e., a solution with minimum channel height).

We now discuss problem formulation for both channel and switch-box routing problems.

3.1.3 Routing Problems

A *channel* is a routing region bounded by two parallel rows of terminals. Without loss of generality, it is assumed that the two rows are horizontal. The top and the bottom rows are also called *top boundary* and *bottom boundary*, respectively. Each terminal is assigned a number representing the net to which that terminal belongs to (see Figure 3.1). Terminals numbered zero are called *vacant terminals*. A vacant terminal does not belong to any net and therefore requires no electrical connection. The net list of a channel is the primary input to most of the routing algorithms.

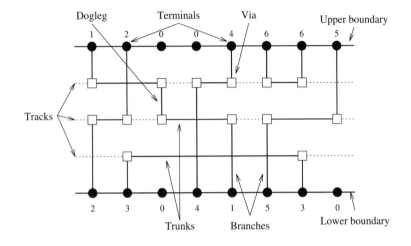

Netlist: 1 2 0 0 4 6 6 5
            2 3 0 4 1 5 3 0

Fig. 3.1   Terminology for channel routing problems.

The horizontal dimension of the routed channel is called the *channel length* and the vertical dimension of the routed channel is called the *channel height*. The horizontal segment of a net is called a *trunk* and the vertical segments that connect the trunk to the terminals are called its *branches*. The horizontal line along which a trunk is placed is called a *track*. A *dogleg* is a vertical segment that is used to maintain the connectivity of the two trunks of a net on two different tracks. A pictorial representation of the terms mentioned above is shown in Figure 3.1.

A channel routing problem is specified by four parameters: channel length, top (bottom) terminal list, left (right) connection list, and the number of layers. The channel length is specified in terms of number of columns in grid-based models, while in gridless models it is specified in terms of λ. The top and the bottom lists specify the terminals in the channel. The top list is denoted by $T = (T_1, T_2, \ldots, T_m)$ and the bottom list by $B = (B_1, B_2, \ldots, B_m)$. In grid-based models, $T_i(B_i)$ is the net number for the terminal at the top (bottom) of the ith column, or is 0 if the terminal does not belong to any net. In gridless models, each terminal $T_i(B_i)$ indicates the net number to which the ith terminal belongs. The left (right) connection list consist of nets that enter the channel from the left (right) end of the channel. It is an ordered list if the channel to the left (right) of the given channel has already been routed.

Given the above specifications, the problem is to find the interconnections of all the nets in the channel, including the connection sets, so that the channel uses the minimum possible area. A solution to a channel routing problem is a set of horizontal and vertical segments for each net. This set of segments must make all terminals of the net electrically equivalent. In the grid-based model, the solution specifies the channel height in terms of the total number of tracks required for routing. In gridless models, the channel height is specified in terms of λ.

The main objective of channel routing is to minimize the channel height. Additional objectives are minimizing the total number of vias used in a multilayer routing solution and minimizing the length of any particular net.

3.1.4 Routing Constraints

In grid-based models, the channel routing problem is essentially the assignment of horizontal segments of nets to tracks. Vertical segments are used to connect horizontal segments of the same net in different tracks and to connect the terminals to the horizontal segments. The problem is very similar in gridless models with the exception that the assignment of horizontal segments is to specific locations in the channel rather than tracks. There are two key constraints that must be satisfied while assigning the horizontal and vertical segments.

1. **Horizontal Constraints:** There is a horizontal constraint between two nets if the trunks of these two nets overlap each other when placed on the same track. For a net N_i, the interval spanned by the net, denoted by I_i, is defined by (r_i, l_i), where r_i is the right-most terminal of the net and l_i is the left-most terminal of the net. Given a channel routing problem, a *horizontal-constraint graph* (HCG) is a undirected graph $G_h = (V, E_h)$, where

   $$V = \{v_i | v_i \text{ represents } I_i \text{ corresponding to } N_i\}$$

   $$E_h = \{(v_i, v_j) | I_i \text{ and } I_j \text{ have nonempty intersection}\}$$

   Note that the HCG is in fact an interval graph as defined in Chapter 2. Figure 3.2(a) shows a channel routing problem, and the associated HCG is shown in Figure 3.2(b).

   The HCG plays a major role in determining the channel height. In a grid-based two-layer model, no two nets that have a horizontal constraint may be assigned to the same track. As a result, the maximum clique in the HCG forms a lower bound for channel height. In the two-layer gridless model, the summation of widths of nets involved in the maximum clique determines the lower bound.

2. **Vertical Constraints:** A net N_i, in a grid-based model, has a vertical constraint with net N_j if there exists a column such that the top terminal of the column belongs to N_i and the bottom terminal belongs to N_j and $i \neq j$.

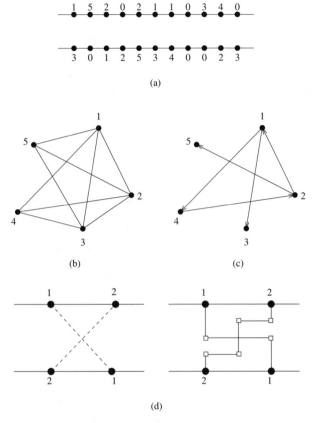

Fig. 3.2   **(a)** A channel routing problem; **(b)** its horizontal-constraint graph; **(c)** vertical-constraint graph; **(d)** a dogleg.

In the case of the gridless model, the definition of vertical constraint is somewhat similar, except that the overlap is between the actual vertical segments rather than the terminals in a column. Given a channel routing problem, a *vertical-constraint graph* (VCG) is a directed graph $G_v = (V, E_v)$, where

$$E_v = \{(v_i, v_j) | N_i \text{ has vertical constraint with } N_j\}$$

It is easy to see that a vertical constraint implies a horizontal constraint. The converse, however, is not true. Figure 3.2(c) shows the VCG for the channel routing problem in Figure 3.2(a).

A directed path in the VCG is one of the deterministic factors of the channel height. If doglegs are not allowed, then the length of the longest path in VCG forms a lower bound on the channel height in the grid-based model. This is due to the fact that no two nets in a directed path may be routed on the same track. Note that if VCG is not acyclic, then some nets must be doglegged. Figure 3.2(d) shows a channel routing problem with a vertical-constraint cycle and how a dogleg can be used to break this cycle.

As discussed in Chapter 2, two interesting graphs related to channel routing problems are the permutation graph and the circle graph. The permutation graph can only be defined for channel routing problems for two-terminal nets and no net has both of its terminals on one boundary (see Chapter 2). Two graphs discussed in this section allow us to consider the channel routing problem as a graph theoretic problem.

The switch-box routing problem is a generalization of the channel routing problem, where terminals are located on all four sides. Switch boxes are formed in two ways. There may be be a four-sided enclosed region within which the routing must be completed or a four-sided region may be formed due to the intersection of two channels. A *switch box* is formally defined as a rectangular region $R(h \times w)$, where h and w are positive integers. Each pair (i, j) in R is a grid point. The ith column and jth row or track are the sets of grid points. The 0th and hth columns are the LEFT and RIGHT boundaries, respectively. Similarly, the 0th and wth rows are TOP and BOTTOM boundaries, respectively. The connectivity and location of each terminal is represented as $LEFT(i) = k$, $RIGHT(i) = k$, $TOP(i) = k$, or $BOTTOM(i) = k$, depending on the side of the switch box it is located on, where i is the coordinate of the terminal along the edge and k is a positive integer specifying the net to which the ith terminal belongs.

Since it is assumed that the terminals are fixed on the boundaries, the routing area in a switch box is fixed. Therefore, the objective of switch-box routing is not to minimize the routing area but to complete the routing within the routing area. In other words, the switch-box routing problem is a routability problem, i.e., to decide if a routing solution is possible. Unlike the channel routing problem, a switch-box routing problem is typically represented by its circle graph (see Chapter 2).

3.2 Classification of Channel Routing Algorithms

There could be many possible ways to classify the detailed-routing algorithms. Classification can be on the basis of the routing models used. Some routing algorithms use grid-based models, while some other algorithms use the gridless model. The gridless model is more practical, since all the wires in a design do not have the same widths. Another possible scheme would be to classify the algorithms based on the strategy they use. Thus, we can have greedy routers or hierarchical routers, for example. We classify the algorithms based on the number of layers used for routing.

1. **Single-Layer Routing Algorithms:** Single-layer routing problems frequently appear as subproblems in multilayer routing problems. In OTC

routing, single-layer channel routing, which is also called *river routing*, is used to route in OTC areas if the fabrication process does not allow vias in OTC areas.

2. **Two-Layer Routing Algorithms:** Two-layer routing problems have been thoroughly investigated, since until recently, due to the limitations of the fabrication process, only two metal layers were allowed for routing. Several hundred papers have been written about two-layer routing problems. In particular, the two-layer channel routing problem is extremely well studied and is considered by many as solved.

3. **Three-Layer Routing Algorithms:** Recently, due to improvements in the fabrication process, a third metal layer is also allowed, but it is expensive compared to the two-layer metal process. Three-layer channel routers are essentially extensions of two-layer routers. These types of routers are of great interest to OTC routing, since most OTC routers are based on the availability of M2 and M3 in the OTC area.

4. **Multilayer Routing Algorithms:** Several multilayer routing algorithms have also been developed recently, which can be used for routing MCMs with up to 32 layers. Multilayer routing is of no significant interest to OTC routing, since we concentrate on routing with three to four layers. A large number of layers is useful for increasing performance even if the zero-routing footprint is achievable with a fewer number of tracks.

3.3 River Routing Algorithms

River routing is a restricted channel routing problem. It has been widely studied for a variety of river routing models.

We are given two sequences of increasing integers $T = (t_1, t_2, \ldots, t_n)$ and $B = (b_1, b_2, \ldots, b_n)$ representing the coordinates of two sets of terminals (pins) on two parallel (horizontal) lines. The distance between these lines, denoted by s, is a positive integer and is a design variable called *separation*. A wire representing the net N_i, for $i = 1, 2, \ldots, n$, must join the terminal at t_i to the terminal at b_i using a continuous rectilinear curve of total length $s + |t_i - b_i|$. This suffices to connect t_i and b_i because wires that are extended beyond the two end points do not help reduce separation. Given a river routing problem, it is desirable to find a solution that minimizes the width of the channel.

The river routing problem arises in OTC routing when one wishes to provide connections between pins in a terminal row $(t_1, t_2, \ldots, t_n)$ and pins on the boundary of the cell $(b_1, b_2, \ldots, b_n)$. This is particularly true for MTM- and CTM-based cell designs when vias are not allowed in OTC areas.

The river routing problem has been widely studied [DKS81, Sie83, LP83], and several objective functions, such as minimization of distance between two terminal rows (also called *separation*) and minimization of total number of jogs, have been considered.

When it is possible to legally route a given instance of river routing, a simple greedy algorithm is sufficient to find the layout with minimum separation. Nets of a river routing problem can be partitioned as left blocks and right blocks. A *left block* is a maximal consecutive sequence of nets $N_i, \ldots, N_j$ such that each net moves from right to left as we go from bottom to top. That is, for all k, $i \le k \le j, a_k < b_k$. Right blocks are defined analogously. It is clear that all the problems of wiring occur within a single block, and there is no interaction between the blocks. A net N_k with $t_k = b_k$ is a special case, belonging to neither a left nor a right block, and is called *straight net*. Evidently, straight nets can be wired vertically across the horizontal lines without interfering with the wiring of any other net. Thus, we shall ignore straight nets and concentrate on the wiring of nets of left and right blocks. Furthermore, it is sufficient to describe the routing of blocks, since all blocks can be routed independently. In order to route a block, a greedy left-most first net strategy is used. The first net is routed as straight as possible, the second net is routed along the contour of the first net, and so on. Figure 3.3 shows an example of single-layer river routing. In the figure, we have ten terminals on the top and ten terminals on the bottom. For the first four nets, $t_k < b_k$. So they belong to the left block, and for the last four nets, $t_k > b_k$ and they belong to the right block. Nets 5 and 6 are straight nets because $t_k = b_k$. We need four tracks to completely route the given problem. So the channel density is 4.

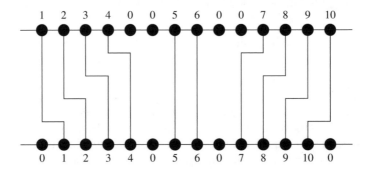

Fig. 3.3   Example of single-layer river routing.

The version of the river routing problem that concerns OTC routing is related to routability. In other words, given a river routing problem and an integer (specifying separation), is the given river routing problem routable in the given separation?

It turns out this problem is quite simple. Let us define *cut* as a line segment that runs from the top terminal row to the bottom terminal row. For a given cut $\overline{pq}$, let $t_1, t_2, \ldots, t_i$ be the top terminals, which are to the left of p, and let $b_1, b_2, \ldots, b_j$ be the terminals to the left of q. Now we define quantities: the congestion of a cut and capacity of a cut. There are exactly $|j - i|$ different nets, which have terminals on both ends of the cut $\overline{pq}$, and they must cross the cut. The quantity $|j - i|$ is referred to as the *congestion* of the cut $\overline{pq}$. The *capacity* of a cut $\overline{pq}$ is equal to the number of nets that can cross $\overline{pq}$. It is easy to see that the capacity of cut $\overline{pq}$ is equal to the larger of two quantities: horizontal separation between p and q, and vertical separation (or simply separation) between p and q.

It is easy to see that if the congestion of a cut exceeds its capacity, then it is impossible to complete the routing. As a result, in order to test routability of a given routing problem, one has to simply verify that the congestion of all cuts is less than or equal to their capacity (see [LP83] for details).

3.4 Two-Layer Channel Routing Algorithms

The two-layer channel routing problem is perhaps the most well studied routing problem. Its similarity with single-layer routing depends on the allowance of vias between metal layers. If vias are not allowed between metal layers, the two-layer routing problem can be seen as an extension of the single-layer routing problem, since two planar sets of nets are generated instead of one. On the contrary, if vias are allowed, the horizontal and vertical segments of each net are routed on different layers. Hence, checking for routability is unnecessary, since all channel routing problems can be completed in two layers. Hence, the main objective function of two-layer channel routers is not just to complete the desired routing, but to do it in a manner that minimizes the necessary routing. In this section, we limit our discussion to those channel routers that allow vias between metal layers and the main objective of the router is minimization of channel height.

In a grid-based channel routing problem, *channel height* can be simply defined as the number of tracks necessary to complete a given routing. For a given grid-based channel routing problem, any solution requires a certain minimum number of tracks. This requirement is called the *lower bound* for the channel routing problem.

The following theorem presents the lower bound for the channel routing problems assuming a two-layer reserved routing model with no doglegs. Let $h_{\max}$ and $v_{\max}$ represent the maximum clique in the HCG and the longest path in VCG, respectively, for a routing instance.

Theorem 10   *The minimum number of tracks required to route a two-layer dogleg-free routing problem is* $\max\{h_{\max}, v_{\max}\}$.

For gridless channel routing problems, the width of nets must be taken into account while computing $v_{\max}$ and $h_{\max}$.

3.4.1 Classification of Two-Layer Channel Routing Algorithms

The channel routing problem has been very widely studied within the VLSI community, leading to many different approaches. We have classified these algorithms based on the approach employed.

1. **Left-Edge Algorithms (LEA):** Left-edge-based algorithms start with sorting the trunks from left to right and then assigning the segments to a track so that no two segments overlap.

2. **Constraint Graph–Based Routing Algorithms:** These routing algorithms use the graph theoretic approach to solve the channel routing problem. The horizontal and vertical constraints are represented by graphs. The algorithms then apply different techniques on these graphs to generate the routing in the channel.

3. **Greedy Routing Algorithm:** The greedy routing algorithm uses a greedy strategy to route the nets in the channel. It starts with the left-most column and works toward the right end of the channel by routing the nets one column at a time.

4. **Hierarchical Routing Algorithm:** The hierarchical router generates the routing in the channel by repeatedly bisecting the routing region and then routing each net within the smaller routing regions to generate the complete routing.

In the following subsections, we present three different routing algorithms which play a significant role in the remainder of the book.

3.4.2 Left-Edge Algorithms

Hashimoto and Stevens [HS71] developed the LEA, which is indeed the first algorithm developed for channel routing. The motivation for development came from routing problems in array-based two-layer PCBs. In such designs, the chips are placed in rows and the area between the rows and underneath the boards are divided into rectangular channels. Although the basic LEA is quite restrictive and is not directly applicable to many practical routing problems, it has served as a start for many practical routers. As a result, it is important to understand the basic idea behind the LEA. In this section, we present the basic LEA and some practical routers based on it.

The LEA is a reserved layer router for channels with no vertical constraints, and it does not allow doglegs. It sorts the intervals formed by the trunks of the nets in ascending order, relative to the x coordinate of the left end points of intervals. It then allocates intervals to each of the tracks, considering them one at a time (following their sorted order) using a greedy method. To allocate an interval to a track, LEA scans through the tracks from the top to the bottom and assigns the net to the first track that can accommodate the net. The allocation process is restricted to one layer, since the other layer is used for the vertical segments (branches) of the nets. The detailed description of the LEA is in Figure 3.4. Figure 3.5 shows a routing produced by the LEA. The terminal permutation is shown in Figure 3.5(a). Figure 3.5(b) shows the horizontal trunks of the nets to be routed, and Figure 3.5(c) shows the routed channel. Net N_1 is assigned to track 1. Net N_2 is assigned to track 2, since it intersects with N_1 and cannot be assigned to track 1. Net N_3 is similarly assigned to track 3. Net N_4 is assigned to track 1, since it does not intersect with N_1.

Algorithm LEA $(\mathcal{N}, \mathcal{I})$

Input: Net list $\mathcal{N}$, associated interval set $\mathcal{I}$
Output: A set of horizontal $\mathcal{H}$ and vertical $\mathcal{V}$ wire segments that satisfy the required connections of netlist $\mathcal{N}$

---

begin
    Form Intervals($\mathcal{N}, \mathcal{I}$);
    Form Horzontal Constraint Graph HCG($\mathcal{I}$,HCG);
    Let d = size of the maximum clique in HCG;
    Let $T = \{T_1, T_2, \ldots, T_d\}$ denote the set of tracks
    Sort intervals in ($\mathcal{I}$) on their left end points;
    for $i = 1$ to n **do**
        for $j = 1$ to d **do**
            if I_i does not overlap with another
                interval in T_j **then**
                assign interval I_i to T_j;
    for $i = 1$ to n **do**
        (* connect the terminals of net N_i to their trunk *)
        Connect left(I_i) with left(N_i) with a vertical segment;
        Connect right(I_i) with right(N_i) with a vertical segment;
end.

Fig. 3.4   Algorithm LEA.

The following theorem establishes the optimality of the LEA and is quite easy to prove by a simple induction argument.

Theorem 11   *Given a two-layer channel routing problem with no vertical constraints, the LEA produces a routing solution with a minimum number of tracks.*

The input to the algorithm is a set of two-terminal nets $\mathcal{N} = \{N_1, N_2, \ldots, N_n\}$. The time complexity of this algorithm is $O(n \log n)$, which is the time needed for sorting n intervals.

To assume that no two nets share a common end point is unrealistically restrictive, and as a result, the LEA is not practical for most channel routing problems.

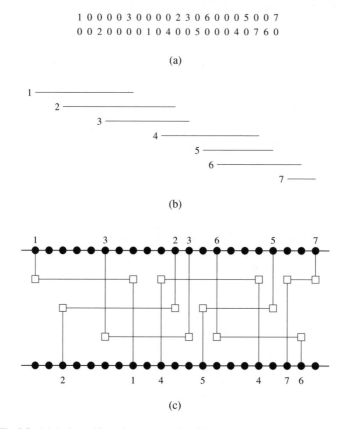

(a)

(b)

(c)

Fig. 3.5   (a) A channel boundary permutation; (b) horizontal segments of the net list; (c) LEA channel routing.

However, the LEA can be used in PCB routing problems with vertical constraints, since there is sufficient space between the adjacent pins to create a *jog*. The LEA is also useful as an *initial* router for channels with vertical constraints. The basic idea of this variation is to create a layout with design rule violations and then use cleanup procedures to remove the violations.

3.4.3 Greedy Channel Router

The LEA does not allow doglegging, thus severely restricting its application to those routing problems in which no vertical constraints are present. If vertical constraints are present, doglegs are necessary. One simple approach is to place a dogleg in the column, such that a net has a terminal in that column. However, this restricts doglegging to only multiterminal nets. Moreover, an optimal solution may require that a net's dogleg be present in a nonterminal column. Based on this observation, a new technique was developed by Rivest and Fiduccia [RF82] using the following approach. A *greedy router* assigns net segments to tracks on a column-by-column basis, beginning from the left-most column of the channel. The router places all the net segments present in a column before proceeding to the next column. Perhaps, the main reason behind the superior performance of this greedy router as compared to the LEA and dogleg routers is that it allows for doglegs to be placed in any column of the channel.

The greedy router uses several different steps in each column. An attempt is made to complete the connection or change tracks of a net (by doglegging) and bring it closer to its target terminal. The following is an explanation of each step, preceded by a procedure name that will be used to represent that step in the detailed description of the algorithm.

1. **CONNECT:** In this step, the algorithm connects any terminal on the boundary of the channel to the trunk segment of its corresponding net. This connection is completed by routing a vertical segment to the track that already contains the net. If a vertical constraint prohibits this connection or if the net is not present on any track, the first empty track is allocated to that net (if no empty track is present, refer to step 5). In other words, a minimum vertical segment is used to connect the trunk to the terminal. For example, net 5 in Figure 3.6(a) in column 3 is connected to its trunk.

2. **COLLAPSE:** The second step attempts to collapse any *split nets* (horizontal segments of the same net present on two different tracks) using a vertical segment as shown for net 8 in Figure 3.6(b). Split nets occur when two terminals of the same net are located on different tracks of the channel within a given column and cannot be immediately connected because of existing vertical constraints. If there are two overlapping sets of split nets, only one can be collapsed in a column.

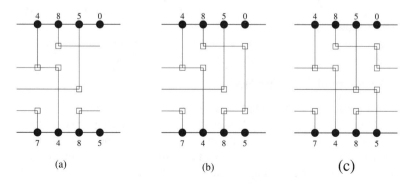

Fig. 3.6   (a) A split net; (b) the collapsed split net; (c) doglegging to reduce range.

3. **REDUCE:** In the third step, the algorithm tries to reduce the range or the distance between two tracks of the same net. This reduction is accomplished by using a dogleg as shown in Figures 3.7(a) and (b). In Figure 3.6 (c), net 8 uses a dogleg to reduce its range.

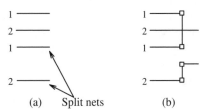

Fig. 3.7 (a) Split nets; (b) doglegging to reduce range.

4. **CONSIDER-NEXT-TERMINAL:** The fourth step attempts to move the nets closer to the boundary which contains the next terminal of that net.

5. **ADD-TRACK:** If a terminal on the boundary of the given column has no available track in which to enter the channel, this step adds an extra track and the terminal is then connected to this new track.

After all five steps have been completed, the trunks of each net are extended to the next column and the steps are repeated. The detailed description of the greedy channel routing algorithm is in Figure 3.8. Note that if there exists a terminal P_i^k $k \in \{t, b\} = 0$, then no electrical connection is necessary (Figure 3.8). If at column i there exists one or more split nets, the function SPLIT-NETS will return TRUE, and function ASSIGNED-TRACK will return TRUE if some terminal P_i^k has not been assigned a track.

The greedy router is one of the best channel routers. It is simple to implement and gives close to optimal results for most channels. The greedy router is more flexible in the placement of doglegs due to fewer assumptions about the topology of the connections. An example of a routing by the greedy channel router is shown in Figure 3.9. Despite it advantages, it does have two drawbacks. One drawback of the greedy router is its excessive use of vias and doglegs in some routing instances. Another drawback is the use of routing area outside the channel. If there exists any split nets remaining after the last column of the channel has been routed, the routing must continue outside the channel's horizontal span.

3.4.4 Yoshimura-Kuh Router

Yoshimura and Kuh [YK82] presented a new channel routing algorithm (YK algorithm), which explicitly addresses the problem of vertical constraints. The major contribution of the YK algorithm is to expose the dependence of the channel height on the constraints. It was the first attempt to analyze the graph theoretic structure of the channel routing problem. The YK algorithm considers both the horizontal- and vertical-constraint graphs and assigns tracks to nets so as to mini-

---

Algorithm GREEDY-CHANNEL-ROUTER ($\mathcal{N}$)

Input: Netlist $\mathcal{N}$
Output: A set of horizontal and vertical wire segments that
satisfy the required connections given by $\mathcal{N}$

---

begin
    $d = $ CHANNEL_DENSITY($\mathcal{N}$);
    (* d is the lower bound of channel density *)
    let $\mathcal{T} = \{T_1, T_2, \ldots T_d\}$ denote the set of tracks;
    for $i = 1$ to m **do**
        (* P_i^t and P_i^b represent the top and bottom
        terminals (pins) at column i respectively *)
        if $P_i^t \neq 0$ **then** CONNECT($P_i^t, i, \mathcal{T}$);
        if $P_i^b \neq 0$ **then** CONNECT($P_i^b, i, \mathcal{T}$);
        if SPLIT-NETS($i, \mathcal{T}$) = TRUE **then** COLLAPSE($i, \mathcal{T}$);
        if SPLIT-NETS($i, \mathcal{T}$) = TRUE **then** REDUCE($i, \mathcal{T}$);
        CONSIDER-NEXT-TERMINAL($i, \mathcal{N}, \mathcal{T}$);
        if $P_i^t \neq 0$ and ASSIGNED-TRACK($i, P_i^t, \mathcal{T}$) = FALSE
          then ADD-TRACK($P_i^t, i, \mathcal{T}$);
        if $P_i^b \neq 0$ and ASSIGNED-TRACK($i, P_i^b, \mathcal{T}$) = FALSE
          then ADD-TRACK($P_i^b, i, \mathcal{T}$);
    while SPLIT-NETS($i, \mathcal{T}$) = TRUE **do**
    $i = i + 1$;
    COLLAPSE($i, \mathcal{T}$);
end.

Fig. 3.8 Algorithm GREEDY-CHANNEL-ROUTER.

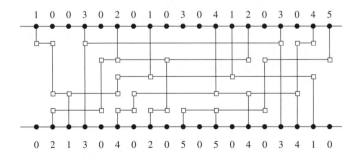

Fig. 3.9 Channel routed using a greedy router.

mize the effect of vertical-constraint chains in the VCG. It does not allow doglegs and cannot handle vertical-constraint cycles. In this section, we briefly review the YK algorithm.

For a channel routing problem with no vertical constraints, the minimum number of tracks required is determined by the maximum clique h_{max} in the HCG, and the LEA produces the optimal results if doglegs are not allowed. But when vertical constraints are allowed, the channel height also depends on the length of the longest path v_{max} in the VCG. The nets that lie on these long paths must be carefully assigned to tracks. When two nets are assigned to the same track without considering the constraint chains, the channel height increases. In the example shown in Figure 3.10(a), a channel height of 5 is obtained. However, a better assignment resulting in a channel height of 3 is shown in Figure 3.10(b). Therefore, in order to minimize the effect of vertical-constraint chains on channel height, two nets may be assigned to the same track only if both of them have small ancestor chains, or both of them have small descendent chains.

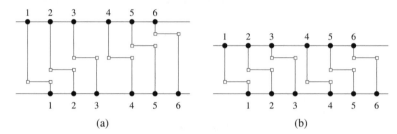

(a)   (b)

Fig. 3.10   **(a)** Channel before net merge; **(b)** channel after net merge.

The YK algorithm partitions the routing channel into a number of regions called *zones* based on the horizontal segments of different nets and their constraints. The basic observation is that a column-by-column scan of the channel is not necessary, since nets within a zone cannot be merged together and must be routed in a separate track. This observation improves the efficiency of the algorithm. The algorithm proceeds from left to right of the channel and merges nets from adjacent zones. The nets that are merged are considered as one composite net and are routed on a single track. In each zone, new nets are combined with the nets in the previous zone. After all zones have been considered, the algorithm assigns each composite net to a track. The key steps in the algorithm are zone representation, net merging to minimize the vertical-constraint chains, and track assignment. Throughout our discussion, we will use the example given in [YK82], since it serves as a benchmark.

   1. Zone Representation: The zones are essentially maximal cliques in the HCG or the interval graph. The interval graph or HCG of the net list in Figure 3.11(a) is shown in Figure 3.11(e). In order to determine zones, let us define $S(i)$ to be the set of nets whose horizontal segments inter-

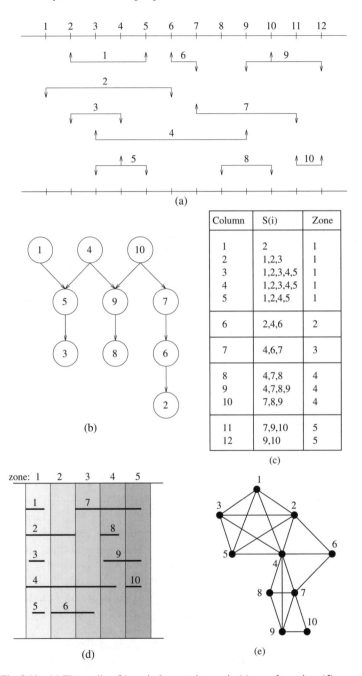

Fig. 3.11   (a) The net list; (b) vertical-constraint graph; (c) zone formation; (d) zone representation; (e) horizontal-constraint graph.

sect column i. Assign *zones* the sequential number to the columns at which $S(i)$ are maximal. These columns define zone 1, zone 2, etc., as shown in Figure 3.11(c). The cardinality of $S(i)$ is called *local density* and the maximum among all local densities is called *maximum density*, which is the lower bound on the channel density. The maximum density for this example is 5. It should be noted that a channel routing problem is completely characterized by the VCG (Figure 3.11(b)) and its zone representation (Figure 3.11(d)).

2. **Merging of Nets:** Let N_i and N_j be the nets for which the following two conditions are satisfied:

(a) There is no edge between v_i and v_j in the HCG.

(b) There is no directed path between v_i and v_j in the VCG.

If these conditions are satisfied, net N_i and net N_j can be merged to form a new composite net. The operation of merging nets N_i and N_j modifies the VCG by shrinking nodes v_i and v_j into node $v_{i,j}$, and updates the zone representation by replacing nets N_i and N_j with net $N_{i,j}$, which occupies the consecutive zones including those of nets N_i and N_j. Finding optimal net pairs for merging is a hard problem, since the future effects of a net merge cannot be determined.

3. **Track Assignment:** Each node in the graph is assigned a separate track.

Let us consider the example shown in Figure 3.11(a). Nets N_6 and N_9 are merged, and the modified VCG, along with the zone representation, is shown in Figures 3.12(a) and (b), respectively. The updated VCG and the zone representation correspond to the net list in Figure 3.11(a), where N_6 and N_9 are replaced by net $N_{6,9}$. The algorithm merges nets as long as two nets from different zones can be merged.

In each iteration, the nets ending in zone z_i are added to the list L, while the nets starting in z_{i+1} are kept in list R. The function MERGE then merges two lists L and R so as to minimize the increase in the longest path length in the VCG. The list L' returned by the function MERGE consists of all the nets merged by the function. These nets are not considered further. Figure 3.13 illustrates how the VCG is updated by the YK algorithm. The length of the longest path in the VCG is 4 and the size of the maximum clique is 5. Hence, any optimal solution takes at least five tracks. In the first iteration, $L = \{N_1, N_3, N_5\}$ and $R = \{N_6\}$. There are three possible net mergings, $N_{1,6}$, $N_{3,6}$, and $N_{5,6}$. Merging N_1 and N_6 creates a path of length 5, merging N_3 and N_6 creates a path of length 4, and merging N_5 and N_6 creates a path of length 4. Therefore, either $N_{3,6}$ or $N_{5,6}$ may be formed. Let us merge N_5 and N_6. Similarly, in the second iteration, nets N_1 and N_7 are merged. In the fourth iteration, N_{10} and N_4 are merged. The final graph is shown in Figure 3.13(e). The track assignment is straightforward. For example, track 1 can be assigned to net $N_{10,4}$. Similarly, tracks 2 and 3 can be assigned to nets

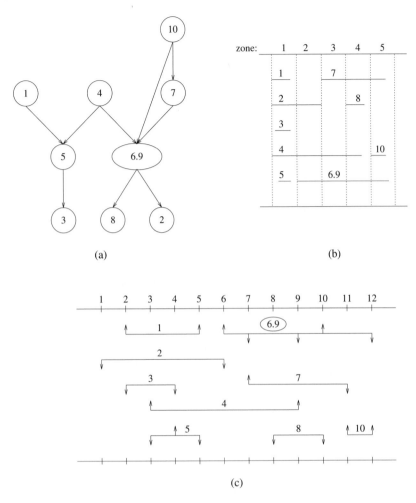

Fig. 3.12   **(a)** Modified VCG; **(b)** modified zone representation; **(c)** channel showing net merge of 6 and 9.

$N_{1.7}$ and $N_{5.6.9}$, respectively. For nets N_2 and $N_{3.8}$, either track 4 or 5 can be assigned.

3.4.5 Experimental Evaluations

The two-layer routing problem has been a topic of intense research in the last decade. While research has led to the development of extremely efficient routers, we have also realized that no algorithm is suitable for all problems and applications. Routing problems differ considerably and problem characteristics must be

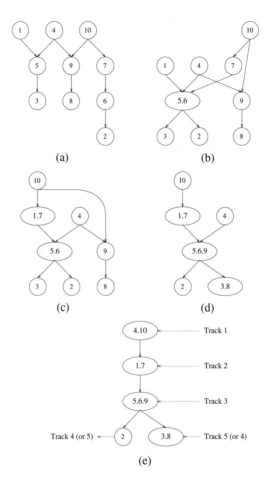

Fig. 3.13 Illustration of YK algorithm: **(a)** VCG; **(b)** VCG after a net merge; **(c)** VCG after two net merges; **(d)** VCG after four net merges; **(e)** final VCG.

carefully studied in order to match them to a routing algorithm. In Table 3.1 we summarize different features of several important two-layer routers.

  Due to the heuristic nature of the algorithms, many benchmark channel routing examples have been proposed. The *Deutsch difficult example* is perhaps the most famous benchmark for the channel routing problem. We reproduce Table 3.2 from [PL88], which summarizes the routing results by different algorithms on the Deutsch difficult example. The table shows that YACR2, a considerably complex router, produces the best routing for the Deutsch difficult example both in terms of tracks and vias. YACR2 is also more complicated to implement as compared to the greedy router. But since the greedy router produces near-optimal results in most practical examples, the complexity tradeoff in favor of YACR2 is usually not well justified. In fact, in most examples, the greedy router produces a solution within one or two tracks of the optimal.

Table 3.1: Comparison of Different Features of Two-Layer Routers

Features	Algorithm						
	LEA	Dogleg	YK	Greedy	YACR2	Hierar-chical	Glitter
Model	Grid-based	Grid-based	Grid-based	Grid-based	Grid-based	Grid-based	Grid-less
Dogleg	Not allowed	Allowed	Allowed	Allowed	Allowed	Allowed	Not allowed
Layer assignment	Reserved	Reserved	Reserved	Reserved	Reserved*	Reserved	Reserved
Vertical constraints	Not allowed	Allowed	Allowed	Allowed	Allowed	Allowed	Allowed
Cyclic constraints	Not allowed	Not allowed	Not allowed	Allowed	Allowed	Not allowed	Allowed

*With some exceptions.

Table 3.2: Results on the Deutsch Difficult Example

Router	Tracks	Vias	Wire Length
LEA	31	290	6526
Dogleg router	21	346	5331
Y-K router	20	403	5381
Greedy router	20	329	5078
Hierarchical router	19	336	5023
YACR2	19	287	5020

3.5 Three-Layer Channel Routing Algorithms

As VLSI fabrication technology advances, three or more layers for signal routing are made feasible. Most of the current gate-array technologies use three layers for routing. For example, the Motorola 2900ETL macrocell array is a bipolar gate array that uses three metal layers for routing. The latest microprocessors such as INTEL's Pentium and DEC's Alpha chip also use three metal layers for routing. As a result, a considerable research effort has recently been directed toward the three-layer channel routing problem.

We classify three-layer routing algorithms into two main categories: the reserved-layer routing algorithm and the unreserved-layer routing algorithm. The reserved-layer model can further be classified into the VHV model and the HVH model. It is easy to derive lower bounds of the channel routing problem in the three-layer reserved-layer routing model in terms of the maximum clique size in the HCG and the longest path in the VCG of the corresponding routing problem.

Theorem 12   *In the VHV routing model, a channel routing problem requires at least $h_{\max}$ routing tracks.*

Theorem 13   *In the HVH model, a channel routing problem requires at least* $\max\{v_{\max}, \frac{h_{\max}}{2}\}$ *routing tracks.*

Theorem 12 follows directly from the fact that in VHV routing the vertical constraints between nets no longer exist. The vertical constraints are eliminated because vertical segments connecting a horizontal segment to the top boundary are routed in the M1 layer, while vertical segments connecting a horizontal segment to the bottom boundary are routed in the M3 layer. In the absence of vertical constraints, the channel height equal to the maximum density can always be realized using the LEA.

Generally, all three-layer routers are based on two-layer routers. The net-merge algorithm by Yoshimura and Kuh [YK82] has been extended by Chen and Liu [CL84]. The gridless router Glitter [CK86] has been extended to Trigger by Chen [Che86]. Cong, Wong, and Liu [CWL87] take a general approach and obtain a three-layer solution from a two-layer solution. Finally, Pitchumani and Zhang [PZ87] partition the given problem into two subproblems and route them into VHV and HVH models. In this section, we discuss one simple three-layer channel routing algorithm.

3.5.1 Three-Layer Routing Based on Two-Layer Routing

Considering the existence of several efficient two-layer routers, it is logical to develop a three-layer one, which obtains a three-layer solution from a two-layer routing solution. In [CWL87], Cong, Wong, and Liu developed an efficient algorithm that systematically transforms a two-layer HV routing solution into a three-layer HVH routing solution. The basic idea of the algorithm is very similar to the YK algorithm [YK82]. In the YK algorithm, nets are merged so that all merged nets forming a composite net are assigned to one track. The objective is to minimize the number of composite nets.

The basic idea of the algorithm is to pair up composite nets. Each composite net pair is assigned to a track. Hence, the objective is to minimize the composite net pairs. In order to find the optimal pair of composite nets that can be merged to form composite net pairs, a directed acyclic graph called *track ordering graph*, TOG $= (V, E)$, is defined. The vertices in V represent the composites (tracks) in a given two-layer solution. The directed edges in $G(S)$ represent the ordering restrictions on pairs of tracks. Composite interval t_i must be routed above composite interval t_j if there exists a net $N_p \in t_i$ and $N_q \in t_j$, such that N_p and N_q have a vertical constraint. The TOG is in fact a VCG between tracks of composite intervals. The objective of the algorithm is to find a track pairing that reduces the total number of such pairs. Obviously, we must have at least $|V|/2$ pairs. It is easy to see that the problem of finding an optimal track pairing of a given two-layer solution S can be reduced to the problem of two-processor scheduling, in which tracks of V are tasks and the TOG is the task precedence graph. Since the two-processor scheduling problems can be optimally solved in linear time [Gab85, JG72], the optimal track permutation can also be found in linear time. Figure 3.14(b) shows the track-ordering graph obtained by the greedy router for the two-layer routing solution, shown in Figure 3.14(a). Figure 3.14(c) shows an optimal scheduling solution for the corresponding two-processor scheduling problem.

Several tracks remain unpaired due to vias in exactly the same locations. These vias are called *adjacent vias*. If adjacent vias can be removed between two nonpaired tracks so that the tracks can be paired together, it would lead to saving two empty tracks. In [CWL87], an attempt is made to move the via aside, and then a maze router is used to reroute the net.

It is obvious that we must minimize the number of adjacent vias between two tracks in order to maximize the tracks that can merge. This is accomplished by the layer reassignment of tracks. It is easy to see that if tracks t_i and t_j are assigned to be routed in the pth track, then the layer on which a particular track gets routed is still to be decided. In other words, for each track we have two choices. However, the choice that we make for each track can affect the number of adjacent vias. This problem can be solved by creating a *layer assignment graph* (LAG), which consists of vertices representing both the possible choices for the track. Thus, each track is represented by two vertices. Four vertices of two adjacent tracks are joined by edges. Thus, each edge represents a possible configuration of two adjacent tracks. Each edge is assigned a weight equal to the number of adjacent vias if the configuration represented by the edge is used. As a result, the problem of finding optimal configuration is reduced to a problem of finding the shortest path in the LAG. The shortest path in the LAG precisely defines the edges, which correspond to layer assignment choices of minimum total adjacent vias. Thus, the problem can be optimally solved in $O(n^2)$ time. Figure 3.14(d) shows the LAG for the problem in Figure 3.14(c).

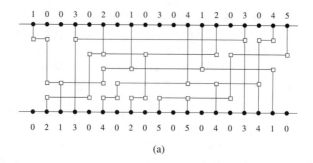

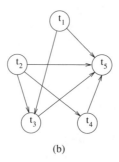

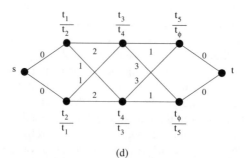

Fig. 3.14 **(a)** A two-layer solution; **(b)** track-ordering graph; **(c)** an optimal scheduling solution; **(d)** a layer assignment graph.

3.6 Multilayer Channel Routing Algorithms

Due to planarization and improved fabrication techniques, more and more routing layers are being made feasible. As a result, there is a need for developing multilayer routing algorithms. Since we do not explicitly use a multilayer router, we present a very short summary of results in this area.

In [Ham85], Hambrusch presented an algorithm for an n-layer channel router. The number of layers, the channel width, the amount of overlap, and the number of contact points are four important factors for routing multiterminal nets in multilayer channels. An insight into the relationship between these four factors is also presented in [Ham85]. In [BBD86], Braun et al. developed a multilayer channel router called *Chameleon*, based on YACR2. The main feature of Chameleon is that it uses a general approach for multilayer channel routing. Stacked vias can be included or excluded, and separate design rules for each layer can be specified. Chameleon consists of two stages: a partitioner and a detailed router. The partitioner divides the problem into two- and three-layer subproblems such that the global channel area is minimized. The detailed router then implements the connections using generalizations of the algorithms employed in YACR2. In [ED86], Enbody and Du presented two algorithms for n-layer channel routing that guarantee successful routing of the channel for n greater than 3.

Because many standard cell designs can be completed without channel areas by using the three-layer metal process by OTC techniques, four-layer channel routing may not be of interest. In the case of full custom, perhaps four layers would be sufficient to obtain layouts with a zero-routing footprint. However, new technologies, such as MCM, require true multilayer capabilities, since as many as 64 layers may be used.

3.7 Terminal Assignment Algorithms

In several OTC routing models, nets have to be routed from terminal rows in the middle or center of the cells to the boundary of the cells. Due to the restrictions of vias, routing must be planar, and therefore nets cannot change their order, but they can change their relative positions. The relative change can be beneficial for OTC routing or channel routing. This problem is generally studied under the topic of pin assignment problems. The problem has various versions depending on how the pin assignment constraints are specified. Many special cases of this problem have been investigated.

In [GCW83], Gopal, Coppersmith, and Wong considered the channel routing problem with movable terminals. Figure 3.15 shows how channel density could be reduced by moving the terminals. Figure 3.15(a) shows a channel that needs three tracks. By moving the pins, the routing can be improved such that it requires one track as shown in Figure 3.15 (b).

In [YW91], the channel pin assignment problem in which the assignment of terminals is subject to linear order position constraints is solved using a dynamic programming formulation by Yang and Wong. In [BPS93b], Bhingarde, Panyam, and Sherwani present an algorithm for assigning the terminals in a channel such that channel density is minimized and nets are routable in a given number of tracks.

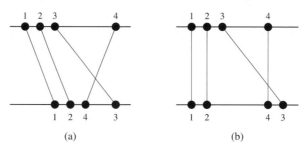

Fig. 3.15   (a) A channel with density = 3; (b) pin reassignment to reduce density to 1.

In this section, we describe pin assignment algorithms for channel routing problems [YW91]. Also, a pin assignment algorithm [BPS93b] specific to a routing problem in CTM and MTM is elaborated.

3.7.1 Terminal Assignment for Minimum Density

Consider a horizontal channel with an imaginary superimposed grid. Terminals can be placed only at the grid points at the top and the bottom of the channel. Let L be the length of the channel and N be the set of nets to be routed. The set of terminals on the top of the channel is denoted by TOP $= \{t_1, t_2, \ldots, t_p\}$, $p \leq L$, and the set of terminals on the bottom of the channel is denoted by BOTTOM $= \{b_1, b_2, \ldots, b_q\}$, $q \leq L$. Let $N = \{N_1, N_2, \ldots, N_n\}$ be a partition of TOP $\cup$ BOTTOM, giving the net list of the channel where N_i is the set of terminals belonging to net i, $1 \leq i \leq n$. The position constraints are specified by $T = T_k \in \{1, 2, \ldots, L\} : \forall 1 \leq k \leq p$, which means that t_k can only be assigned to positions in T_k, $1 \leq k \leq p$. Similarly the bottom constraints are specified by $B = B_k \in \{1, 2, \ldots, L\} : \forall 1 \leq k \leq p$. $R_T = (\text{TOP}, <_T)$ is a partial order set (poset) defining the order constraints for the terminals on the top of the channel; i.e., if $t_i <_T t_j$, then t_i must be assigned to the left of t_j. Similarly, $R_B = (\text{BOTTOM}, <_B)$ specifies the order constraints for the terminals on the bottom of the channel. An instance of this channel pin assignment (CPA) problem is represented as an 8-tuple $\Phi = (L, \text{TOP}, \text{BOTTOM}, N, T, B, R_T, R_B)$.

A solution to Φ is an ordered pair of functions $\pi = (f, g)$, where f: TOP $\to \{1, 2, \ldots, L\}$ and g: BOTTOM $\to \{1, 2, \ldots, L\}$ such that

$$f(t_k) \in T_k, \forall k, 1 \leq k \leq p$$
$$g(b_k) \in B_k, \forall k, 1 \leq k \leq q$$

$$t_i <_T t_j \Rightarrow f(t_i) < f(t_j), \forall t_i, t_j \in \text{TOP}$$
$$b_i <_B b_j \Rightarrow g(b_i) < f(b_j), \forall b_i, b_j \in \text{BOTTOM}$$

Both f and g are injective functions.

The CPA problem with position and order constraints was shown to be NP-hard in general, and the problem of finding whether a solution exists for a CPA problem with a density less than or equal to a fixed integer d was shown to be NP-complete, even if all the nets are two-terminal nets having one terminal on each side of the channel. Hence, it is very unlikely that this problem can be solved efficiently. The algorithm described here is optimal for an important case of the CPA problem with position and order constraints, where both R_T and R_B are linearly ordered sets. For convenience, the channel is augmented with column 0 and two trivial nets $N_0 = \{t_0\}$ and $N_0' = \{b_0\}$, with $t_0 <_T t_1, b_0 <_B b_1$ and $T_0 = \text{TOP}, B_0 = \text{BOTTOM}$ introduced. A solution to Φ is called an (i, j, k) solution if it assigns exactly $t_0, t_1, \ldots, t_i, b_0, b_1, \ldots, b_j$ to the first k columns of the channel. The number of nets with their left-most terminals to the left of (or on) a column and their right-most terminals to the right of (or on) that column is called the *density* at that column. The maximum of the densities at the first k columns of a channel is called the *k-density* of the channel.

A function $d : \{0, 1, \ldots, p\} \times \{0, 1, \ldots, q\} \times \{0, 1, \ldots, L\} \rightarrow I^0 \cup \{+\infty\}$, where I^0 denotes the set of nonnegative integers, is called the *density function* with respect to Φ. If for all values of i, j, k, Φ has an (i, j, k) solution, then $d(i, j, k)$ is equal to the minimum of the *k-densities* of all (i, j, k) solutions of Φ, else $d(i, j, k) = +\infty$. The main idea of the algorithm is as follows. Given Φ, the density function with respect to Φ is computed using dynamic programming, and the optimal solution is reconstructed by a backtracking procedure.

The (i, j, k) solutions can be classified into the following four types according to the pin assignment at column k as illustrated in Figure 3.16.

- **Type 0:** No terminal is assigned to either end points of column k as shown in Figure 3.16(a).
- **Type 1:** Only t_i is assigned to the top end point of column k as shown in Figure 3.16(b).
- **Type 2:** Only b_j is assigned to the bottom end point of column k as shown in Figure 3.16(c).
- **Type 3:** Both t_i and b_j are assigned to column k as shown in Figure 3.16(d).

Let us define sets $R_1(i, j)$, $R_2(i, j)$, and $R_3(i, j)$, as follows:

$R_1(i, j)$: A set of nets with one terminal in $\{t_1, t_2, \ldots, t_{i-1}, b_1, b_2, \ldots, b_j\}$ and another terminal in $\text{TOP} \cup \text{BOTTOM} - \{t_1, t_2, \ldots, t_i, b_1, b_2, \ldots, b_j\}$ and the net containing t_i if it is not trivial.

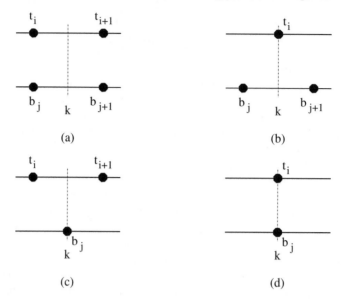

Fig. 3.16 Four types of (i, j, k) solutions: **(a)** type 0; **(b)** type 1; **(c)** type 2; **(d)** type 3.

$R_2(i, j)$: A set of nets with one terminal in $\{t_1, t_2, \ldots, t_i, b_1, b_2, \ldots, b_{j-1}\}$ and another terminal in TOP $\cup$ BOTTOM $- \{t_1, t_2, \ldots, t_i, b_1, b_2, \ldots, b_j\}$ and the net containing b_j if it is not trivial.

$R_3(i, j)$: A set of nets with one terminal in $\{t_1, t_2, \ldots, t_{i-1}, b_1, b_2, \ldots, b_{j-1}\}$ and another terminal in TOP $\cup$ BOTTOM $- \{t_1, t_2, \ldots, t_i, b_1, b_2, \ldots, b_j\}$ and the nets containing t_i and b_j if they are not trivial.

The algorithm finds $R_1(i, j)$, $R_2(i, j)$, and $R_3(i, j)$ for $0 \le i \le p, 0 \le j \le q$. Next, it calculates the *crossing numbers at (i, j, k)* for all $0 \le i \le p, 0 \le j \le q, 0 \le k \le L$, which are defined as

$$x(i, j, k) = \begin{cases} +\infty & \text{if } k \notin T_i \\ |R_1(i, j)|, & \text{otherwise} \end{cases}$$

$$y(i, j, k) = \begin{cases} +\infty & \text{if } k \notin B_j \\ |R_2(i, j)|, & \text{otherwise} \end{cases}$$

$$z(i, j, k) = \begin{cases} +\infty & \text{if } k \notin T_i \cap B_j \\ |R_3(i, j)|, & \text{otherwise} \end{cases}$$

The procedure to compute the crossing numbers $x(i, j, k)$, of type 1 is given in Figure 3.17. It can be easily seen that the complexity of this procedure is $O(pqL)$. Crossing numbers $y(i, j, k)$, of type 2 and $z(i, j, k)$ of type 3 can be computed by similar procedures.

The actual algorithm is described as follows. The input to the algorithm is an instance $\Phi = (L, \text{TOP}, \text{BOTTOM}, N, T, B, R_T, R_B)$ of the linear CPA problem,

Procedure: COMPUTE_CROSS (1)

Input: TOP, BOTTOM, 1
Output: $x(i, j, k)$

```
begin
    for i = 0 to p do
        for j = 0 to q do
            begin
                COMPUTE (|R₁(i, j)|)
                for k = 0 to L do
                    x(i, j, k) = +∞
            end;
    for i = 0 to p do
        for j = 0 to q do
            for k = 0 to L do
                x(i, j, k) = |R₁(i, j)|
end.
```

Fig. 3.17   The procedure to compute the crossing numbers of type 1.

and the output is an optimal solution to Φ, or an indication that no such solution exists. The algorithm consists of four phases.

The first phase is the initialization phase in which the density function is initialized for certain values of i, j, k. The array $d(i, j, 0)$ for $0 \le i \le p, 0 \le j \le q$ is initialized to $+\infty$ since no solution exists for $k = 0$. However $d(0, 0, 0) = 0$, since no terminals are assigned in zero columns. Also, $d(0, 0, k) = 0$, for $1 \le k \le L$, since no terminals are assigned on the top and bottom within k columns. In the second phase, the crossing numbers are computed using the procedure COMPUTE_CROSS. In the third phase, $d(p, q, l)$ is computed using dynamic programming. At each column k, the four types of (i, j, k) functions are computed for $0 \le i \le p, 0 \le j \le q$. For instance, the type 1 function denoted by D_1 is taken as the maximum of $d(i - 1, j, k - 1)$ and $x(i, j, k)$. Hence, it is the maximum of the density up to $k - 1$ columns and the local density at the kth column when i terminals are assigned at the top and j terminals assigned at the bottom. In addition, we require that the ith terminal is assigned at the top of the kth column and no terminal is assigned at the bottom of the kth column. Similarly D_2(type 2) and D_3(type 3) are computed. The value of $d(i, j, k)$ is the minimum of D_1, D_2, D_3 and $d(i, j, k - 1)$. In the fourth phase, the optimal solution, if it exists, is constructed using backtracking. At each column k, starting from column L, it is found from which of the four (i, j, k) solutions $d(i, j, k - 1), D_1, D_2$, or

D_3, $d(i, j, k)$ is derived. For instance, if $d(i, j, k)$ is derived from D_1, then the top of the kth column is allotted to terminal t_i and the bottom of the kth terminal is not allotted to any terminal. Thus, all the terminals are allotted to various columns so as to minimize the channel density.

The algorithm is formally described in Figure 3.18. It can be easily seen that the time complexity of phase 1 is $O(pq + L)$ and that of phase 2 is $O(pqL)$ as explained earlier. In phase 3, the three nested **for** loops have a complexity of $O(pqL)$, and the backtracking procedure in phase 4 takes $O(L)$ time. Therefore, the overall time complexity of this algorithm is $O(pqL)$.

3.7.2 Terminal Assignment for Improved OTC Routing

In this section we consider the relationship between river routing and OTC routing. In some cell models, such as the MTM and CTM, the terminals are located in the cell areas and not on the boundaries. When vias are not allowed in OTC areas, the terminals have to be brought to the boundary in a planar fashion. This problem is referred to as *boundary terminal assignment problem* (BTAP). The cell boundary is essentially considered as a row of possible terminal locations. Each net terminal is assigned a terminal location on the boundary, considering river routing constraints while minimizing the channel density. A river router is used to complete the connections, as specified by the terminal assignment, in a planar fashion in the M2 layer of the T and B areas. In the following, we present the algorithm developed in [BPS93b].

Given:

1. L, the total number of terminals available in a row.
2. TOP $= \{t_1, t_2, \ldots, t_p\}$, $p \leq L$, the set of terminals in the lower terminal row of the top cell row.
3. BOT $= \{b_1, b_2, \ldots, b_q\}$, $q \leq L$, the set of terminals in the upper terminal row of the bottom cell row.
4. TOP_BOU $= \{u_1, u_2, \ldots, u_L\}$, an ordered set of positions on the lower boundary of the top cell row.
5. BOT_BOU$= \{l_1, l_2, \ldots, l_L\}$, an ordered set of positions on the upper boundary of the bottom cell row.
6. k_3, the number of tracks in the B area of the top cell row.
7. k_1, number of tracks in the T area of the bottom cell row.
8. $N = \{n_1, n_2, \ldots, n_{N\_max}\}$, a net list that is a partition of TOP $\cup$ BOT.

The BTAP is to find two injective functions f and g such that:

1. $f(t_k) \in$ TOP_BOU.
2. $g(b_k) \in$ BOT_BOU.
3. If $U = \{a_1, a_2, \ldots, a_p\}$ is a set of nets, where $a_i = \{t_i, f(t_i)\}$, then U is river routable in k_3 tracks.

```
Algorithm LINEAR-CPA()
```

```
Input: Φ = (L, TOP, BOTTOM, N, T, B, R_T, R_B)
Output: Optimal solution to Φ
```

```
begin
    (* Phase 1: initialize *)
    for i = 0 to p do
        for j = 0 to q do
            d(i, j, 0) = +∞;
    for k = 0 to L do
        d(0, 0, k) = 0;
    (* Phase 2: Computer corossing numbers for all three types*)
    COMPUTE-CROSS(1);
    COMPUTE-CROSS(2);
    COMPUTE-CROSS(3);
    (* Phase 3: Dynamic programming phase *)
    for k = 1 to L do
        for i = 0 to p do
            for j = 0 to q do
                (* type 1 solution *)
                D₁ = max{d(i − 1, j, k − 1), x(i, j, k)};
                (* type 2 solution *)
                D₂ = max{d(i, j − 1, k − 1), y(i, j, k)};
                (* type 3 solution *)
                D₃ = max{d(i − 1, j − 1, k − 1), z(i, j, k)};
                d(i, j, k) = min{d(i, j, k − 1), D₁, D₂, D₃};

    if d(p, q, l) = +∞ then
        return φ is not feasible;
    else
    (* Phase 4: backtracking for constructing optimal solution *)
        i = p; j = q;
        for k = L down to 1 do
            if d(i, j, k) = D₁ then
                f(i) = k; i = i − 1;
            else if d(i, j, k) = D₂ then
                g(j) = k; j = j − 1;
            else if d(i, j, k) = D₃ then
                f(i) = k; g(j) = k;
                i = i − 1; j = j − 1;
        return π = (f, g);
end.
```

Fig. 3.18   The optimal channel pin assignment algorithm.

4. If $L = \{c_1, c_2, \ldots, c_q\}$ is a set of nets, where $c_i = \{b_i, g(b_i)\}$, then L is river routable in k_1 tracks.

5. If $N' = \{n'_1, n'_2, \ldots, n'_{N\_\max}\}$ is another set of nets, where $n'_i = \{f(t)|$ $t \in n_i, t \in \text{TOP}\} \cup \{g(t)| \, t \in n_i, t \in \text{BOT}\}$, then the density of nets in N' is minimized.

Theorem 14   *Problem BTAP can be optimally solved in $O(pqL)$ time.*

Proof: Nets in set U have to be planar. This imposes linear order constraints on the function f; i.e., in the set TOP sorted on a column position of the terminals, if terminals t_i and t_j are adjacent to each other such that $\text{COL}(t_i) < \text{COL}(t_j)$, then $\text{COL}(f(t_i)) < \text{COL}(f(t_j))$. Note that there are only $O(p)$ order constraints on f. Nets in set U have to be river routable in k_3 tracks, since this imposes position constraints T_i on $f(t_i)$ for each $t_i \in$ TOP. The position constraints corresponding to t_i are dependent on the value of k_3 and the number of nets starting from the positions on the left and right sides of t_i. Let $t_l \in$ TOP be the kth terminal on the left side of t_i, and $t_r \in$ TOP be the kth terminal on the right side of t_i. Let the function call $\text{COL}(t)$ return the column range of a terminal $t \in$ TOP $\cup$ BOU. Let the function call $\text{NUM}(c_1, c_2)$ return the number of terminals from set TOP that have a column position in $[c_1, c_2]$. We define $T_i = \{y_{il}, \ldots, y_{ir}\} \subseteq$ TOP_BOU, such that $y_{il} = u_p$ and $y_{ir} = u_q$, where

$$p = \begin{cases} \text{COL}(t_l) + k_3 & \text{if } t_l \text{ exists} \\ 1 + \text{NUM}(1, \, \text{COL}(t_i) - 1) & \text{if } t_l \text{ does not exist} \end{cases}$$

$$q = \begin{cases} \text{COL}(t_r) - k_3 & \text{if } t_r \text{ exists} \\ L - \text{NUM}(\text{COL}(t_i) + 1, L) & \text{if } t_r \text{ does not exist} \end{cases}$$

Note that the total number of position constraints on f is $O(p)$. Figures 3.19(a) and (c) show valid river routings. In Figure 3.19(a), t_l exists for t_2, and $y_{2l} = u_{11}$. Figure 3.19(b) shows that the net starting at t_5 is unroutable if the net starting at t_2 is routed on the left of u_{11}. In Figure 3.19(c), t_l does not exist for t_6, and $y_{6l} = u_3$. Figure 3.19(d) shows that the net starting at t_4 is unroutable if the net starting at t_6 is routed on the left of u_3. Similarly, the river routability constraints on L impose order and position constraints on g. With the transformation of the river routability constraints on U and L into the order and position constraints on f and g, the BTAP is transformed to CPA [BPS93b]. Generating order and position constraints on f requires sorting TOP on the column positions of terminals; i.e., it requires $O(p \log p)$ time. Similarly, to generate position constraints on g requires $O(q \log q)$ time. Thus, BTA can be transformed into CPA in $O(p \log p + q \log q)$ time. Since CPA can be optimally solved in $O(pqL)$ time, the theorem follows.

$\square$

In [WHSS92], a faster algorithm with the primary objective of eliminating vertical constraints has been proposed for boundary terminal assignment. How-

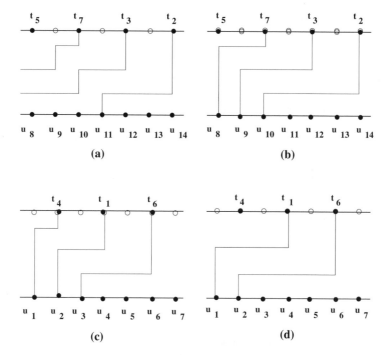

Fig. 3.19   River routing constraints.

ever, this algorithm requires the presence of a large number of vacant terminals. Also, it may not produce optimal channel heights.

3.8 Summary

Our approach to three-dimensional routing draws heavily from channel-based approaches. Channel routing problems have been extensively studied in VLSI. Channel routing plays a critical role in detailed routing. In this chapter, we have reviewed the basic terminology and concepts of channel routing. We have presented basic graphs, such as horizontal- and vertical-constraint graphs, which are used to explore channel routing problems. We have classified the channel routing algorithms according to the number of routing layers. In the case of a single layer, the problem is known as the river routing problem, which is well studied and understood. The two-layer problem is NP-complete, and as a result several heuristic algorithms have been presented. Perhaps the best algorithm is the greedy channel router, and it produces a solution that is close to optimal in majority of the cases. For three-layer routing, we have examined an algorithm that obtains a solution from a two-layer solution. We have also reviewed multilayer routing algorithms. Another problem of concern in routing is pin or terminal assignment. We have presented two algorithms: the first one minimizes the channel density, while the other is used to find a planar routing solution in many routing problems.

PROBLEMS

3-1. Develop an algorithm for planar routing in a channel in the presence of obstacles.

3-2. The river routing techniques discussed in this chapter have channels with straight boundaries. However, in layouts using target-based cell designs, the channels are irregular, as shown in Figure 3.20. Develop a planar routing algotrithm to route the channels with irregular boundaries.

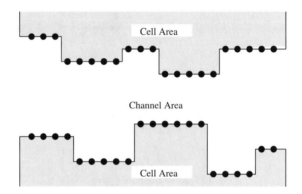

Fig. 3.20   A channel with irregular boundaries.

3-3. Develop a greedy routing algorithm to route a channel with irregular boundaries.

3-4. A typical standard-cell design consists of two metal layers in OTC areas and three metal layers in the channel for routing, as shown in Figure 3.21(a). Develop an algorithm for routing in this environment.

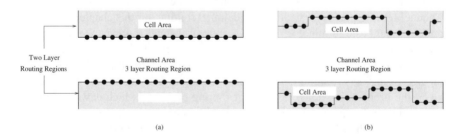

Fig. 3.21   An over-the-cell channel routing problem.

3-5. Consider the OTC irregular boundary channel routing problem shown in Figure 3.21(b). Two metal layers are available for routing in OTC areas and three metal layers are available in the channel. Develop an efficient algorithm to solve this routing problem.

3-6. In OTB routing problems with two metal layers for routing, the routers are allowed to switch routing layers only at certain specific regions. The routing regions are as shown in Figure 3.22. Develop an efficient routing algorithm to solve this problem.

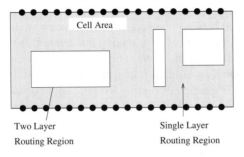

Two Layer   Single Layer
Routing Region   Routing Region

Fig. 3.22   An over-the-block routing problem.

3-7. When vias are not allowed in OTC areas and terminals are nonaligned, the OTC routers have the following problem:
(a) Routing the nets connecting the upper cell row to the upper boundary.
(b) Routing the nets connecting the lower cell row to the lower cell boundary.
(c) Routing the intrarow nets.
The problem is illustrated in Figure 3.23. Develop an efficient routing technique for this problem.

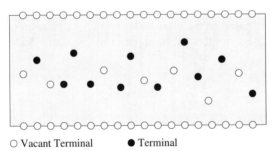

○ Vacant Terminal   ● Terminal

Fig. 3.23   A double-sided planar routing problem.

3-8. Based on the optimal CPA algorithm, develop an efficient pin assignment technique to maximize the independent set for a given channel permutation.

3-9. Develop a CPA algorithm to maximize the maximum bipartite set.

3-10. Develop a CPA technique to minimize vertical constraints.

BIBLIOGRAPHIC NOTES

The general single-layer routing problem was shown to be NP-complete in [Ric84]. In [BP83a, DKS87, LP83, SD81, Tom81], several restricted single-layer routing problems have been solved optimally. River routing was defined in

[DKS87] and refined by Leiserson and Pinter [LP83]. Several extensions of the general river routing algorithm have been proposed [JP89, TH90]. In [JP89], the river routing algorithm is extended to handle multiple parallel channels. River routing is called *feed-through river routing*, because wires must pass through gaps that are to be created between the components in each row. In [TH90], Tuan and Hakimi presented a variation of river routing that minimizes the number of jogs.

In [HS71], Hashimoto and Stevens first introduced the channel routing problem. Another column-by-column router has been proposed by Kawamoto and Kajitani [KK79] that guarantees routing with the upper bound on the number of tracks equal to the density plus 1, but additional columns are needed to complete the routing. Ho, Iyenger and Zheng developed a simple but efficient heuristic channel routing algorithm [HIZ91]. The algorithm is greedy in nature and can be generalized to switch boxes and multilayer problems.

Some other notable efforts to solve the switch-box problem have been reported by Hamachi and Ousterhout [HO84]. This approach is an extension of the greedy approach for channel routing. In [Joo86], Joobbani proposed a knowledge-based expert system called WEAVER for channel and switch-box routing. In [LHT89], Lin, Hsu, and Tsai presented a switch-box router based on the principle of evolution. In [GH89], Gerez and Herrmann presented a switch-box router called PACKER. PACKER is based on a stepwise reshaping technique. WEAVER [Joo86] is an elaborate rule-based expert-system router that often produces excellent-quality routing at the cost of excessive computation time. SILK [LHT89] is a switch-box router based on the simulated evolution technique. A survey and comparison of switch-box routers has been presented by Marek-Sadowska [Mar92].

4

Routing Models

Effective routing depends on several factors. The two most important factors are the routing model and the routing algorithm. In this chapter, we will concentrate on routing models. In particular, we will consider chip and MCM routing models.

The chip routing model specifies in detail the routing resources, such as the number of tracks available in the routing areas, the location of these tracks, and the permissibility of vias over active areas. Thus, the routing model specifies exactly the shape and size of the routing areas. At the chip level, the routing model is based on the cell and block model and the fabrication process. The purpose of the cell and block model is to guide the designer in creating the layout of cells and blocks. Designers must follow the cell and block model, since the routing model is defined by it. The routing algorithms are designed to optimally utilize the routing areas as defined by the routing model. There is a close relationship between routing models and the associated routing algorithms. A "bad" routing model may fragment the routing area, and, as a result, it may lead to inefficient routing. A "good" routing model can aid in the development of an efficient routing algorithm, and it can lead to optimal utilization of OTC and OTB routing areas. Since a "good" routing model is essential in the development of a fast router and better utilization of the routing area, it is critical to study the different parameters defining a routing model.

The MCM routing model specifies in detail the usage of layers and terminal locations, among other parameters. Since the MCM routing problem is complex, an effective routing model helps in simplifying the routing task, as well as improving the utilization of the substrate. In fact, the routing model also guides the placement process to produce routable MCM designs.

In this chapter, we will first discuss the chip-level routing model. We will describe the paramaters involved in the chip-level routing model and cell and block routing models. We next discuss cell layouts using different cell models. Finally, we present a routing model for MCMs.

4.1 Significance of Chip-Level Routing Models

The die size required for a circuit layout is mainly dependent on the routing model. A routing model specifies the following:

1. The shape of the routing region in each layer.
2. The number of metal layers.
3. Terminal locations.
4. Presence of equipotential terminals.
5. Restriction on usage of vias.

The selection of the routing model with an emphasis on achieving a zero-routing footprint leads to an increase in cell widths (see Figure 4.1(a)). On the other hand, the selection of the routing model with an emphasis on cell model does not generate layouts with a zero-routing footprint (see Figure 4.1(b)). A "good" routing model contributes to the reduction in die size by:

1. Reducing the width of the layout.
2. Achieving a zero-routing footprint.

The reduction in layout width is obtained by reducing the cell widths and the zero-routing footprint is achieved by providing the necessary flexibilities for the routing algorithms to complete the entire routing in OTC areas, as shown in Figure 4.1(c).

The standard cells are designed such that all the cells in a library have identical cell heights. Furthermore, cell designers attempt to optimize the cell width for a required cell performance. However, minimum-width cells may not always lead to minimum chip areas, since the chip area is determined by the product of the maximum row width (W_r) and the total chip height (H_c). The width of a row is the sum of widths of all the cells placed in that row and W_r is the maximum among all the row widths. The chip height H_c is given by

$$H_c = m \times h_{\text{cell}} + \sum_{i=1}^{m-1} h_i,$$

where m is the total number of cell rows in the layout, h_{cell} is the cell height, and h_i is the height of the ith channel. Since the cell height is fixed for a cell library, the objective of cell designers has been to design cells with minimum cell widths, thereby reducing row widths. Even though these designs may have minimal row widths, they may not always lead to minimum chip areas, since the chip areas

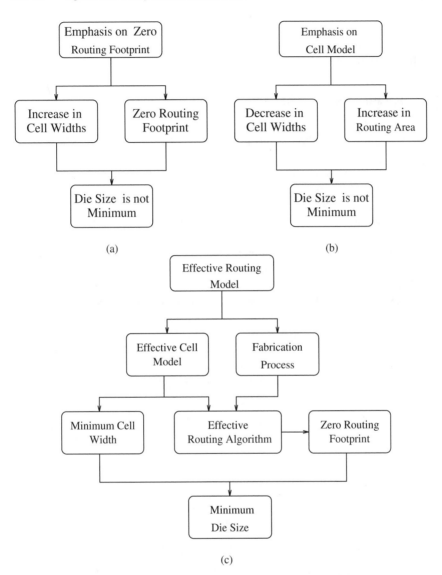

Fig. 4.1   Die size dependent on routing model: **(a)** emphasis on zero-routing footprint; **(b)** emphasis on cell model; **(c)** effective routing model.

are also dependent on the layout height. The minimum layout height is obtained when all the interconnections are completed in the OTC area; i.e., the channels are eliminated. The layout height is influenced by the routing model. The objective of the cell designer in minimizing the chip height is to provide flexibility to the routing algorithms in using the OTC areas for routing.

In the next section, we discuss the cell and block design parameters that affect the utilization of the OTC and OTB area for routing and then we will analyze various cell models based on these parameters.

4.2 Parameters for Chip-Level Routing Model

Let us examine the parameters that influence cell design. First, let us study the basic structure of the CMOS cell. In a typical CMOS standard-cell design, the p-type and n-type transistors are placed in two rows, as shown in Figure 4.2. The poly lines are routed vertically across the two diffusions to form the CMOS transistors. The two diffusions are not placed adjacent to each other on the same row. This is influenced by the MOSIS rules, which require the p-type and n-type diffusions to be separated by at least 10λ. Hence, the row width increases if all the cells in the row are placed 10λ apart. The transistor count in a cell depends on the functionality of the cell. For example, a two-input NAND gate cell can be designed using two CMOS transistors, and a four-input NOR gate cell can be designed using four CMOS transistors. Now the power and ground lines, terminals, and feedthroughs have to be placed. These parameters play a critical role in determining the cell width and efficiency of OTC routing. Some of the parameters, such as locations of terminals, V_{DD} and V_{SS} lines, and feedthroughs are related to the cell model, while the other parameters, such as number of metal layers and the permissibility of the vias in OTC and OTB areas, are entirely dependent on the fabrication process. The classification of various parameters is shown in Figure 4.3.

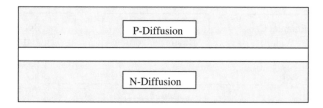

Fig. 4.2   Diffusion rows in a typical standard-cell design.

4.2.1 Parameters Based on Cell Models

The layout-related parameters are dependent on the location of terminals, V_{DD} and V_{SS} lines, and feedthroughs as discussed below.

- **Location of Terminals:** Among several other parameters, the location of terminals is an important parameter which decides the effectiveness of utilization of OTC metal layers for routing. As illustrated in Figure 4.4, the location of terminals specifies the following:
  1. Terminal layer.
  2. Terminal placement.

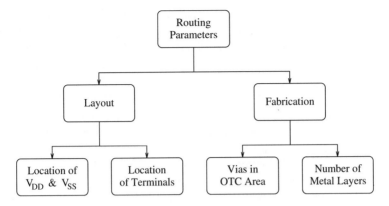

Fig. 4.3 Classification of routing parameters.

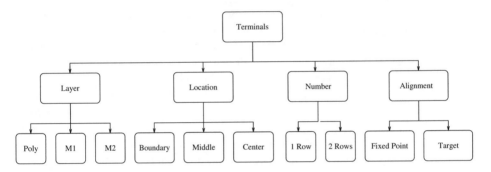

Fig. 4.4 Parameters based on location of terminals.

3. Number of equipotential terminals.

4. Alignment/shape of the terminals.

The terminal layer indicates the layer in which the terminals are located. The terminals may be located in poly, M1, or M2 layers. Terminal placement corresponds to the location of terminals with respect to the cell boundaries. Historically, one terminal row was located at each cell boundary to enable channel routing, since no routing layers were available for OTC routing [CPL90, HSS91b]. The terminals in a cell row may be placed at the center of the cell or at the cell boundaries, or may be located over the diffusion areas. When the terminals are placed in the center, we have only one terminal row, whereas when the terminals are placed at the boundaries, we may have two rows of equipotential terminals in each column. Usage of more than two rows of equipotential terminals in a cell leads to an increase in cell widths and cell design complexities. The terminals in a cell may or may not be aligned with each other. Even though misaligned terminals may ease the cell design to some degree, they do complicate the routing task. It is easier to develop routers for routing designs with aligned terminals. Routing problems with arbitrarily located terminals in cells may necessitate the use of less developed area routing techniques as compared to well-understood and well-developed channel routing techniques, which can be used in the presence of aligned terminals. When the terminals are aligned, there may be one or more rows of equipotential terminals.

- **Location of V_{DD} and V_{SS}:** Power and ground lines normally run along the cell widths. Since the feedthroughs are perpendicular to the V_{DD} and V_{SS} lines, they cannot be routed on the same layer. Hence, there are three options available for deciding the location of these lines. The first option is to route them in the center, adjacent to each other. In this case, it is not always possible to route them in the M1 layer, since the center portion of the M1 layer may be used for the local interconnects; hence, they should be routed in the M2 layer. This segments the entire M2 layer horizontally. The second option is to route one of them in M1 and the other in M2, overlapping with each other along one of the cell boundaries. This also segments the entire M2 layer horizontally. The third option is to route one of them along the upper cell boundary and the other along the lower cell boundary, both in the M1 layer. One advantage of this scheme is that it does not segment the M2 layer, and as a result it is the most popular scheme, since less segmentation leads to more effective use of the routing resources.

  If the terminals are located on the cell boundaries and if the feedthroughs are routed in M1, then V_{DD} and V_{SS} lines have to be in the M2 layer. Furthermore, when vias are not allowed in OTC areas, the M2 layer between V_{DD} and V_{SS} lines is not available for routing. Thus, V_{DD} and V_{SS} lines need to be very close to each other. Therefore, in this case, the most suitable location of V_{DD} and V_{SS} lines would be in the M2 layer, at the center of the cell, since it allows equal OTC area for the nets in the upper as well as lower channel.

  If the terminals are located in the center, in M2, then the nets have to be brought to the boundaries using the M2 layer. However, if V_{DD} and V_{SS} lines are in M2, they will obstruct the routes of these nets. The possibility of routing V_{DD} and V_{SS} in the middle, in the M1 layer, is also ruled out, since the M1 layer in the middle of cells is used for intracell routing. Thus, V_{DD} and V_{SS} lines may be routed in M1 along the cell boundaries.

- **Location of Feedthroughs:** The feedthroughs can be in either M1 or M2 layers. If they are in the M2 layer, the OTC area in the M2 layer is horizontally divided, and if they are routed in M1, the OTC area remains connected. The advantage of routing them in M2 is that the nets can be directly routed in the channel without having to use extra vias to change layers. The advantage of routing the feedthroughs in M1 is that the OTC area in M2 is not partitioned. The location of feedthroughs is based on the location of V_{DD} and V_{SS} lines. When the V_{DD} and V_{SS} lines are routed in the center, in the M2 layer, the feedthroughs must be routed in the M1 layer. However, when the V_{DD} and V_{SS} lines are routed in M1 at the boundaries, the feedthroughs may be routed in the M2 layer.

4.2.2 Parameters Based on Fabrication Process

The fabrication technology used in processing of chips dictate several parameters as discussed below:

- **Number of Layers:** Currently, two and three metal layers are allowed for routing in standard-cell design layouts and cell libraries are available for both technologies. Even though three-layer technology is expensive when compared with the two-layer technology, the third metal layer provides more routing space, resulting in layouts with a smaller area and shorter wire lengths. Therefore, three metal layers are used when performance is the most important criteria for design. When design specifications demand dedicated layers for power, ground, or clock lines, we opt for a fabrication process with more than three metal layers. But this is seldom used due to high cost factors.

- **Permissibility of Vias Over the Cell:** In some existing process technologies, vias are not allowed between the metal layers over the active diffusion areas. One of the reasons for this restriction is that the active surfaces are not planar, but have hills and valleys [Hna87]. If a via is introduced in one of the valleys over the active elements, the oxide layer, which tends to be thinner at the edges, breaks, thus causing a short circuit between the M1 and the active element. Also, it may introduce an open circuit in M2 at the point of the via. Recent technologies allow vias over the cell, since they use the planarization techniques to eliminate hills and valleys [Fan93, Hna87]. Depending on the process technology used for fabricating the chip, vias may or may not be allowed in OTC areas. When vias are not allowed, OTC metal layers can only be used for planar routing.

The permissibility of vias in OTC regions also influences the location of V_{DD} and V_{SS} lines. When vias are not allowed in OTC areas and the terminals are in the center, the V_{DD} and V_{SS} lines have to be routed at the boundaries in the M1 layer. However, when vias are allowed in OTC areas, and the terminals are in the center, the V_{DD} and V_{SS} lines may be routed in either the M1 or M2 layer.

4.3 Standard-Cell Routing Models

The routing parameters determine the effectiveness of the utilization of the OTC area in the standard-cell designs. The key parameters shown in Figure 4.4 are the terminal alignment and terminal placement. Other parameters shown in Figure 4.4 are based on terminal placement. For ease of OTC routing, the terminals are normally placed in the M2 layer. The OTC routing techniques are based on the alignment of terminals. The presence of aligned terminals leads to the usage of well-studied and well-understood channel routing algorithms, while the presence of nonaligned terminals leads to the usage of less understood area routers.

When the terminals are located in the center, then we have a single row of aligned terminals and the V_{DD} and V_{SS} lines are routed at the boundaries in the M1 layer. However, if the terminals are routed at the boundaries in the M2 layer, the V_{DD} and V_{SS} lines are routed in the center, in the M2 layer. Since the M1 layer is used extensively for intracell routing, the V_{DD} and V_{SS} lines are normally routed in the M2 layer. When the terminals are located in the middle (i.e., in the middle of upper and lower halves of the cell) in the M2 layer, the V_{DD} and V_{SS} lines can be located either in the center or at the boundaries in the M1 layer. The placement of V_{DD} and V_{SS} lines in the center leads to an obstruction in intracell routing in M1. Therefore, the power and ground lines are located at the boundaries in the M1 layer.

Based on these routing parameters, the standard-cell designs are broadly classified as *terminal row–based cell models* and *target-based cell models*. Each class has several subclasses. Their classification is shown in Figure 4.5 and are discussed below.

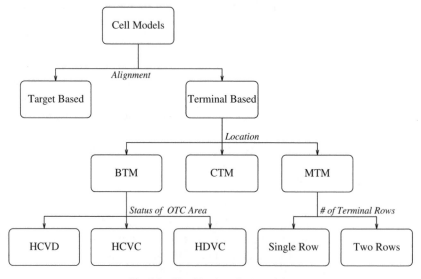

Fig. 4.5   Classification of cell models.

4.3.1 Terminal-Based Cell Models

The cell models with a fixed location for the terminals are terminal-based cell models. The location of the terminals could be either at the boundary or at the center, or at any fixed offset from the cell boundary. For ease of routing, the terminals are horizontally aligned. The set of aligned terminals in a cell row is called the *terminal row*. Standard-cell models may have either one terminal row or two equipotential terminal rows. The V_{DD} and V_{SS} lines are located either at the boundaries or in the center, depending on the cell design style. The location of terminals and V_{DD} and V_{SS} lines plays a very important role in the utilization of the OTC area for routing and will be discussed in Chapters 5 through 8. Based on these parameters, we currently have three different terminal-based cell design styles: *boundary terminal model* (BTM), *center terminal model* (CTM), and *middle terminal model* (MTM), as discussed below.

4.3.1.1 Boundary terminal models

This is the traditional cell model. It was introduced when only two metal layers were available for routing. The BTM-based cell libraries are used extensively in the industry. This is a traditional cell model introduced basically for channel routing. The concept of OTC routing was nonexistent, and hence the terminals were brought to the boundary. The traditional BTM cells have equipotential terminals at the top and bottom cell boundaries, in the poly layer. The power and ground lines are routed in M1 layer and are located just below the top terminal row and just above the bottom terminal row, respectively [Hei88]. The feedthroughs are routed in the M2 layer. The M2 and M3 OTC areas are completely free of any intracell routing. Based on the cell model considerations, Cong et al. presented three subclasses of BTM [CPL90].

1. **The Horizontally Connected, Vertically Divided Model:** In this model, cell terminals are on the M2 layer and feedthroughs on the M1 layer. Power and ground lines are routed on the M2 layer at the middle of the cell row. Clearly, the OTC routing region for each row of cells spans horizontally over the entire row of cells. However, it is divided into two parts vertically in M2 by the power and ground busses at the middle of the cells. This OTC routing model is called the *horizontally connected, vertically divided model* (BTM-HCVD), since the OTC routing layer, M2, is connected in the horizontal direction and disconnected in the vertical direction. (See Figure 4.6.) The BTM-HCVD contains the 2BTM-HCVD for the two-layer process and the 3BTM-HCVD-V for the three-layer process when vias are not allowed over the cell, and the 3BTM-HCVD+V when vias are allowed over the cell.

2. **The Horizontally Connected, Vertically Connected Model:** In this model, cell terminals and feedthroughs are both on the M1 layer. The power bus is on the M2 layer in the upper channel, just above the upper terminals, and the ground bus is on the M2 layer in the lower channel, just below the lower terminals. Clearly, the entire M2 layer over the cells is available for OTC routing. Therefore, this OTC routing model is called

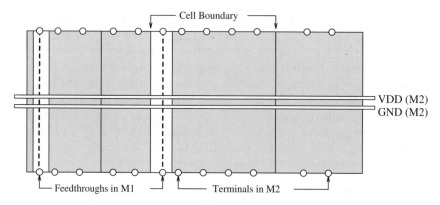

Fig. 4.6   HCVD-BTM.

the *horizontally connected, vertically connected* (HCVC) model. (See Figure 4.7.) The BTM-HCVC contains the 2BTM-HCVC for the two-layer process and the 3BTM-HCVC-V for the three-layer process when vias are not allowed over the cell, and the 3BTM-HCVC+V when vias are allowed over the cell. Since the preferred routing model in the three-layer environment is HVH, the 3BTM-HCVC-V and 3BTM-HCVC+V models have power and ground busses in M1.

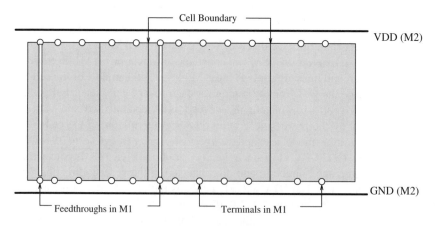

Fig. 4.7   HCVC-BTM.

3. **The Horizontally Divided, Vertically Connected Model:** In this model, both cell terminals and feedthroughs are on M2. Power and ground busses are on the M1 layer at the upper and lower edges of the cells. Clearly, the OTC routing region for each row of cells is divided horizontally (by feedthroughs) into many smaller regions. There is no vertical separation

within regions. This OTC routing model is called the *horizontally divided, vertically connected* (HDVC) model, since the OTC routing region is disconnected in the horizontal direction and connected in the vertical direction. (See Figure 4.8.) The BTM-HDVC contains the 2BTM-HDVC for the two-layer process and the 3BTM-HDVC-V for the three-layer process when vias are not allowed over the cell and the 3BTM-HDVC+V when vias are allowed over the cell. Table 4.1 summarizes the features of the three physical models.

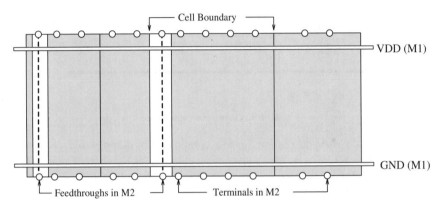

Fig. 4.8   HDVC-BTM.

Table 4.1:   BTM Designs for Over-the-Cell Routing

Model	Terminals	FTs	V_{DD}, GND busses	OTC routing
HCVD	M2	M1	M2, over cells	M2
HCVC	M1	M1	M2, in channels	M2
HDVC	M2	M2	M1, within cells	M2

4.3.1.2 Center terminal models

When more layers are available, it is necessary to develop a new class of cell model. A class of standard-cell models that has terminals in the center of the cell has been proposed in [WHSS92]. The three-layer model from this class is very efficient for processes that permit the use of vias in OTC areas. In the CTM, the terminals are located in M2, in the middle of the cell. The power and ground rails are in M1 near the top and bottom cell boundaries, respectively. (See Figure 4.9.) The connections within the cell are completed in M1. Thus, M2 is only blocked by terminals, and M3 is completely unblocked. OTC routers may use two rectangular regions, one above the terminal row and the other below the terminal row for routing in M2 and M3 layers. Vias may or may not be allowed in OTC areas, depending on the process. The CTM contains the 2CTM for the two-layer process and the 3CTM-V for the three-layer process when vias are not allowed over the cell, and the 3CTM+V when vias are allowed over the cell.

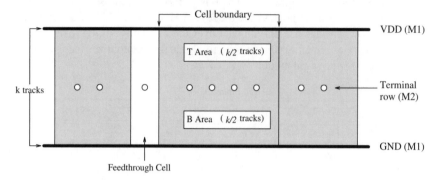

Fig. 4.9   Center terminal model.

In the 3CTM+V, the routing can be accomplished by doing two-layer channel routing in the OTC area and three-layer channel routing in the channel area. When vias are not allowed over the cell, the M2 layer is used for bringing the terminals to the cell boundary using river routing.

4.3.1.3 Practical cell designs in CTM

The cell designs in the CTM require the terminals to be located in the exact center of the cell height. In CMOS cell designs, minimum separation between n-diffusion and p-diffusion as given by MOSIS design rules is 10λ. Hence, we have at least a 10λ gap between diffusions, and the terminals can be located in this separation area if it is in the center of the cell height. But in some CTM designs, due to the presence of intracell routing, positioning the terminals exactly in the center may lead to an increase in the cell widths and also complicate the cell design task. In order to overcome this rigidity in the location of terminals, the cell designers are given a range of tracks in which to place the terminals. As shown in Figure 4.10(a), a range of p tracks from a total of k tracks in the OTC area are allocated for the location of the terminals. All the terminals are considered to be vertical M2 strips, with a uniform height of p tracks, irrespective of their place of origin. Figure 4.10(b) shows an instance of a practical CTM. As seen from the figure, we have only $k - p$ tracks in the M2 layer of the OTC area for routing. The larger the value of p, the larger is the number of choices for the location of the terminals. But an increase in the value of p decreases the effective OTC routable area. Hence, the selection of the most suitable value for p is a tradeoff between the flexibility given to the cell designer (in terms of locations of the terminals) and the resulting OTC routable area.

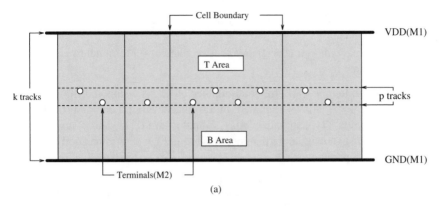

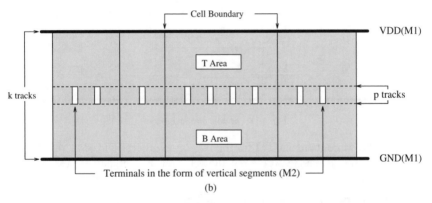

Fig. 4.10   Practical cell designs in the CTM: **(a)** range for the terminals; **(b)** the practical model for the CTM.

4.3.1.4 Middle terminal models

MTMs differ from the BTM and CTM in terms of terminal locations. In the MTM, the terminals are arranged in two rows, one row is located k_1 tracks below the upper cell boundary and another is located k_3 tracks above the lower cell boundary. Both terminals in a column of a cell are equipotential. In the MTM, the terminals are available in the M2 layer, and the power and ground rails are routed in the M1 layer near the top and bottom cell boundaries, respectively. Intracell routing is completed in poly and M1 layers, and as a result, the M2 layer is available for OTC routing. As opposed to two OTC rectangular regions in the CTM, cells in the MTM have three rectangular regions in the M2 layer, as discussed below (see Figure 4.11):

1. An area with k_1 tracks between the upper cell boundary and upper terminal row. We refer to this area as the top area, or T area.

2. An area with k_2 tracks (C area) between the lower terminal row and lower cell boundary.

3. An area with k_3 tracks (B area) between the lower terminal row and upper terminal row.

For a cell with height 150λ, the total number of tracks available in OTC area is 23. Setting $k_1 = k_3$ allows alignment of terminals even if a cell in a row is used in rotated fashion. Typical values of k_1, k_2 and k_3 are 6, 11, and 6, respectively. However, more extensive experimentation is required to determine good empirical values of these parameters.

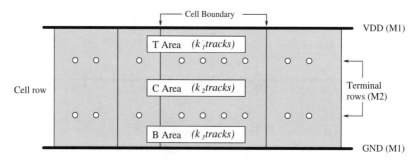

Fig. 4.11   Middle terminal model.

The MTM contains 2MTM for the two-layer process and 3MTM-V for the three-layer process when vias are not allowed over the cell, and 3MTM+V when vias are allowed over the cell.

4.3.1.5 Practical cell designs in MTM

The requirement of having fixed locations for the terminals in MTM for the entire cell library is rather rigid and may lead to an increase in cell widths for some cells. Hence, the cell designer is given a range of tracks in both the upper and lower halves of the cell to locate the terminals. As shown in Figure 4.12(a), we have p tracks in the upper and lower halves of the cell. The terminals in the practical MTM are considered short vertical segments in the M2 layer (see Figure 4.12(b)). The height of these segments is p tracks, and they are located k_1 tracks below the upper cell boundary and k_3 tracks above the lower boundary. Therefore, total OTC tracks k is given by $k = k_1 + k_2 + k_3 + 2p$. For routing flexibility, the value of p should be small, and for cell design flexibility, these values should be large. Hence, we have a tradeoff between the OTC routable area and the cell design flexibility similar to the practical CTM. This gives a flexible cell design in terms of terminal locations.

4.3.2 Target-Based Cell Models

The terminals in the target-based cell (TBC) models are in the form of long vertical segments in the M1 layer, called *targets*. Each target has a unique

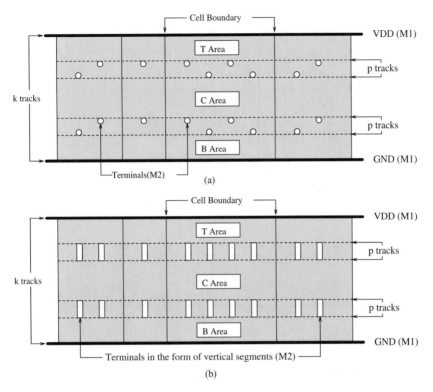

Fig. 4.12   Middle terminal model: **(a)** range for the terminal model; **(b)** a practical MTM model.

x-location. The height of the target is measured in terms of the number of tracks it overlaps. The targets in a cell row may have different heights and may be located at nonuniform offsets from the cell boundaries. The V_{DD} and V_{SS} lines are also in the M1 layer and they are located at the top and bottom cell boundaries, as shown in Figure 4.13. The fabrication process used for processing TBC designs should have at least three metal layers and should permit the use of vias in OTC areas. The exact location of the terminal in TBC designs is determined by the routing algorithm. The routing algorithm selects the most suitable location on each target and places a via there. This via location is called the *contact point*. These contact points are determined by the column-wise densities in each channel, and hence the contact points in a cell row may be nonaligned. The routing algorithms and the optimal contact point selection techniques are discussed in Chapter 8.

A new layout style as shown in Figure 4.14 was proposed in [YHH93a]. This layout style enables either an automatic synthesizer or a layout designer to use the second metal layer for intracell routing. In this design style, the transistors are placed in two parallel rows, with p-type transistors placed in one row and

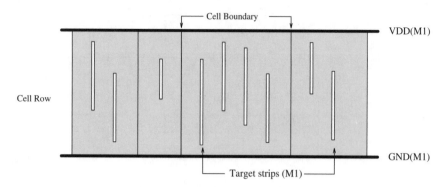

Fig. 4.13 Target-based cell designs.

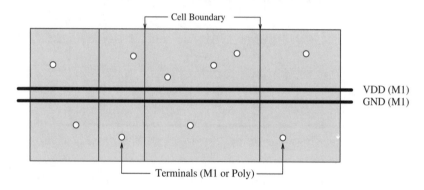

Fig. 4.14 Target-based cell designs using 2-metal intracell routing.

the n-type transistors in the other row. The V_{DD} and V_{SS} lines are routed in M1 between the diffusion rows. (See Figure 4.14.) The OTC area is divided into three regions:

1. The upper region (U region) is the area above the p-type-channel metal oxide semiconductor (PMOS) diffusion row.

2. The middle region (M region) is the area between the PMOS and NMOS diffusion rows.

3. The lower region (L region) is the area below the NMOS diffusion row.

Each pair of aligned p-type and n-type transistors are connected using a vertical polysilicon wire across the V_{DD} and V_{SS} lines. Each pair of aligned drain sources are connected using M2 segments. The gate signal targets are in poly, while the drain-source targets are in M1. Although this layout style aims at a reduction in cell areas, the usage of M2 segments for intracell routing may necessitate the requirement of complex routers.

4.4 Block Routing Models

There are two kinds of blocks that need to be considered in the MBC design style. Reused blocks are blocks that are partial or full layouts of some existing designs. New blocks are specifically designed for performance reasons.

4.4.1 Routing Parameters

There are two significant factors affecting the utilization of the OTB area, as discussed below.

1. **Location of Terminals:** It is reasonable to assume that all reused blocks have terminals on the boundary. This is due to the use of channels for routing. For new blocks, terminal location depends on the number of layers available for routing over the block. For both types of blocks, the routing is affected by the permissibility of vias.

   - If no routing is allowed over the block, then it is reasonable to assume that terminals are located on the boundary. Routing in OTB regions may not be allowed, since the metal used in routing may change the electrical characteristics of the block.

   - If a single layer is allowed for OTB routing, then the terminals may be located anywhere on the block. In this case, the routing must be planar, and essentially all the terminals must be "brought" to the boundaries.

   - If more than one layer is allowed for OTB routing, it is reasonable to assume that terminals may be located inside the block boundaries, in M2.

2. **Permissibility of Vias:** If more than one layer is allowed for routing in OTB regions, then the critical factor in the routing algorithms is the use of vias.

   - **Vias Not Allowed Over the Block:** If vias are not allowed in OTB regions, then these regions can be used for routing several planar (independent) sets of nets.

   - **Vias Allowed Over the Block:** If vias are allowed in OTB areas, then the problem of routing is similar to a multilayer switch-box problem if terminals are located on the boundary. However, if terminals are located inside the block, then the routing problems closely resemble the thin-film MCM routing problem.

     The use of vias may be unrestricted, that is, vias may be placed anywhere over the block, or we may restrict the placement of vias to certain regions. In the latter case, vias must be moved or "combed" to these regions.

4.5 Cell Layouts Using Different Cell Models

In this section, let us analyze the layout characteristics of some example cells in each of the four models discussed above, from a layout designer's point of view. For analysis, let us consider three gates: two-input NOR, two-input NAND, and a high-impedance buffer. The cell layouts are obtained using MAGIC Ver. 6.3 on a Sun Sparc 1+ workstation.

4.5.1 Two-Input NAND Gate Cell

The circuit schematic of a two-input NAND gate cell is as shown in Figure 4.15(a). The CMOS3 cell library [Hei88] uses the layouts in the traditional BTM style. The two-input NAND shown in Figure 4.15(b) has a cell height of 150λ and a width of 36λ [Hei88]. The terminals are in poly, and the V_{DD} and V_{SS} lines are in M1. The layout of such a gate is very simple. The gate poly of the transistors can be directly extended beyond the diffusions and the M1 power and ground lines, to form the input terminals V_{in1} and V_{in2}. For the output terminal V_{out}, originating in the M1 layer, a vertical poly segment outside the diffusion areas is used to generate equipotential terminals at the top and bottom boundaries. In this example (see Figure 4.15(c)), the width of the BTM cell is not affected by the input terminal locations, but in order to route the output terminal to the boundary, the cell width has to be increased.

The layout in the CTM requires the terminals to be located in the M2 layer, in the center, as shown in Figure 4.16(a). Since the MOSIS design rules specify the separation gap between the p-diffusion and n-diffusion to be at least 10λ, and in this example the diffusions are separated toward the center of the cell height, the terminals can be conveniently located in this isolation gap. The CTM requires the usage of two poly contacts, a via for each input terminal, and a via for the output terminal in this example.

The MTM layouts require the terminals to be located in M2, in the middle of the upper and lower halves of the cell. The location of terminals in M2 necessitates the usage of vias over the diffusion areas. However, some fabrication processes restrict the usage of vias in OTC areas. When such fabrication processes are being used, the areas under the terminals must be clear of diffusion. However, when the fabrication process permits the usage of vias in OTC areas, the terminals can be directly placed at the specified locations without any changes to the diffusion masks. The layout of a two-input NAND in the MTM is shown in Figure 4.16(b). The power and ground lines are in M1 at the cell boundaries. The equipotential terminals are vertically aligned, and since the terminals are in M2, they must be spaced at least 4λ apart, in accordance with the MOSIS design rules.

The layouts in TBC design style do not have any fixed locations for the targets, but have V_{DD} and V_{SS} lines located in the M1 layer at the boundaries. Since the height and location of the targets are independent for each terminal, the cell designer has the flexibility in selecting the most suitable location and

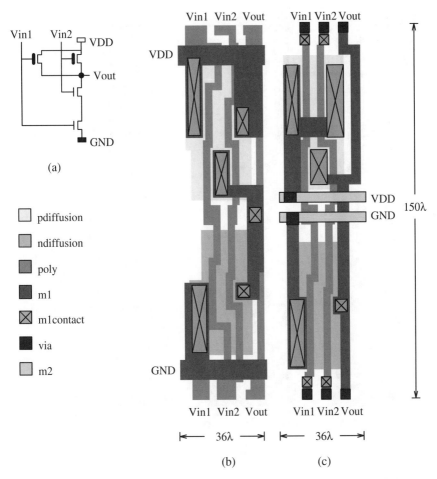

Fig. 4.15   Two-input NAND gate cell: **(a)** schematic circuit; **(b)** traditional BTM;
(c) HCVD-BTM.

maximum permissible height of each target. The layout of the two-input NAND
gate in the TBC is shown in Figure 4.16(c). The targets V_{in1}, V_{in2} and the output target V_{out}, are stretched to their maximum heights without any design rule
violations.

The objective of a TBC designer is to minimize cell widths while maximizing
the target heights. The minimization in cell widths leads to a reduction in the
corresponding row widths and an increase in the target heights gives flexibility for
the routers to minimize the channel heights. Since the total chip area is composed
of areas occupied by the cells and routing area, the area occupied by the cells can
be reduced by designing cells with minimum cell widths, and the routing area is
reduced by providing flexibility to the routers in the form of long targets.

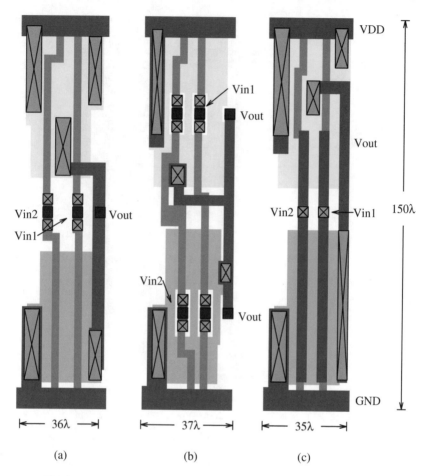

Fig. 4.16 Layout of a two-input NAND gate cell: **(a)** CTM; **(b)** MTM; **(c)** TBC.

4.5.2 High-Impedance Buffer Cell

The circuit schematic of a high-impedance buffer cell is shown in Figure 4.17(a). The layout of this gate in the traditional BTM [Hei88] is shown in Figure 4.17(b). In this cell, we have only two terminals, an input and an output. Both of these terminals are routed to the boundary in the poly layer, outside the diffusion areas, on either side of the diffusions. However, as shown in Figure 4.18, the cell layouts in the CTM, MTM, and TBC do not require this additional area for routing terminals to the boundary. Hence, for the high-impedance buffer cell, the width in the BTM is greater than the corresponding widths in other cell models.

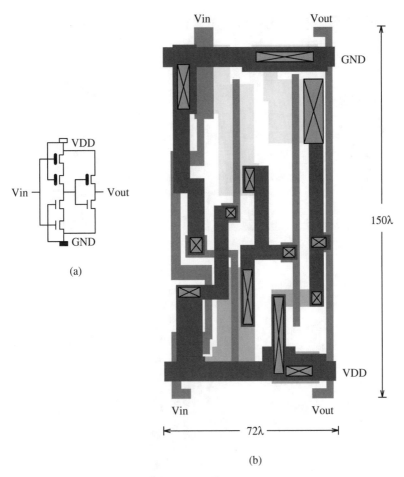

Fig. 4.17   Layout of a high-impedance buffer: **(a)** schematic; **(b)** BTM.

4.5.3 Two-Input NOR Gate Cell

The schematic circuit layout of a two-input NOR gate cell is as shown in Figure 4.19(a). The gate has two input terminals and one output terminal. The area required to route the terminals to the boundary in the BTM is used by the other models to locate the terminals in M2. Figures 4.19(b) and (c) show the gate in the BTM and CTM, respectively. The layouts of the gate in the MTM and TBC are shown in Figures 4.20(a) and (b), respectively. The layout of the two-input NOR gate is based on two CMOS transistors. The p-transistors are connected in series, while the n-transistors are connected in parallel. The simplicity in the intracell routing and the presence of only three terminals leads to identical cell widths in all four models.

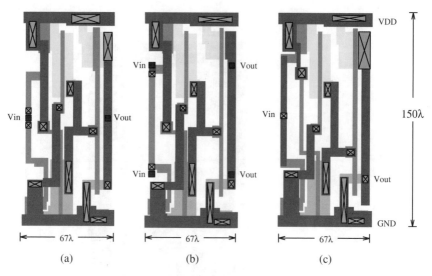

Fig. 4.18   Layout of a high-impedance buffer using: **(a)** CTM; **(b)** MTM; **(c)** TBC.

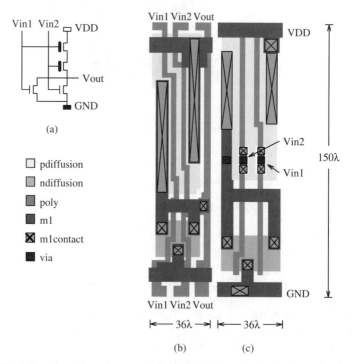

Fig. 4.19   Layout of a two-input NOR gate using: **(a)** schematic circuit; **(b)** BTM; **(c)** CTM.

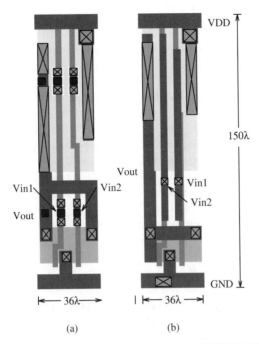

Fig. 4.20   Layout of a two-input NOR gate using: **(a)** MTM; **(b)** TBC.

4.6 Comparison of Different Cell Models

In this chapter, we first discussed the cell design parameters. Then we familiarized ourselves with various cell models and their physical layouts. Our main objective is the reduction of die size. The die size is reduced by developing cells with the twin objectives of reducing cell width and providing flexibility for the routers in the utilization of the OTC area for routing. Now let us compare the cell models for cell widths, and after developing routing algorithms for all the cell models in the subsequent chapters, we will compare the utilization of the OTC areas for routing.

4.6.1 Cell Width Analysis

In the BTM, the terminals are located at the top and bottom cell boundaries in the poly layer. As shown in Figure 4.15(a), the gate poly is routed from the top cell boundary to the bottom cell boundary, forming p-type and n-type transistors. These poly routes can be directly extended beyond the power and ground lines to form the input terminals. However, the cell outputs are usually available in the M1 layer, and they have to be brought to the boundary in the poly layer. Hence, the output terminals have to be routed to the boundary in the poly layer, outside the diffusion areas. This may tend to increase the cell width in the BTM cells. The problem of routing the terminals to the cell boundaries is eliminated in the CTM, since the terminals are located in the center, in M2. However, the cell width in the CTM may be larger as compared to the BTM for circuits with dense intracell connections due to two additional polycontacts and a via for each input terminal. In the MTM, two equipotential terminals have to be located in the M2 layer, one in the middle of the upper half of the cell and the other in the middle of the lower half. The intracell connections required to connect VDD and GND lines to the diffusions and the routing of the output terminals necessitate the use of the M1 layer. The terminals have to be located at least 3λ from these connections. In addition, the MTM requires the equipotential terminals to be aligned. This requirement leads to larger cell widths for some MTM cells with a large number of terminals. However, the TBC designs have flexibility in the location of targets. The targets are in the M1 layer and the exact location of the contact on these targets is determined by the routing algorithm. Even though the MOSIS design rules specify the minimum width and separation for M1 lines to be 3λ, the M1 targets in TBC designs require the minimum width and separation for M1 lines to be 4λ in order to accommodate vias in adjacent columns on the same track.

A list of cell-library elements is shown in Table 4.2. For ease of reference, the cell numbers correspond to the numbers in the CMOS3 cell library [Hei88]. The comparisons of the cell widths for these elements are shown in Table 4.3. The widths of some of the cells in the MTM are larger than the cells in the BTM and CTM, and some MTM cells also have smaller widths than the corresponding BTM cells. The cells in the TBC have the minimum width when compared with other cell models. The variation in cell widths between the cell models can be attributed to the following factors:

- **Intracell Routing:** Intracell routing over the diffusion areas is accomplished in the M1 layer, and routing between the diffusions is accomplished in both poly and M1 layers. For circuits with dense intracell routing, the exact terminal locations may lead to an increase in the cell width. The MTM requires the terminals to be vertically aligned and located at a specific location with respect to the cell boundaries. In the CTM, the terminals are located at the exact center. Hence, the cells in the CTM and MTM may be wider in circuits with dense intracell routing. The TBC design has an advantage over other cell models, since there is no fixed location for terminals. Hence, the cell widths in the TBC are always smaller than the corresponding widths in the other cell models.

- **Number of Terminals:** The cell widths also depend on the total number of terminals in the cell. This factor becomes critical in the MTM and CTM, since the terminals in these models have to be located in the M2 layer. For small cells (cell widths, typically, $<50\lambda$) with a large number of terminals (typically >5), the cells in the MTM and CTM may be wider than the corresponding cells in the BTM or TBC, since the TBC has the terminals in M1 and has the flexibility of locating them. The BTM has terminals in the poly, and the separation and thickness requirements for the poly are smaller than those of the M2 lines. Therefore, the total terminal count in a cell is not critical in determining the widths of BTM and TBC cells, whereas it is critical in determining cell widths in the MTM and CTM.

Table 4.2: Cell-Library Elements

Sl. No.	Cell No.	Cell Name	T	Cell Description
1	1120	2NOR	3	Two-input NOR gate
2	1130	3NOR	4	Three-input NOR gate
3	1140	4NOR	5	Four-input NOR gate
4	1220	2NAND	3	Two-input NAND gate
5	1230	3NAND	4	Three-input NAND gate
6	1310	INV	2	Inverter
7	1320	NINV	2	Noninverting buffer
8	1370	TRANS	3	Transmission gate
9	1420	NORLATCH	4	NOR latch
10	1430	PULLUP	1	Pull-up
11	1440	PULLDOWN	1	Pull-down
12	1550	HIB	2	High-impedance buffer
13	1560	DELAY	2	Delay gate
14	1570	DFLIP	4	D flip-flop
15	1660	2NAND-AND	4	Two-input NAND and AND gate
16	1670	3NAND-AND	5	Three-input NAND and AND gate
17	1680	4NAND-AND	6	Four-input NAND and AND gate
18	1740	4OR	5	Four-input NOR gate
19	1760	2OR-NOR	4	Two-input NOR and OR gate
20	1770	3OR-NOR	5	Three-input OR and NOR gate
21	1850	FULL ADDER	5	Two-bit full adder
22	2310	EX-OR	3	Exclusive-OR gate
23	2350	EXNOR	3	Exclusive-NOR gate

T = No. of terminals.

The cell width is also dependent on other electrical parameters such as fanout, timing specifications, and so on. However, the variations in cell widths between the cell models for a given set of electrical specifications are mainly due to the two parameters discussed above. Table 4.4 shows the number of terminals for some example cells and the change in cell width with respect to the BTM. From Table 4.4, it can be seen that the cell widths in TBC designs are always minimum when compared with the other cell models, and the variation in cell widths in the MTM and CTM are dependent on the number of terminals and on intracell routing.

Table 4.3:   Cell Width Comparison

Sl. No.	Cell	BTM	CTM	MTM	TBC
1	2NAND-AND	49	44	55	47
2	2NAND	36	36	37	35
3	2NOR	36	36	36	36
4	2OR-NOR	48	43	46	45
5	3NAND-AND	60	56	66	56
6	3NAND	48	45	54	41
7	3NOR	48	48	46	42
8	3OR-NOR	60	55	60	56
9	4NAND-AND	72	68	73	66
10	4NOR	60	54	56	54
11	4OR	73	67	68	67
12	DELAY	84	81	81	75
13	DFLIP	156	154	175	153
14	EX-OR	74	69	75	68
15	EXNOR	72	67	70	66
16	FULL ADDER	168	156	163	156
17	HIB	72	67	67	67
18	INV	36	33	31	31
19	NINV	36	35	37	32
20	NORLATCH	60	56	66	60
21	PULLDOWN	36	29	29	18
22	PULLUP	36	28	25	18
23	TRANS	72	72	72	72
	TOTAL	1336	1245	1320	1210

4.6.2 Gate-Delay Analysis

In the previous sections, the emphasis has been on the physical layout features. In this section, let us study the delay characteristics of the cells in the various cell models. For analysis, the layouts are extracted from MAGIC and simulated using SPICE. The switching speed of a CMOS gate is dependent on the time taken to charge and discharge the load capacitance C_L [WE85]. A change in the input logic level results in an output transition that either charges C_L toward V_{DD} or discharges C_L toward V_{SS} (input ground level). The load capacitance C_L is determined by adding the capacitances at the inputs of the next gate, routing capacitance, and the output capacitance of the gate under consideration.

There are several factors that contribute to the delay of the gate [WE85], as explained below:

1. The rise time (t_r) is the time required for the signal to rise from 10% to 90% of V_{DD}.

2. The fall time (t_f) is the time required for the signal to fall from 90% to 10% of V_{DD}.

Table 4.4:   % Change in Cell Widths with Respect to the BTM

Sl. No.	Cell	T	W_{BTM}	% Change in Cell Width		
				CTM	MTM	TBC
1	2NAND–AND	4	49	−10.20	12.20	−4.08
2	2NAND	3	36	0.00	2.77	−2.77
3	2NOR	3	36	0.00	0.00	0.00
4	2OR-NOR	4	48	−10.40	−4.10	−6.25
5	3NAND-AND	5	60	−6.60	10.00	−6.60
6	3NAND	4	48	−6.25	12.50	−14.50
7	3NOR	4	48	0.00	−4.10	−12.50
8	3OR-NOR	5	60	−8.33	0.00	−6.66
9	4NAND-AND	6	72	−5.55	1.38	−8.33
10	4NOR	5	60	−10.00	−6.66	−10.00
11	4OR	5	73	−8.20	−6.80	−8.20
12	DELAY	2	84	−3.50	−3.50	−10.70
13	DFLIP	4	156	−1.20	12.17	−1.90
14	EX-OR	3	74	−6.70	1.35	−8.10
15	EXNOR	3	72	−6.90	4.16	−8.30
16	FULL ADDER	5	168	−7.10	−2.90	−7.10
17	HIB	2	72	−6.94	−6.94	−6.94
18	INV	2	36	−8.30	−13.8	−13.80
19	NINV	2	36	−2.70	2.70	−11.10
20	NORLATCH	4	60	−6.66	10.00	0.00
21	PULLDOWN	1	36	−19.40	−19.40	−50.00
22	PULLUP	1	36	−22.20	−30.50	−50.00
23	TRANS	3	0.00	0.00	0.00	0.00

T = No. of terminals.
W_{BTM} = Cell width in the BTM.

3. The delay time is the time difference between t_1 and t_2, where t_1 (t_2) is the time taken by input (output) to change from the initial level to 50% of the final level. Delay time is qualitatively defined as the time taken for a logic transition from input to output.

Figure 4.21 shows the delay characteristics of a typical CMOS gate.

The delay of a single gate is dominated by the output rise and fall times. The delay is approximately given by $t_{dr} = t_r/2$ and $t_{df} = t_f/2$ [WE85]. The average gate delay (t_{av}) for rising and falling transitions is then calculated as

$$t_{av} = \frac{t_{dr} + t_{df}}{2} = \frac{t_r + t_f}{4}$$

Table 4.5 shows the circuit delays for a four-input OR gate in different cell models. It clearly can be seen that the gate delays in all the models are similar. Figure 4.22 shows the comparative delay performance of a four-input OR gate cell in four

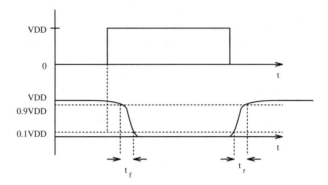

Fig. 4.21   Switching characteristics for CMOS gate.

Table 4.5:   Delay Comparisons for a Four-Input OR Gate

Cell Model	Total Area	Delay
BTM	$150\lambda \times 73\lambda$	1.0025 ns
CTM	$150\lambda \times 67\lambda$	0.9113 ns
MTM	$150\lambda \times 68\lambda$	0.9375 ns
TBC	$150\lambda \times 67\lambda$	0.8975 ns

models for different load capacitances. The gate delays in all the cell models are identical.

4.6.3 Layout Comparisons

In this section, we describe key advantages and disadvantages of each cell model in terms of cell layouts. The effectiveness of each cell model in using the OTC area for routing is discussed in detail in the subsequent chapters.

1. **Boundary Terminal Model:** The BTM-HCVD has two advantages. First, the terminals are in the M2 layer, and hence extra (or larger) vias are not needed to connect the nets routed in the M2 layer of the OTC area. Second, the OTC routing region horizontally spans the entire row of cells. This long span is advantageous when compared with the BTM-HDVC model in which the OTC routing region is divided horizontally into smaller regions. The BTM-HCVD model also has shortcomings. Since power and ground lines are in M2, some vias are required to bring them to M1, in addition to diffusion contacts required for connecting them to shared sources and drains. Also, extra vias are required to bring the terminals of feedthroughs up to the M2 layer for OTC routing and channel routing using the HV or

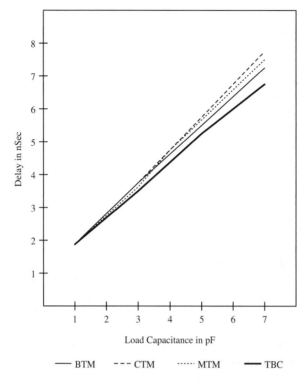

Load Capacitance in pF

—— BTM   --- CTM   ····· MTM   —— TBC

Fig. 4.22 Delay performance of different cell models.

HVH model. As power and ground lines are laid in the center, their proximity to each other induces coupling effects.

The BTM-HCVC model has two advantages. First, the entire M2 layer is available for OTC routing, which not only maximizes the opportunity for OTC connections, but also allows OTC connections of terminals between the upper and lower terminal rows of a cell row. Second, the terminals of the cells and the feedthroughs are all on the M1 layer, so they can be connected to branches without using extra vias. The BTM-HCVC model also has several shortcomings. Since cell terminals and OTC connections are on different layers, extra vias are needed to bring the terminals up to M2. In three-layer BTM-HCVC models, since the preferred model for channel routing is HVH, extra vias are needed for connecting the branches to the terminals. Cells based on the BTM-HCVC model will probably be wider than the cells of the other models because of the need to bring power and ground signals into the cells. In the layout based on BTM-HCVC cells, power and ground lines take up an additional two tracks in each channel.

The BTM-HDVC has two advantages. First, since terminals, feed-throughs, branches, and OTC connections are on the same layer, no extra vias are required to connect them. Second, since the poser and ground lines are in M1, they can easily be connected to diffusion. However, the OTC area of cells based on the BTM-HDVC is horizontally divided by the feedthroughs, which makes it difficult to use the OTC area. Also, some M1 routing area within the cells is lost, since power and ground lines are routed in the M1 layer within the cell.

2. **Center Terminal Model:** The separation gap between the p-diffusion and n-diffusion can be used to bring the terminals to the M2 layer. There-fore, no etching in diffusion is required to bring the terminals to the M2 layer. Since M1 is extensively used for intracell routing over the diffusion areas, avoiding the terminal locations over the diffusion areas minimizes the cell widths. Since the terminals are located between the diffusions, the intracell routing in the poly is restricted.

3. **Middle Terminal Model:** The terminals need not be extended all the way to the boundaries or to the center. The requirement of aligned terminals over the middle of the diffusions calls for an etching in the diffusion areas. Vias are to be used in these areas. Generating aligned vias in the active areas increases the cell widths. This is a serious disadvantage for developing the cells that have many terminals and very much less width.

4. **Target-Based Cells:** This model provides maximum flexibility to the designer. The terminals need not be extended all the way to the boundaries or to the center. Further, the terminals need not be aligned, or have uniform heights. The TBC is most suitable for cell-level design, since it imposes very few restrictions on the designer. The TBC designs have long vertical strips for the targets. This will increase the terminal capacitances.

4.7 MCM Routing Model

The MCM routing model specifies the number of layers and terminal locations. Before explaining the parameters that determine the routing model, a brief intro-duction to different types of MCMs is presented.

MCMs have been introduced as an alternative packaging approach to com-plement the advances taking place in IC technology. At present, there are three different types of MCMs available; they are MCM-L, MCM-C, and MCM-D.

MCM-L is the oldest technology available. MCM-L is essentially an ad-vanced PCB on which bare IC chips are mounted using chip-on-board (COB) tech-nology. The well-established PCB infrastructure can be used to produce MCM-L modules at low cost, making them an attractive electronic packaging alternative for many low-end MCM applications with low interconnect densities. MCM-L becomes less cost-effective at higher densities, where many additional layers are required. For cost effectiveness, MCM technology must increase the functionality of each layer instead of adding more layers. MCM-L is considered a suitable technology for applications requiring a low-risk packaging approach and where most of the steps have already been automated.

MCM-C (ceramic) refers to MCMs with substrates fabricated with cofired ceramic or glass-ceramic techniques. These have been used for many years, and MCM-C has been the primary packaging choice in many advanced applications requiring both performance and reliability. Ceramic substrates, having excellent thermal conductivity and low thermal expansion, have also been used to serve as the package. While interconnect densities are in the range of 200 to 400 cm/cm^2, this is still not enough for high-end applications.

MCM-D (deposited) technology is closest to IC technology. It consists of substrates that have alternating deposited layers of high-density thin-film metals and low-dielectric materials such as poly or silicon dioxide. MCM-D technology is an extension of conventional IC technology and is developed specifically for high-performance applications demanding superior electrical performance and high interconnect density. This technology, being relatively new, has neither a cost-effective manufacturing infrastructure nor a high-volume application that serves as a driving force. We refer to MCM-D as thin-film MCMs, while all other MCMs are referred to as general MCMs.

4.7.1 Parameters for Routing Model

The MCM routing model is composed of a *fabrication model* and a *terminal model*. The fabrication model determines the number of layers available for routing, while the terminal model is based on the chip attachment techniques adopted, as shown in Figure 4.23 and discussed below.

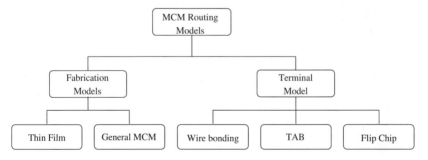

Fig. 4.23   MCM routing model.

1. **Terminal Model:** The chip attachment technique is specified by the terminal model. The two popular techniques are wire bonding, TAB, and flip-chip.

2. **Fabrication Model:** The fabrication model specifies the number of metal layers available for routing. In a thin-film MCM, 4 to 5 layers are allowed, while thick-film MCM, allow up to 63 (or more) layers for routing.

4.7.1.1 Parameters based on terminal model

The terminal locations are specified by the terminal model. The terminal locations are based on the chip attachment techniques. Bare chips are attached to the MCM substrates in the following three ways:

1. **Wire Bonding:** In this chip attachment technique, the back side of a chip (nondevice side) is attached to the substrate and the electrical connections are made by attaching very small wires from the I/O pads on the device side of the chip to the appropriate points on the substrate. The wires are attached to the chip by thermal compression. Wire bonding is shown in Figure 4.24(a). The terminals appear in rectangles forming one or two rows.

2. **Tape Automated Bonding:** TAB is a relatively new method of attaching chips to a substrate. It uses a thin polymer tape containing metallic circuitry. The connection pattern is simply etched on a polymer tape. As shown in Figure 4.24(b), the actual path is simply a set of connections from inner leads to outer leads. The inner leads are positioned on the I/O pads of the chips, while the outer leads are positioned on the connection points on the substrate. The tape is placed on top of the chip and substrate and pressed. The metallic material on the tape is deposited on the chip and substrate to make the desired connections. The terminals usually appear in a rectangle on the outline of the chip.

3. **Flip-Chip Bonding:** Flip-chip bonding uses small solder balls on the I/O pads of the chip to both physically attach the chip and make required electrical connections (see Figure 4.24(c)). This is also called *face-down bonding* or *controlled-collapse chip connections*. Terminals may be placed randomly on the chip surface, so they appear in a distributed fashion on the substrate.

Obviously, using chips that have terminals located evenly on the chip surface and connected to the MCM substrate using flip-chip bonding helps to distribute routing congestion evenly, whereas using the chips with terminals on the boundary requires a pin distribution phase while routing, which may consume two to three routing layers.

4.7.1.2 Parameters based on fabrication model

The fabrication model specifies the number of layers available for routing. The MCMs are classified into two types as discussed below.

1. **Thin-Film MCMs:** The term *thin film* does not refer to a specific range of film thickness, but implies the use of IC processes to achieve high-density

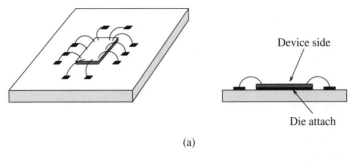

(a)

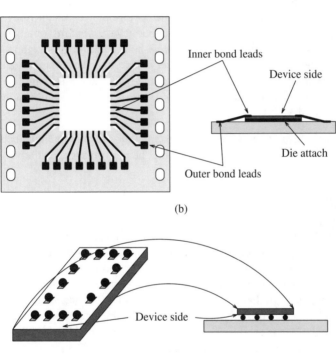

(b)

(c)

Fig. 4.24   Die attachment techniques: **(a)** wire bonding; **(b)** tape automated bonding;
(**c**) flip-chip bonding.

patterns in conductor and dielectric layers, roughly 2 to 25 μm thick. There
are several ways to incorporate thin-film multilayer interconnections into
multichip packaging structures. In one of the methods, the thin films may
be patterned on a blank substrate mounted in a second-level package such
as a perimeter-leaded flat pack. Electrical connections (such as wire bonds)
are made between the substrate and the package, and then the package is
sealed. The shapes and size of the substrate and other different design

parameters depend on the manufacturer. Some of the design parameters for thin-film technologies are substrate thickness of 5 μm, line pitch of 50 to 125 μm, bond pad pitch of 100 μm, and ability to use a maximum of six layers.

2. **General MCMs:** General MCMs or *thick film* is a technology in which specially developed pastes are deposited and patterned by screen printing onto a ceramic substrate. The paste can be formulated to produce any number of different passive electrical components such as conductors, resistors, inductors, and capacitors. The substrate in the thick-film-based technology is a passive inorganic material used to provide a mechanical platform for the thick-film circuit. Well-defined design rules exist for thick-film technologies. We can either have a *standard-* or a *custom*-built thick-film substrate. For the *custom*-built thick films, substrate sizes are greater than 7 in $\times$ 7 in, the number of layers can be several dozen, and the minimum via diameter, minimum line width, and minimum line spacing are 0.005 in. The large number of layers is the most significant feature of general MCMs.

The objective of MCM routing is to complete the chip interconnections in a minimum number of layers while satisfying the performance requirements.

4.8 Summary

In this chapter, we have investigated different factors that influence the cell and block and MCM layouts. We have presented several cell models based on location of terminals, location of ground and power busses, and permissibility of vias. We have compared the cell models for cell widths and delay performance. It can be seen that the cell models behave in an identical manner in terms of delay performance. However, the cell widths vary for different models. The parameters responsible for this variation in cell widths are the intracell routing and number of terminals in a cell. The die size is essentially determined by the cell widths and the flexibility offered by routing models to the routing algorithms in utilization of OTC areas for routing. TBCs have minimum cell widths. Therefore, chip designs using TBCs have minimum row widths. However, the area of a chip is determined by both the row width and the chip height. The chip height depends on the flexibility rendered to the routing algorithms by the cell model in effective utilization of OTC areas for routing. Hence, the performance of each cell model has to be analyzed at a global level, considering the routing flexibility and restrictions imposed by the routing models. In the next chapter, we introduce the concept of OTC routing and discuss various problems pertaining to OTC routing using the cell models illustrated in this chapter.

For MCM routing, we have presented two factors that influence routing effectiveness. These are the fabrication process, which determines the number of layers, and the die attachment technique, which determines the terminal location on the substrate.

PROBLEMS

4-1. In the middle terminal cell model, the equipotential terminals are required to be aligned in a column. This requirement tends to increase the cell widths. What is the reduction in cell widths if the equipotential terminals in a column are not required to be aligned?

4-2. In target-based cell designs, the targets are of nonuniform heights and appear at different offsets from the boundaries. This may necessitate the use of complex routers to route large designs based on the TBC. However, the task of routing can be simplified by generating TBCs consisting of uniform heights and placed at fixed offsets from the boundary. What is the percentage increase in cell widths if this flexibility is incorporated?

4-3. In middle terminal model, we have two equipotential terminals in each column. What is the reduction in cell width if we have only one terminal per column, either in the upper half or in the bottom half of the cell row?

4-4. The target-based cell model has vertical strips of M1 as targets. Design TBCs with L-shaped targets and compare the cell widths.

4-5. The standard cell-based layouts employing OTC routing techniques require interconnections, including the high-frequency clock signals, to be routed in OTC areas. What are the EMC and EMI effects due to clock lines on the cell performance?

BIBLIOGRAPHIC NOTES

The use of vias in OTC areas leads to step coverage and oxide undercut problems and is discussed in [Hna87]. The step coverage problem refers to how well the interconnection material will adhere to the nearly vertical sides of the contact holes in the insulating oxide that reaches down to the previous connecting layer. These problems can be overcome using new planarization techniques. In [Ell89, Hna87], the various planarization techniques for surface planarization are discussed in detail.

One disadvantage of planarizing in general is that automatic alignment depends on surface topography for light reflection. A planar dielectric will nearly remove the steps normally used for alignment of the M2 mask [Hna87].

New planarization techniques, the use of antireflection coatings, and plasma etching can be found in [Fan93, Hor91].

Cong, Preas, and Liu [CPL90] presented three physical models based on the boundary terminal cell model that effectively uses the OTC areas for routing.

Various algorithms have been presented for layout optimization and generation of functional cells [MH90, LM88, BSC91]. The focus in these papers is on automatic cell layout generations.

Lin et al. [YHH93b] presented an automatic layout generation system called LiB for SSI cells used in the CMOS VLSI design. LiB takes a transistor-level circuit schematic in the SPICE format and outputs a mask layout in CIF. LiB can be used as a cell-library builder or as a subsystem of a random logic module generator.

The same authors proposed another layout style in [YHH93a], which uses two metal layers for intracell routing. The power and ground lines are placed in M1 in the middle of the cell, and M2 segments are used to join vertically aligned drain sources.

Cohen and Shechory [CS93] present a text-driven system for assembling a data path block layout conforming to a double-metal-layer technology, featuring abutted standard cells and OTC routing. Fuji et al. [FMMY92] presented a cell model for processes having four layers for routing. The terminals are located around the horizontal center line in the center. [LPH92] considers cell height reduction by routing over the cell. The concept of OTC routing is applied to the global routing problem in leaf cell synthesis.

A new CMOS standard cell library is proposed in [Lin84] for complex and higher speed semicustom VLSI circuits using 3-μm technologies. The library is designed to be compatible with p- and n-well CMOS processes, one- and two-layer metallization, and suitable single- or double-sided cell auto-layout software.

5

Basic Problems in Routing

Our objective is to develop efficient, optimal, and approximation algorithms for routing problems in OTC, thick-film MCMs, and thin-film MCMs. These problems are complex due to the different cell models used, fabrication process, and other factors.

Our approach to solving the above-stated complex problems is to decompose them into conceptually simpler and smaller problems and develop efficient algorithms for solving these relatively simpler problems. We refer to these "simpler" problems as the *basic problems*.

Several basic problems have already been identified, and efficient algorithms have been proposed for solving them. As a result, the complex problems can be decomposed into smaller problems which have already been studied and solved efficiently. Even though the routing approaches for designs using different models and fabrication processes are not similar, they can be modeled as one or more of the known basic problems. Therefore, the key step in the development of new routing algorithms for any cell model and fabrication process is to decompose the given problem into one or more of the basic problems. This leads to the development of efficient algorithms for both OTC and MCM routing problems.

In this chapter, we first present a historical perspective of OTC and MCM routing. We then review the required routing terminology, and finally define the basic problems in OTC and MCM routing. The solutions to these problems are discussed in the subsequent chapters.

5.1 Historical Perspective of OTC and MCM Routing

The concept of OTC routing was introduced by Deutsch and Glick in 1980 [DG80]. They presented an algorithm that produces single-layer planar routing over the cells for I^2L and LST^2L logic arrays. A gate array router presented in [Kro83] used horizontal and vertical OTC routing channels to increase cell density. In [SS87], an OTC router called a *permeation* router was presented. Gudmundsson and Ntafes [GN87] studied the problem of choosing net segments to be connected in the channel after routing over the cell.

In [CL88], a symbolic model for OTC channel routing was presented together with the algorithms for each stage of the entire routing process. In [CPL90], three physical models were presented for using the OTC area for routing in standard-cell designs. In [HSS91b], the concept of using vacant terminals for the BTM was introduced for routing over the cell for the two-layer process. This concept was extended for the three-layer process in [HSS91a]. In [LPHL91], it was shown that only the removal of critical nets from the channel leads to reduction in channel density. An improved heuristic algorithm for OTC channel routing was presented in [DNB91]. In [CH90], a pseudopin assignment technique was presented for single-layer OTC routing. A router that formulates an OTC channel routing problem as a 0-1 integer linear programming problem was developed in [PP92]. Larmore et al. [LGW92] defined a new sliced layout architecture using OTC routing on the second metal layer for compilation of arbitrary schematics into layouts for CMOS technology. In [WHSS92], a new cell model, the CTM was introduced to take advantage of OTC routing. In [NSHS92], a high-performance OTC router for the BTM was presented. This router uses $45°$ segments for routing in OTC areas. In [HC92], a dynamic programming-based algorithm was presented that uses the concept of pin permutation for improving OTC channel routing.

In [FMMY92], a new cell model using three metal layers for intracell routing was introduced, which shortens the length of polysilicon wires so as to achieve high-performance circuits.

In [SH93], an OTC router was presented for the BTM for the three-layer process when vias are permitted in OTC areas. In [BPS93c], a new cell model, the MTM, was presented for efficient utilization of OTC areas. In [BPS93b], an OTC router for the MTM was presented when vias are not allowed on OTC areas. In [BPS93a], an OTC router for the MTM was presented when vias *are* allowed on OTC areas.

The MCM routing problem is an immense three-dimensional general area routing problem; in this problem, routing can be carried out almost everywhere in the entire multilayer substrate. The line width is much smaller and the routing result is much denser in MCM routing as compared to those of conventional PCB routing. Therefore, traditional PCB routing tools are often inadequate in dealing with MCM designs.

Hanafusa, Yamashita, and Yasuda [HYY90] presented a three-dimensional maze routing algorithm for multilayer ceramic PCBs. Ho et al. [HSVW90] presented a method for multilayer MCM routing in which the routing layers are divided into several x-y layer pairs. Nets are first assigned to the x-y layer pairs and then two-layer routing is carried out for each.

Khoo and Cong [KC92] presented a multilayer MCM router called SLICE. It computes the routing solution on a layer-by-layer basis and carries out planar routing in each layer. Efficient routers have been proposed for silicon-on-silicon–based MCM technology [DDS91]. Since the number of routing layers is usually small (two layers for signal routing in most cases) in this technology, many techniques for IC routing, such as hierarchical routing and rubber-band routing, can be applied to yield good solutions. Recently, Yu, Badida, and Sherwani [YBS93] proposed a "true" three-dimensional approach to MCM routing.

5.2 Routing Terminology

A typical standard cell layout is shown in Figure 5.1. The entire layout is assumed to be superimposed by a $w\lambda \times s\lambda$ grid, where w and s are the minimum width and isolation requirements for metal wires, as specified by the design rules (fabrication process). Horizontal grid lines are called *tracks* and the vertical grid lines are called *columns*. The cells are placed in rows. Let $\mathcal{R} = R_1, R_2, \ldots, R_K$ be the set of all cells, such that R_i denotes the cells in the ith row. The areas between the cell rows, called *channels*, are used for routing. A layout with K rows has $K + 1$ channels. For ease of notation, top I/O pads and bottom I/O pads are considered cell row R_0 and R_{K+1}, respectively. A channel between two cell rows R_i and R_{i+1} for $1 \leq i \leq (K - 1)$ is denoted by C_i. For a standard-cell layout with K rows, the set of channels $\mathcal{C} = C_0, C_1, \ldots, C_K$. *Channel height* ($h_i$) denotes the total number of tracks in a channel C_i.

Let L denote the total number of columns in a layout, or width of the layout. The terminal positions in a cell row are denoted by the column numbers. If a terminal does not exist on a column of a cell row, then that column is called a *vacant terminal*. k denotes the total number of tracks available in the OTC area of a cell row for routing (cell height).

The upper boundary of a cell row is called the *top boundary* and the lower boundary is called the *bottom boundary*. The channel adjacent to the top boundary is called the *upper channel* and the channel adjacent to the bottom boundary is called the *lower channel*. The cell row adjacent to the upper boundary of a channel is called *upper cell row* and that adjacent to the lower boundary is the *lower cell row*. A complete set of nonequipotential terminals in a cell row is called a *terminal row*. Based on the design style, a cell row may have one or two terminal rows. If a cell row has two terminal rows, then the upper terminal row is called the *UPPER row*, denoted by $\mathcal{U} = u_1, u_2, \ldots, u_L$, and the lower terminal row,

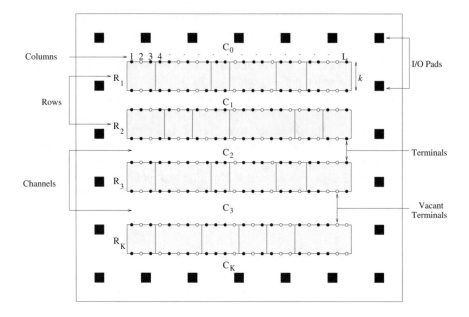

Fig. 5.1   Basic terminology in routing.

called *LOWER* row, is denoted by $\mathcal{L} = l_1, l_2, \ldots l_L$. If the cell row has only one terminal row, then the nomenclature for the terminal row is defined in reference to the channel. The terminal row in the upper cell row is denoted by $\mathcal{U}$ and the terminal row in the lower cell row by $\mathcal{L}$. If the UPPER or LOWER row is located on the top or bottom boundary respectively, then the terminal rows are defined in reference to the channel. The terminal row on the upper boundary of the channel is called the *TOP* row, and is denoted by $\mathcal{T} = t_1, t_2, \ldots t_L$. Similarly, the terminal row on the lower boundary of the channel, called the *BOT* row, is denoted by $\mathcal{B} = b_1, b_2, \ldots, b_L$.

The interconnections required between the terminals are specified by a *net list*, $\mathcal{N} = \{N_1, N_2, \ldots, N_L\}$, where n is the total number of nets. Let $N_i = t_{i_1}, t_{i_2}, \ldots, t_{i_{m_i}}$ for all $1 \leq i \leq n$, where m_i is the total number of terminals in a net i. The nets may span between two or more terminals. The nets with only two terminals are called *two-terminal nets*, and nets with more than two terminals are called *multiterminal nets*. One popular routing technique involving multiterminal nets is to decompose each multiterminal net into several two-terminal nets. The algorithms are usually developed for two-terminal nets, and most of them can be easily extended to multiterminal nets.

The location of the terminal row is specified by the cell model. In the BTM, the terminal rows are located at the top and bottom boundaries, as shown in Figure 5.2(a), they are denoted by $\mathcal{T}$ and $\mathcal{B}$. In the MTM, the terminal rows are located in the middle of the upper and lower halves of the cell areas and are denoted by $\mathcal{U}$ and $\mathcal{L}$, respectively. For ease of routing, a pair of terminal rows called *pseudoterminal* rows are assumed to be located at the cell boundaries. The pseudoterminal row located on the upper boundary of a channel is denoted by $\mathcal{T}$, and the pseudoterminal row located at the lower boundary of the channel is denoted by $\mathcal{B}$, as shown in Figure 5.2(b). In the CTM, the terminals are located in the center of the cell. Since we have only one terminal row, we denote the terminal row in reference to the channel. $\mathcal{L}$ represents the terminal row in the upper cell row, and $\mathcal{U}$ refers to the terminals in the lower cell row. As shown in Figure 5.2(c), CTM cell rows have $\mathcal{T}$ and $\mathcal{B}$ pseudoterminal rows. For ease of reference, some of the notations are tabulated below.

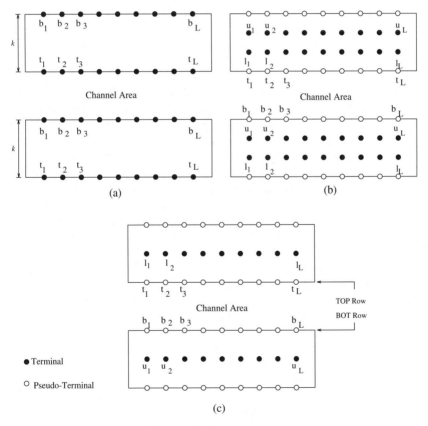

Fig. 5.2   Basic terminology in OTC routing (contd.): **(a)** BTM cell model; **(b)** MTM cell model; **(c)** CTM cell model.

OTC Routing Terminology

h	Channel height
k	Cell height
K	Total number of cell rows
L	Total number of columns in a cell row
$\mathcal{T} = t_1, t_2, \ldots, t_L$	TOP row
$\mathcal{B} = b_1, b_2, \ldots, b_L$	BOT row
$\mathcal{U} = u_1, u_2, \ldots, u_L$	UPPER row
$\mathcal{L} = l_1, l_2, \ldots, l_L$	LOWER row
$\mathcal{N} = N_1, N_2, \ldots, N_n$	Net list
$\mathcal{R} = R_1, R_2, \ldots, R_K$	Cell rows
$\mathcal{C} = C_0, C_1, \ldots, C_K$	Channels

An MCM routing problem is specified by a set of nets $\mathcal{N}$ to be routed over the substrate of MCM. To *route* a net is to interconnect its terminals by electric wires. There are two kinds of terminals of the nets. The electrical connections between the chips and substrate are made by attaching wires from the I/O pads on the device side of the chip to the appropriate points on the top-most layer on the substrate, called the *chip layer*. These points are called *die terminals*. Die terminals are used to make interconnections from bare dies to bare dies or from bare dies to the MCM. The terminals located in the periphery area of the chip layer are used to make interconnections from the MCM to the outside world and are called *I/O terminals*. Beneath the chip layer, there are several routing layers available for placing the electric wires. A three-dimensional grid is superimposed on the chip layer and routing layers in an MCM such that the signal wires can only be laid out along the grid lines to meet the design rule requirement (refer to Figure 5.3). An XYZ Cartesian coordinate system is set up such that the Z axis is perpendiculiar to the chip layer, the X and Y axes are parallel to the boundaries of the substrate, and the origin is at a corner of the MCM on the chip layer (see Figure 5.3). A *point* is an intersection among some grid lines. Assume that each terminal of the net list is located at some point on the chip layer.

Suppose that the size of the substrate is $l \times w$, and the number of routing layers is h. Then the cube on the bottom of the substrate, in which the signal wires are to be laid out, is called a *routing space*. Clearly, the size of the routing space is $l \times w \times h$. The substrate is partitioned into a set of tiles. Each *tile* is a rectangular area of the substrate. The cube on the bottom of the tile is called a *tower*. Therefore, the partitioning of the substrate into tiles is equivalent to the partitioning of the routing space into towers. We use these two notations interchangeably. A boundary plane of a tower is called a *face* of the tower, while a boundary of a tile is referred to as an *edge* of the tile. A vertical line on the face of a tower is called a *pillar*. The terminologies associated with a tower are shown in Figure 5.4.

Given a net $N_i = (t_{i_1}, t_{i_2}, \ldots, t_{i_{m_i}})$ consisting of terminals $t_{i_1}, t_{i_2}, \ldots,$ and $t_{i_{m_i}}$, we define a three-dimensional Steiner tree for n as a tree in the three-dimensional grid interconnecting $t_{i_1}, t_{i_2}, \ldots,$ and $t_{i_{m_i}}$, which are called *demand points*, and

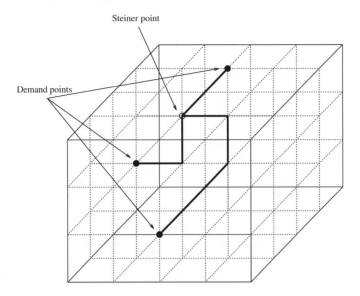

Fig. 5.3   Grid, coordinate system, and three-dimensional Steiner tree.

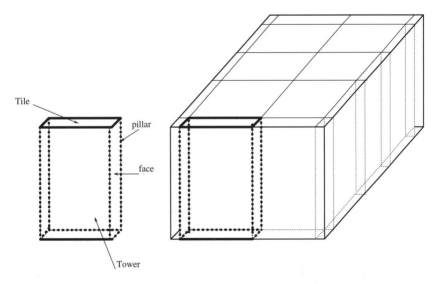

Fig. 5.4   A tower and terminologies associated with a tower.

some arbitrary points, which are called *Steiner points* (see Figure 5.3). The MCM routing problem can be formally defined as finding a three-dimensional Steiner tree for each net in the net list such that no three-dimensional Steiner trees of two different nets overlap or intersect. The Steiner trees for these nets are also called the *routes* of the nets. A point that is on the route of a net and the face of a tower is called a *terminal* of the net in the tower. The routing of a net can be decomposed into two steps. The first step is to find out the towers that the net passes through and the location of the terminals in the faces of the towers that the net passes through. The second step is to route the net within each tower it passes through. The routing of a net within a tower is to find a three-dimensional Steiner tree interconnecting all the terminals of the net in that tower.

5.3 Basic Routing Problems

OTC and MCM routing algorithms are custom developed for a given routing model. Hence, we have independent algorithms for various routing environments. However, all the routing problems can be decomposed into one or more of the basic routing problems discussed below.

1. **Single-Sided Planar Routing (SSPR) Problem:** Given a row of terminals $\mathcal{T} = t_1, t_2, \ldots, t_L$; a net list $\mathcal{N} = \{N_1, N_2, \ldots, N_n\}$, where $N_i = (t_{i_1}, t_{i_2})$, $1 \leq i \leq n$, $t_{i_1} \in \mathcal{T}$ and $t_{i_2} \in \mathcal{T}$; and a fixed number of tracks k.

   Find a planar routing solution for a subset of nets, $\mathcal{N}' \subseteq \mathcal{N}$, in k tracks on one side of the terminal row such that $|\mathcal{N}'|$ is maximized.

   The weighted version of the problem is called the *weighted single-sided planar routing* (WSSPR) problem. For each net $N_i \in \mathcal{N}$, $w(N_i)$ specifies its weight. WSSPR is the same as SSPR, except that we need to find a subset $\mathcal{N}'$ such that

$$\sum_{\forall N_i \in \mathcal{N}'} w(N_i)$$

   is maximized.

   The SSPR problem finds applications in solving OTC routing problems in the BTM, CTM, and MTM for planar OTC routing. The algorithm for solving SSPR is given in Chapter 6. A valid routing solution for SSPR is illustrated in Figure 5.5(a). All the terminals belonging to a single net have been denoted by a unique number. For example, the net between the terminals marked 1 is called net 1, the net between terminals marked 2 is called *net* 2, and so forth. An efficient solution includes routing all the nets except net 3 and net 7. Since routing net 3 necessitates the elimination of nets 1, 2, 5, 6 from the planar routing solution, net 3 is not routed. Net 7 is not routed because it would require an additional track, and only two tracks are available in the OTC area for routing ($k = 2$).

2. **Two-Sided Planar Routing (TSPR) Problem:** Given two rows of terminals $\mathcal{T} = t_1, t_2, \ldots, t_L$ and $\mathcal{B} = b_1, b_2, \ldots, b_L$; a fixed number of tracks k between them; and a net list $\mathcal{N} = \{N_1, N_2, \ldots, N_n\}$, such that $N_i = t_{i_1}, t_{i_2}$, where $1 \leq i \leq n$ and $t_{i_1}, t_{i_2} \in \mathcal{T}$ or $t_{i_1}, t_{i_2} \in \mathcal{B}$.

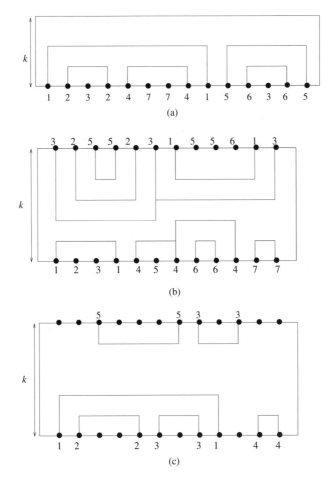

Fig. 5.5   Basic OTC routing problems: **(a)** SSPR ($k = 2$); **(b)** TSPR ($k = 4$); **(c)** ETSPR
($k = 3$).

Find a planar routing solution for a subset of nets, $\mathcal{N}' \subseteq \mathcal{N}$, in k
tracks on one side of each terminal row, between the terminal rows, such
that $|\mathcal{N}'|$ is maximized.

The weighted version of this problem is called the *weighted two-sided
planar routing* (WTSPR) problem. For each net $N_i \in \mathcal{N}$, $w(N_i)$ specifies
its weight. WSTPR is the same as TSPR, except that we need to find a
subset $\mathcal{N}'$ such that

$$\sum_{\forall N_i \in \mathcal{N}'} w(N_i)$$

is maximized.

The TSPR problem requires a selection of nets to be routed between
the terminals. The net list comprises two subsets of nets. Each subset

consists of nets that start and finish on the same row. The solution to this problem is required to solve OTC routing problems in the CTM and MTM, and the algorithms to solve these problems are given in Chapters 6 and 7. A valid routing solution is shown in Figure 5.5(b). The TSPR problem is considerably harder than the SSPR problem. Note that the nets from both sides may "entangle." That is, the number of tracks used by nets from either side is not constant for the entire cell row. In the example shown in Figure 5.5(b), the first six columns of the second track from the bottom cell boundary are used to route the net from the upper terminal row and then the same track is used to route net 4 from the bottom lower terminal row. If this sharing of tracks between the nets belonging to the upper and lower terminal rows is not allowed, then the TSPR problem can be essentially decomposed into two SSPR problems.

3. **Equipotential Two-Sided Planar Routing (ETSPR) Problem:** Given two rows of equipotential terminals $T = t_1, t_2, \ldots, t_L$ and $B = b_1, b_2, \ldots, b_L$, such that t_i and b_i are equipotential for all $i = 1, 2, \ldots, n$; a fixed number of tracks k between them; and a net list $\mathcal{N} = \{N_1, N_2, \ldots, N_n\}$, such that $N_i = t_{i_1}, t_{i_2}$, where $t_{i_1}, t_{i_2} \in T$ or $t_{i_1}, t_{i_2} \in B$.

Find a planar routing solution for a subset of nets, $\mathcal{N}' \subseteq \mathcal{N}$, in k tracks on one side of each terminal row, between the terminal rows, such that $|\mathcal{N}'|$ is maximized.

The weighted version of this problem is called *weighted equipotential two-sided planar routing* (WETSPR) problem. For each net $N_i \in \mathcal{N}$, $w(N_i)$ specifies its weight. WETSPR is the same as TSPR, except that we need to find a subset $\mathcal{N}'$ such that

$$\sum_{\forall N_i \in \mathcal{N}'} w(N_i)$$

is maximized.

The ETSPR problem is an open problem. However, some variations of this problem are solved in Chapters 6 and 7. A valid solution to the ETSPR routing problem is shown in Figure 5.5(c).

4. **Boundary Terminal Assignment (BTA) Problem:** Given the following:
A terminal row $\mathcal{L} = l_1, l_2, \ldots, l_L$ in the upper cell row
A terminal row $\mathcal{U} = u_1, u_2, \ldots, u_L$ in the lower cell row
A netlist $\mathcal{N} = N_1, N_2, \ldots, N_n$, such that $N_i = t_{i_1}, t_{i_2}$, where $t_{i_1}, t_{i_2} \in \mathcal{L} \cup \mathcal{U}$
A pseudoterminal row $T = t_1, t_2, \ldots, t_L$ on the bottom cell boundary of the upper cell row
A pseudoterminal row $B = b_1, b_2, \ldots, b_L$ on the top cell boundary of the lower cell row
Fixed planar routing areas of k_1 tracks each between $\mathcal{L}$ and T and between $\mathcal{U}$ and B

A nonplanar channel routing area of variable height between the terminal rows T and B

For each terminal in L that belongs to a net in N, assign a terminal in T, and for each terminal in U that belongs to a net in N, assign a terminal in B such that there exist planar routing solutions in k tracks between T and U and between B and L, and the resulting channel density between T and B is minimum among all such assignments.

A routing solution to this problem is illustrated in Figure 5.6(a). The basic routing idea in the fixed-height planar routing areas is to jog the nets toward the destination terminals. The algorithms to solve this problem were presented in Chapter 3. The BTA is encountered in solving planar OTC routing problems of the CTM and MTM, as discussed in Chapters 6 and 7.

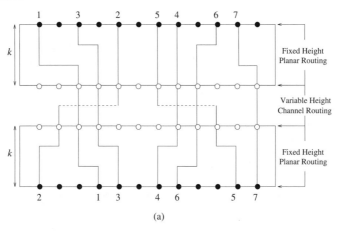

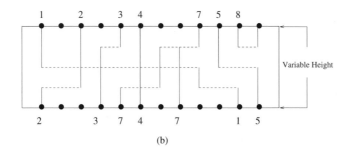

Fig. 5.6   Basic OTC routing problems (contd.): **(a)** BTA ($k = 2$), **(b)** CCR.

5. **Conventional Channel Routing (CCR) Problem:** Given two rows of terminals $T = t_1, t_2, \ldots, t_L$ and $B = b_1, b_2, \ldots, b_L$; a net list $N =$

$\{N_1, N_2, \ldots, N_n\}$, where each net $N_i = t_{i_1}, t_{i_2}$ and $t_{i_1}, t_{i_2} \in (\mathcal{T} \cup \mathcal{B})$; and a fixed number of routing layers. Find a routing solution for $\mathcal{N}$ such that the channel height is minimized.

The CCR problem (or channel routing problem) is well studied and several algorithms have been presented. (Refer to Chapter 3.) The CCR problem is a general routing problem in layouts using any of the cell models BTM, CTM, MTM, and TBC. An example showing a routing solution for the CCR problem is shown in Figure 5.6(b). Since the number of tracks in the channel is not fixed, there always exists a routing solution. However, the objective is to generate a solution with a minimum number of tracks in the channel.

6. **Nonuniform Layer Channel Routing (NLCR) Problem:** Given two terminal rows $\mathcal{L} = l_1, l_2, \ldots, l_L$ and $\mathcal{U} = u_1, u_2, \ldots, u_L$; a net list $\mathcal{N} = N_1, N_2, \ldots, N_n$, where $N_i = t_{i_1}, t_{i_2}$, where $t_{i_1}, t_{i_2} \in (\mathcal{U} \cup \mathcal{L})$; and a routing area consisting of two fixed-height OTC regions with two-layers available for routing and a channel with three routing layers.

Find a routing solution for $\mathcal{N}$, such that the channel height is minimized.

The NLCR problem finds applications in solving OTC routing problems of CTM, MTM, and TBC cell models when vias are permitted in OTC areas. The routing area between the terminal rows consists of two 2-layer OTC routing regions, and one 3-layer channel between the cell rows, as shown in Figure 5.7. Algorithms to solve the NLCR problem are presented in Chapter 8.

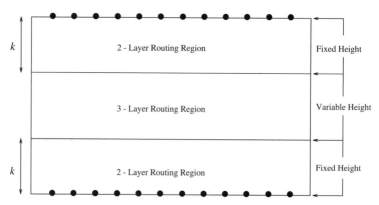

Fig. 5.7   Nonuniform layer channel routing problem.

7. **Terminal Assignment (TA) Problem:** Given three terminal rows $\mathcal{L} = l_1, l_2, \ldots, l_L$ and $\mathcal{U} = u_1, u_2, \ldots, u_L$; $\mathcal{T} = t_1, t_2, \ldots, t_L$; a net list $\mathcal{N} = N_1, N_2, \ldots, N_n$, where $N_i = t_{i_1}, t_{i_2}$, where $t_{i_1} \in \mathcal{U}, t_{i_2} \in \mathcal{T}$; and a routing area consisting of k layers.

For each net N_i, assign a terminal $t_i \in \mathcal{T}$ such that the number of net crossings is minimized.

The TA routing problem is shown in Figure 5.8. The TA finds applications in MCM routing problems, where the net crossings are minimized in order to minimize the number of vias used. Algorithms to solve the TA routing problem are given in Chapter 10.

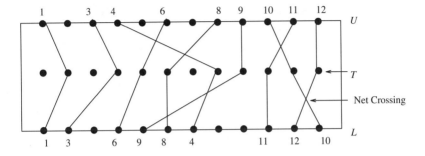

Fig. 5.8   Terminal assignment routing problem.

8. **Layer Assignment (LA):** Given (1) $V = \{V_1, V_2, \ldots, V_k\}$, where k is the number of layers and V_i is the set of nets that can only be routed on layer i for all $1 \leq i \leq k$; (2) $V' = \{V_{12}, V_{23}, \ldots, V_{(k-1)k}\}$, where $V_{(i-1)i}$ is the set of nets that can be routed either on layer $i - 1$ or i; (3) $S = \{S_1, S_2, \ldots, S_k\}$, where $S_i \subseteq V_{(i-1)i} \cup V_i \cup V_{i(i+1)}$, $V_{01} = \phi$, and $S_1 \cap S_2 \cap \cdots S_k = \phi$.

Find the LA such that

$$\sum_{i=1}^{k} |S_i|$$

is maximized.

The LA finds applications in MCM routing problems. Algorithms to solve the LA routing problem are given in Chapter 10.

5.4 OTC Routing Problems for the BTM

The BTM has two equipotential terminal rows, one located at the top boundary and the other located at the bottom boundary, as shown in Figure 5.9(a). The terminal rows are denoted with reference to the channel. The bottom terminal row for the upper cell row, called *upper terminal row*, is denoted by $\mathcal{T}$, and the top terminal row for the lower cell row of a channel, called *lower terminal row*, is denoted by $\mathcal{B}$. Each cell row has k tracks in the OTC area. Since the terminals are located at the boundary, the routing in the OTC area is accomplished by routing the nets from the boundary into the cell area and back again to the boundary. The BTM class consists of several models, as discussed in Chapter 4. Each of these

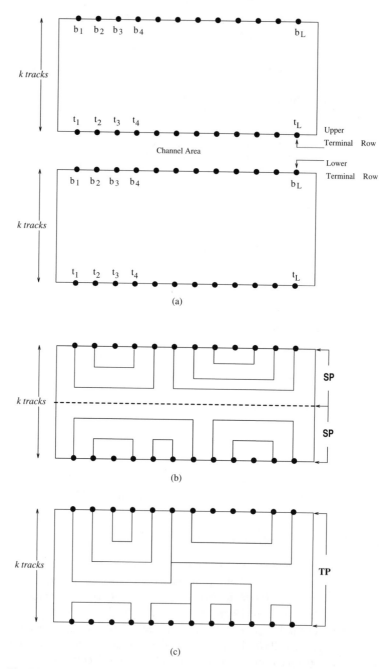

Fig. 5.9 (a) Cell rows in the BTM; (b) a valid solution for the BTM-HCVD; (c) a valid solution for the BTM-HCVC.

models has a unique routing problem associated with it. Let us now formulate the routing problems for all the cell models in the BTM class for both two-layer and three-layer processes.

5.4.1 Two Layer Problems

In this section, let us discuss various OTC routing problems associated with the BTM in a two-metal-layer fabrication process environment. The BTM comprises four different layout styles, and the routing problems vary for each of the models, as discussed below.

1. **Horizontally Connected, Vertically Divided Model:** In this model, for each cell row R, since the OTC routing region is divided vertically into two subregions because of power and ground lines, the OTC routing for top and bottom terminals can be carried out independently. However, the height of each region is limited by half cell heights.

   The OTC routing problem for the two-layer BTM-HCVD problem is a special case in ETSPR and is referred to as ETSPR-1. ETSPR-1 can be solved by restricting ETSPR to a fixed number of tracks for routing from each terminal row. The height is fixed at $k - 2/2$, where k is the total number of tracks in the OTC area and two tracks are required for routing power and ground lines. Figure 5.9(a) illustrates the routing regions in a BTM-HCVD-based cell row and a valid routing solution for a two-layer BTM-HCVD problem is illustrated in Figure 5.9(b).

2. **Horizontally Connected, Vertically Connected Model:** In this model, top and bottom terminals of a cell row R share the OTC routing area. Both the rows have to be routed simultaneously so that the common routing region will be used efficiently. Furthermore, the number of tracks in the OTC routing region is limited by the height of the cells. This problem is equivalent to the ETSPR problem.

3. **Horizontally Divided, Vertically Connected Model:** In this model, since the OTC area is divided horizontally into several subregions by feedthroughs, OTC routing can be accomplished only within a subregion. However, the OTC area within a subregion is not segmented and hence the area can be shared by both the top and bottom terminal rows. Therefore, the OTC routing problem for each subregion is exactly equivalent to the ETSPR problem.

4. **Traditional BTM:** In this model, the entire OTC area in the M2 layer between the terminal rows is unblocked. Therefore, the OTC routing problem in the traditional BTM is equivalent to the ETSPR problem.

5.4.2 Three-layer problems

The use of the three-layer process leads to yet another routing parameter, permissibility of vias in OTC areas. The techniques adopted to solve the three-layer OTC routing problems when vias are restricted or allowed in OTC areas are completely different and are discussed below.

1. **Vias Not Allowed in OTC Areas:** When vias are not allowed in OTC areas, the routing in the M3 layer is required to be planar. Therefore, for all the models in the BTM class, the additional advantage of the three-layer process when compared to the two-layer process is the availability of an additional layer for planar OTC routing. The advantages of this model over other models are that the terminals are located at the boundary, and hence for all the models of BTM, the routing in the third layer is equivalent to the ETSPR problem. The OTC routing in M2 for various models in the BTM is similar to the corresponding problem in the two-layer process. For example, consider the routing areas in OTC regions and channels for the HCVD model, shown in Figure 5.10. The terminals are located in the M2 layer. The M2 layer is vertically segmented by the power and ground lines in the middle of the cell row. The M3 layer has terminals located at the boundaries and is completely unblocked. Therefore, OTC routing in the M3 layer is equivalent to ETSPR.

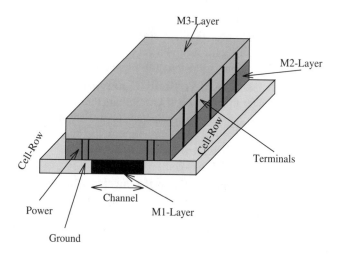

Fig. 5.10   Routing regions in the BTM-HCVD, for three-layer process.

2. **Vias Allowed in OTC Areas:** The advantage of using vias in OTC areas is that the routing in OTC areas need not be planar; that is, the nets can switch layers in OTC areas. This is a major advantage over the previous case when vias were not allowed in OTC areas. When vias are allowed OTC areas, the OTC routing problems in various models in the BTM class are special cases of the NLCR problem. These are stated below and solved in Chapter 8.

(a) **BTM-HCVD Model:** The routing area consists of two OTC routing regions of a fixed-height $k - 2/2$ M2 OTC area and a completely unblocked M3 layer for routing. There is a three-layer channel area for routing between the cell rows. This problem is called the NLCR-1.

(b) **BTM-HCVC Model:** The routing problem is a special case of the NLCR problem, called NLCR-2.

(c) **BTM-HDVC:** This is composed of multiple NLCR-2 problems.

5.5 OTC Routing Problems for the CTM

In the CTM, the terminal row is located in the center of the cell in the M2 layer. The power and the ground lines are located at the boundaries in the M1 layer. The terminal row in the upper cell row is called the *UPPER row* and the terminal row in the lower cell row is called *LOWER row*. When vias are not allowed in OTC areas, the terminals are brought to the boundary in a planar fashion. The pseudoterminals at the boundary are called the *TOP* or *BOT terminal row* for the upper or lower cell row, respectively. The terminal rows are referred to with reference to the channel. For each channel, the upper cell row consists of the terminal row $\mathcal{L}$, and the lower cell row consists of the terminal row $\mathcal{U}$. The upper boundary of the channel area consists of the pseudoterminal row $\mathcal{T}$, and the lower boundary of the channel consists of the pseudoterminal row $\mathcal{B}$.

5.5.1 Two-Layer Problems

The OTC routing area consisting of k tracks is split into two equal regions, each with $k/2$ tracks. The power and ground lines do not obstruct OTC routing, since they are in the M1 layer. In this model, the entire OTC routing layer is required to route the terminals to the boundary, and hence, unlike the BTM, routing of independent sets in OTC areas is not possible. The routing area in the two-layer CTM consists of three regions: the upper OTC region consisting of $k/2$ tracks, the middle variable height channel region (with two routing layers), and a lower OTC region consisting of $k/2$ tracks. Thus, as illustrated in Figure 5.11, the routing problem for two-layer CTM is equivalent to the BTA problem.

5.5.2 Three-Layer Problems

The use of the three-layer process leads to yet another routing parameter, permissibility of vias in OTC areas. The techniques adopted to solve the three OTC routing problems when vias are restricted or allowed in OTC areas are completely different and are discussed below.

1. **Vias Not Allowed in OTC Areas:** When vias are not allowed in OTC areas, the routing in the M3 layer is required to be planar. Therefore, for the CTM, the additional advantage of the three-layer process when compared to the two-layer process is the availability of an additional layer for planar

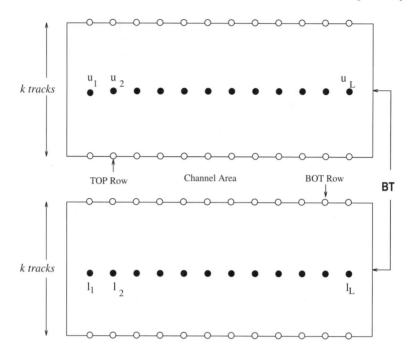

Fig. 5.11   Two-layer routing problem in CTM.

OTC routing. The OTC routing problem in the three-layer CTM can be split into two subproblems. First, the terminals have to be routed to the boundary in a planar fashion in M2. This problem is exactly equivalent to the BTAP. Second, since the terminals are now available at the boundaries, the routing in the OTC areas is equivalent to the TSPR problem, and the routing in the channels is a conventional channel routing problem.

2. **Vias Allowed in OTC Areas:** When vias are allowed in OTC areas, the routing problem between two adjacent cell rows is a classical NLCR problem.

5.6 OTC Routing Problems for the MTM

In the MTM, the terminals are in the M2 layer in the middle of the upper and lower halves of the cell rows (see Figure 5.12(a)). The upper half of the cell row consists of the terminal row $\mathcal{U}$ and the lower half consists of the terminal row $\mathcal{L}$. The pseudoterminal rows are located at the top and bottom boundaries and are denoted with reference to the channel. The pseudoterminal row at the top or bottom boundary of a channel is denoted by $\mathcal{T}$ or $\mathcal{B}$, respectively.

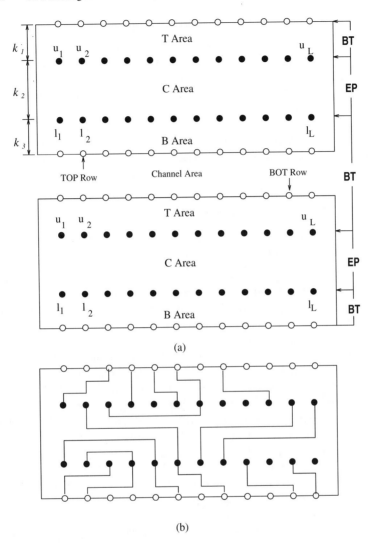

Fig. 5.12   (a) Cell rows in MTM, (b) two-layer MTM routing problem.

The OTC area in MTM is divided into three regions:

1. **T Area:** The area between the upper terminal row and the top cell boundary, consisting of k_1 tracks.

2. **C Area:** The region between the two terminal rows, conisting of k_2 tracks.

3. **B Area:** The area between the lower terminal row and the lower cell boundary, consisting of k_3 tracks.

5.6.1 Problems in the Two-Layer MTM

The MTM has terminals in the M2 layer, and the power and ground lines are located in the M1 layer. The advantage of the MTM over the CTM is that the intrarow connections are accomplished in the C area and the interrow connections are completed in the T and B areas. The terminals in each column are equipotential. The OTC routing problem in the C area is exactly equivalent to the ETSPR problem, while the routing between adjacent cell rows is equivalent to the BTAP, as shown in Figure 5.12(a). A valid OTC routing solution for two-layer MTM over a single cell row is shown in Figure 5.12(b). The intrarow routing is accomplished in the C area, and the river routing technique is adopted for routing the nets from the terminals to the boundaries.

5.6.2 Three-Layer Problems

In the three-layer process, vias may or may not be allowed in OTC areas for routing. The techniques adopted to solve two-layer and three-layer problems are completely different, as discussed below.

1. **Vias Not Allowed in OTC Areas:** When vias are not allowed in OTC areas, the routing in the M3 layer is required to be planar. Therefore, for the MTM, the additional advantage of the three-layer process when compared to the two-layer process is the availability of an additional layer for planar OTC routing. The OTC routing problem in the three-layer MTM can be divided into two subproblems. First, the OTC routing in the M2 layer and then the OTC routing in the M3 layer. The routing in the M2 layer is exactly as discussed in the previous section and comprises the BTAP and ETSPR problems. Second, since the M3 layer is completely unblocked and the terminals are at the boundaries, the OTC routing problem in the M3 layer is exactly equivalent to the two-sided planar routing problem. Therefore, the OTC routing problem for the three-layer MTM when vias are not allowed in OTC areas comprises BTAP, ETSPR, and TSPR problems and conventional channel routing problems in the channel.

2. **Vias Allowed in OTC Areas:** There are several methods for solving this problem. One simple method is to divide the problem into two subtasks. The first task involves routing between the terminal rows $\mathcal{U}$ and $\mathcal{L}$ of the cell row, converting it into a modified channel routing problem with a fixed channel height of k_2 tracks. The second task is to route between $\mathcal{L}$ of the upper cell row and $\mathcal{T}$ of the lower cell row. This problem is exactly equivalent to the NLCR problem.

5.7 OTC Routing Problems for the TBC

The TBC designs have targets in the form of long vertical strips in the M1 layer. The routing algorithm locates the exact position of the terminal on each target.

The power and ground lines are also in the M1 layer, at the top and bottom cell boundaries. Since the terminals are located by the routing algorithm based on columnwise net densities, the terminals in a cell row are not aligned. The TBCs are designed for processes using at least three metal layers for routing, and vias are allowed in OTC regions. There are several techniques to solve the TBC routing problem. One simple method is to transform the TBC routing problem into a CTM routing problem. The three-layer CTM routing problem when vias are allowed in OTC areas is similar to the TBC routing problem, the only difference being the shape of the terminal rows. In the CTM, the terminals are aligned, and hence it has rectangular routing boundaries. However, the TBC has irregular OTC routing boundaries, as shown in Figure 5.13. Therefore, the routing problem in TBC is a modified NLCR problem called NLCR-3. The modification required is the shape of the OTC routing boundaries.

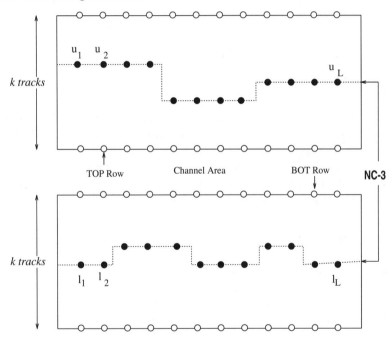

Fig. 5.13   OTC routing problem in the TBC.

5.8 Routing Problems for the MCM

Based on the terminal model used, the MCMs may either have uniformly located terminals or scattered terminals, as shown in Figures 5.14(a) and (b), respectively. The number of available layers for routing depends on the fabrication model. The routing approach in the scattered terminal model is an area routing approach, while the MCMs with uniformly located terminals can use the concept of *tile routing*, discussed in Chapter 10.

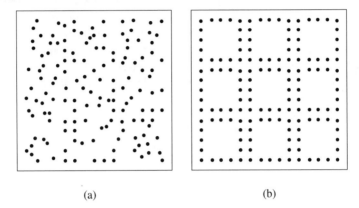

(a)   (b)

Fig. 5.14   Routing problems in the MCM.

5.9 Summary

The development of efficient routers critically depends on identification of the problem, decomposing the problem into solvable and simple subproblems, and developing efficient algorithms for each subproblem.

In this chapter, we have analyzed different cell models and shown that OTC and MCM routing problems are variations of some basic problems. Efficient algorithms for these basic problems would lead to efficient OTC routers.

PROBLEMS

5-1. Consider a design layout with both MTM and CTM cells. Assume a two-layer process and develop a routing problem.

5-2. Decompose the above problem into one or more basic routing problems.

5-3. Formulate a routing problem for a layout using two layers and arbitrarily located terminals.

5-4. Consider an HCVC-based standard cell layout using the three-layer fabrication process, with vias in OTC areas. Formulate a routing problem for the entire layout, such that the total density in the channels is minimized.

5-5. Formulate a routing problem for a layout based on the cell models that use L-shaped terminals and use the two-layer process.

5-6. Formulate OTB routing problems in four-layer environment for the following cases:
(a) Terminals are located on the boundaries of the blocks.
(b) Terminals are randomly placed inside the blocks.

5-7. Formulate the MCM routing problem for the following cases:
  (a) Stacked vias are allowed.
  (b) Stacked vias are not allowed.

BIBLIOGRAPHIC NOTES

Cong and Liu [CL88] formulate the SSPR problem. The problem is solved in three stages. In the first stage, nets are routed in the OTC areas on either side of the channel. In the second stage, net segments are selected for interconnecting multiterminal nets routed on either side of the channel, and in the third stage, the terminals corresponding to the net segments are connected.

Cong, Preas, and Liu consider the TSPR problem in [CPL90] in order to solve the HCVC routing problem. The complexity of the problem is unknown and an approximate solution is developed. The terminals on either boundary are considered to be nonequipotential. However, when the terminals are equipotential, the TSPR problem becomes an ETSPR problem. Bhingarde, Panyam, and Sherwani develop an approximate solution for ETSPR in [BPS93b] in order to generate a planar routing solution for the same-row nets in the C area of the MTM.

The BTAP problem requires river routing in the cell rows on either side of the channel such that the density in the channel is minimized. Seigel [Sie83] provides an excellent reference for river routing algorithms. Tuan and Hakimi [TH90] present an algorithm for river routing with a small number of jogs. A similar problem is considered by Yang and Wong [YW91] for channel pin assignment. Wu et al. consider the BTAP for planar routing in OTC areas of the CTM to "bring" the terminals to the boundaries.

Several algorithms have been developed for solving the CCR problem, as discussed in Chapter 3. The NLCR problem is considered in solving the three-layer routing problems of the CTM, MTM, and TBC-based layouts [BPS93a, KBPS94, BMPS93].

6

Routing Algorithms for the Two-Layer Process

The fabrication process in which two metal layers are available for routing is currently the most widely used process for VLSI chips. Due to its maturity, it is very cost-effective. Many cell libraries are available that are designed specially for the two-layer process. The two-layer process uses two metal routing layers designated metal1 (M1) and metal2 (M2). In addition to these metal layers, poly or a local interconnnect layer may also be used for localized routing. However, the use and effectiveness of these layers varies widely from one process to another. Both of these layers cannot be used for routing long wires due to their undesirable delay characteristics. We do not consider the effect of these layers in routing algorithms, since we assume that these layers are used for local routing. The effect of these layers has been considered during cell design. In the channels, both metal layers are used for intercell routing. M1 is normally used for intracell routing, and therefore M2 is used in OTC areas for intercell routing. Obviously, vias cannot be used in OTC areas, since the routing is accomplished using only one metal layer. As a result, many OTC routing problems for the two-layer process are in fact planar routing problems.

The standard cells use the M1 layer for both power and intracell routing. The power routing comprises routing VDD and VSS lines horizontally across the cells. The intracell routing is accomplished both over and between the P and N diffusion areas. Since only one metal layer is normally used for intracell routing, it is used for both horizontal and vertical routing. The number of vertical and horizontal tracks routed in the M1 layer for intracell routing depends on the cell functionality and cell model. With the exception of BTM-T, all the cell models require the terminals to be located in a metal layer.

The TBC designs have terminals in the form of long vertical strips in M1. Since the M1 layer is heavily used within the cell, OTC routing in the M1 layer is not considered. The CTM, MTM, and some BTMs locate the terminals in the M2 layer, which necessitates the use of the M1 layer to route to the boundary.

The total usable M1 area for OTC routing would average between 20 to 40%. However, the M1 routing area is available only as randomly placed small segments. Therefore, only area routing approaches such as maze routers can effectively utilize the M1 layer for OTC routing.

In this chapter, we present OTC routing algorithms for all applicable cell models. We will discuss routing algorithms for the BTM, CTM, and MTM. The TBC model is not relevant to the discussion in this chapter, since it assumes M2 and M3 layers in OTC areas. It should be noted that the BTM was developed before the introduction of OTC routing. Hence, when the concept of OTC routing was introduced in [DG80], several researchers presented various routing techniques for efficient OTC routing using the BTM [SS87, GN87, CL88, CPL90, HSS91b, HSS91a, LPHL91, DNB91]. The BTM is the traditional and well-studied cell model. On the other hand, the CTM and MTM are recent cell models developed specifically for OTC routing and very few routers have been developed for these models [WHSS92, BPS93b, BPS93a].

The rest of this chapter is organized as follows. First, we classify the algorithms on the basis of the algorithmic techniques used. Second, we present three different algorithms for the BTM based on maximum independent sets in overlap graphs, vacant terminals, and integer linear programming. Third, we present algorithms for the CTM and complete the chapter with approximation and optimal algorithms for the MTM.

6.1 Classification of BTM Routing Algorithms

The problem of OTC routing in the BTM is usually solved by dividing it into the following three steps:

1. Selecting nets for routing over the cells
2. Choosing net segments for routing in the channel
3. Routing in the channel

The net selection problem for two layers essentially boils down to the selection of two sets of net segments. However, the actual selection depends on the routing perspective.

If we consider the problem from the channel routing perspective and route one channel at a time, we should select two sets of nets. One set is routed in the upper OTC area, while the other is routed in the lower OTC region. The nets that are not selected are routed in the channel area. There are two routing layers in the channel, and there is a single routing layer over the cells for intercell connections. Clearly, the set of nets selected for OTC routing must be planar.

On the other hand, if we route one cell row at a time, then we must select two sets of nets, one from the upper channel and one from the lower channel. The nets thus selected should form a planar set routable in k tracks in the OTC area. This perpective is of special interest in BTM-HCVC-type models.

OTC routers for the BTM can be classified on the basis of the net selection algorithm used for selecting nets in step 1. We classify the algorithms in four categories:

1. Heuristic algorithms

2. MIS-based algorithms

3. Approximation algorithms

4. Integer linear programming–based algorithms

In the following, we briefly discuss all the algorithms in each of these categories. We then select a few algorithms from each category and present them in detail. Purely heuristic algorithms will not be discussed in detail.

6.1.1 Heuristic Algorithms

Several algorithms have been developed that can be classified as heuristic algorithms. We will briefly mention some of them.

In [Kro83, SS87], heuristic algorithms are presented for OTC routing. In [GN87], the problem of choosing net segments for routing in the channel is considered. Gudmundsson and Ntafos [GN87] assume that nets have been selected for OTC routing and have indicated that results on small examples produced good results if maximum cardinality matching is used to select nets for OTC routing. Since selection of nets gives rise to several sets of locally interconnected terminals (called superterminals), the problem is to find a set of connections among these superterminals. They suggested a left-to-right scan of the channel in order to identify a column, or a short segment, that can be used to interconnect two adjoining superterminals. This process is carried out for all the superterminals of a net.

In [DNB91], Das et al. use a greedy approach for OTC routing. During each iterative step, a net segment is chosen for OTC routing that has the largest congestion and span, and as a result the algorithm attempts to reduce the channel density due to nets remaining in the channel. In addition, selection of nets for OTC areas is carried out simultaneously for both top and bottom. A heuristic cost function is used to guide the search for the best candidate net in each iteration.

6.1.2 MIS-Based Algorithms

In [CL90], the basic three-step approach stated above for OTC routing is followed. However, their main contribution of Coney and Liu is that they formulated the net selection problem in a very natural way as the problem of finding an MIS of a circle graph. Since the latter problem can be solved optimally in quadratic time, an efficient optimal algorithm is obtained for the first step. Also, the second step is formulated as the problem of finding a minimum-density spanning forest of a graph. The minimum-density spanning forest problem is shown to be NP-hard, and an efficient heuristic algorithm is presented that produces very satisfactory results. A greedy channel router [RF82] is used for the third step. The most significant contribution of this paper is to formulate the OTC routing problem from an algorithmic point of view.

In [CPL90], three variations of BTM are presented (see Chapter 4 for details). The first model, HCVD, allows k tracks in both top and bottom OTC areas. As a result, the k-DMIS algorithm for overlap graphs was developed to select two planar subsets with density k. The details of that algorithm are given in Chapter 3. In the second model, HCVC, the entire M2 layer in the OTC area is available for routing. The second model is discussed in the approximation algorithms below. The third model, HDVC, partitioned the M2 vertically by feedthroughs into several smaller regions. Each region is routed independently, using the approximation algorithm developed for the HCVC model. Experimental evaluation on benchmarks from Xerox PARC and Physical Design Workshop indicated that HCVD was the most area-efficient among the three models presented. Detailed descriptions of MIS base algorithms are given in Section 6.2.

6.1.3 Approximation Algorithms

There are four routing algorithms in this category [HSS91b, CPL90, HSS91a, RS93]. The first one develops two key ideas: the use of vacant terminals and an approximation algorithm for net selection for the two-layer problem. The routing algorithm in [HSS91a] is a generalization of the first and develops an approximation algorithm for the three-layer case. In [RS93], a better approximation algorithm is presented for the two-layer case.

In [HSS91b], the concept of vacant terminals is used to increase the number of nets that can be routed in OTC areas. A terminal is called *vacant* if it is not used in any interconnection. They classify nets into several different types according to their terminal locations with respect to the vacant terminals. The algorithm includes two new steps in addition to the four steps of the basic OTC routing algorithm discussed above:

1. Net classification
2. Vacant terminal assignment
3. Net selection
4. OTC routing
5. Channel segment assignment
6. Channel routing

It is shown that the vacant terminal assignment problem is NP-complete. It is also shown that finding a maximum-weighted subset of nets that may be routed over the cells is NP-complete. For step 3, the M2IS algorithm for circle graphs is used, which guarantees a solution of at least 75% of the optimal result. Another unique feature of the approach is that nets are selected for OTC routing on the basis of their relationship to the length of the longest path (v_{max}) in the VCG, as well as the size of the maximum clique (h_{max}) in the HCG. Since the channel height depends on both h_{max} and v_{max}, this approach leads to a better OTC routing solution. Experimental results on industrial benchmarks show significant reduction in channel height and number of vias.

In [CPL90], three variations of BTM are presented (see Chapter 4). In one of the models, HCVC, the entire M2 layer is available for OTC routing. As a result, nets from two channels on either side of a cell row can be routed in the OTC area. However, this set of nets must be planar and routable in a fixed number of tracks. The complexity of finding such a set is open. (This problem has recently been resolved. See bibliographic notes at the end of the chapter.) An approximation algorithm is developed for this purpose, which finds the required subset with approximation $\rho = t/d$, where t is the number of tracks in the OTC area and d is the density of all the nets eligible for routing in the OTC region.

In [RS93], it is shown that the 0.75 approximation algorithm in [HSS91b] is only applicable if all nets are routable on both top and bottom OTC areas. As discussed above, among all nets routable in the OTC area, only a subset of nets is routable in the top OTC area, while another subset is routable in the bottom area. In [RS93], all three types of nets are considered and a 0.60 approximation algorithm is presented for a selection of nets.

6.1.4 Integer Linear Programming–Based Algorithms

In this category, two algorithms have been developed and are briefly discussed below. The first algorithm is discussed in detail in Section 6.4.

In [LPHL91], an integer linear programming (ILP)–based algorithm is proposed to reduce the channel density. The basic idea is to select only those nets that contribute to the minimization of the channel density. Pairwise net constraints ensure that the set of nets routed in the OTC area is planar. For each clique or zone in the HCG, a constraint is formed to minimize its density. LINDO is used to solve the ILP, and experimental evaluation shows that the approach is effective.

Another ILP-based algorithm is developed in [PP92]. The router is capable of handling a limited number of tracks in the OTC area. The resulting ILP is solved using interior point methods. Experimental results indicate that router performance is similar to that of the router presented in [LPHL91].

However, one major concern common to all ILP-based algorithms is the computational inefficiency of the algorithms. Due to the exponential time complexity of the algorithms used to solve integer linear programs, the proposed OTC algorithms are not feasible for large examples.

6.2 MIS-Based Algorithm for the BTM

In [CL90], Cong and Liu presented an algorithm for OTC channel routing. This was the first router, which provided the graph theoretic formulations used in many of the subsequent routers.

6.2.1 Algorithm Overview

This algorithm is developed for the HDVC model, and, as a result, the first step is to connect terminals on each side of the channel using the OTC routing area on that side. The same procedure is carried out for each side (upper or lower) of the channel independently. Let $t_{i,j}$ denote the terminal of net N_i at column j. In a given planar routing on one side of the channel, a *hyperterminal* of a net is defined to be a maximal set of terminals that are connected by wires in the OTC routing area on that side. For example, for the terminals in the upper side of the channel in Figure 6.1, $\{t_{5,4}, t_{5,6}, t_{5,11}\}$ is a hyperterminal of net 5. Obviously, when the routing within the channel (the third step) is to be done, all the hyperterminals of a net need to be connected instead of connecting all the terminals of the net, because the terminals in each hyperterminal have already been connected in the OTC routing area. Intuitively, the fewer the hyperterminals obtained after routing over the cells, the simpler the subsequent channel routing problem. Therefore, the first step of the problem can be formulated as routing a row of terminals using a single routing layer on one side of the row such that the number of hyperterminals is minimum.

Column: 1   2   3   4   5   6   7   8   9   10 11

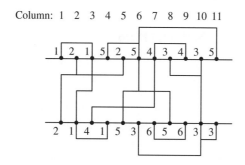

Fig. 6.1   A valid OTC routing solution.

After the completion of the OTC routing step, the second step is to choose net segments to connect the hyperterminals that belong to the same net. A net segment is a set of two terminals of the same net belonging to two different hyperterminals. For example, consider the channel shown in Figure 6.2. Net N_1 has two hyperterminals on each side of the channel and hence there are four possible net segments that can be used to connect these two hyperterminals (indicated by dashed edges between the hyperterminals of net N_1), while only one of them is needed to complete the connection. Similarly, net N_2, which has two terminals

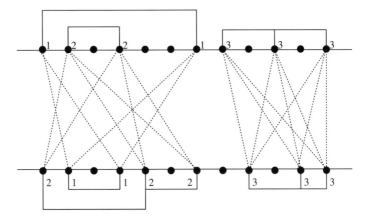

Fig. 6.2   Possible net segments for connecting two hyperterminals.

on the upper terminal row and three terminals on the lower terminal row, has a total of six possible net segments that can be used to connect the hyperterminals. Thus, the second step of the problem is to choose net segments to connect all the hyperterminals of each net such that the resulting channel density is minimum.

After the net segments for all the nets are chosen, the terminals specified by the selected net segments are connected using the routing area in the channel. The problem is now reduced to the conventional two-layer channel routing problem. A greedy channel router [RF82] is used for this step. Other two-layer channel routers may also be used.

6.2.2 Net Selection for OTC Routing

The first step of the OTC channel routing problem is to route a row of terminals using a single routing layer on one side of the channel such that the resulting number of hyperterminals is minimized. This problem is solved by using the dynamic programming method in $O(c^3)$ time, where c is the total number of columns in the channel. This problem is called the *multiterminal single-layer one-sided* (MSO) routing problem. Figure 6.3(a) shows an instance of the problem for the upper side of the channel in Figure 6.1. A valid routing solution is a set of nonintersecting wires that connect terminals in the same net on one side of the channel such that all the wires lie on one side of the channel. For example, Figure 6.3(b) is a valid routing solution for the instance in Figure 6.3(a).

It was shown that an MSO routing problem can be transformed into its two-terminal version (SPO) by decomposing each m-terminal net into $m(m-1)/2$ two-terminal nets. It was also shown that minimizing the number of hyperterminals is equivalant to maximizing two terminals in the SPO routing problem. As a result, the SPO routing problem can be modeled as an MIS in a circle graph and can be solved in $O(L^2)$ time.

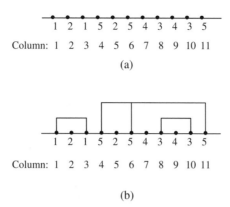

Fig. 6.3   (a) An instance of the MSO routing problem; (b) one of its valid solutions.

The MSO routing problem can also be solved directly, as follows. Given an instance I of an MSO routing problem, let $I(i, j)$ denote the instance resulting from restricting I to the interval $[i, j]$. Let $\mathcal{S}(i, j)$ denote the set of all the possible routing solutions for $I(i, j)$. Let

$$M(i, j) = \max_{S \in \mathcal{S}(i, j)} \left\{ \sum_{k \geq 2} (k - 1)d_p(S) \right\}$$

where $d_p(S)$ is the number of hyperterminals of degree p in S.

The solution for $M(i, j)$ is constructed by using dynamic programming. We assume that all solutions within $[i + 1, j]$ are already known. If the terminal at column i does not belong to any net, then clearly, $M(i, j) = M(i + 1, j)$. Otherwise, assume that the terminal at column i belongs to net N. Let $x_1, x_2, \ldots, x_s$ be the column indices of other terminals that belong to net N in interval $[i, j]$. Then it is easy to verify that

$$M(i, j) = \max\{M(i + 1, j), \max_{1 \leq l \leq s} \{M(i + 1, x_l) + M(x_l, j)\}\}$$

It is easy to see that this recurrence relation leads to an $O(L^3)$ time dynamic programming solution to the MSO routing problem.

6.2.3 Channel Segment Selection

After the completion of OTC routing, a set of hyperterminals is obtained. The terminals in each hyperterminal are connected together by OTC connections. The next problem is to choose a set of net segments to connect all the hyperterminals of each net such that the channel density is minimized. This problem can be transformed into a special spanning forest problem, as discussed below.

For an instance I of the net segment selection problem, the connection graph $CG(I) = (V, E)$ is defined to be a weighted multigraph. Each node in V represents a hyperterminal. Let h_1 and h_2 be two hyperterminals that belong to the same net N_i. For every terminal t_{i_j} in h_1 and for every terminal t_{i_k} in h_2 there is a corresponding edge (h_1, h_2) in E, and the weight of this edge $w((h_1, h_2))$ is the interval $[j, k]$ (assume that $j \le k$; otherwise, it will be $[k, j]$). Clearly, if h_1 contains p_1 terminals and h_2 contains p_2 terminals, then there are $p_1 \times p_2$ parallel edges connecting h_1 and h_2 in CG. Furthermore, corresponding to each net in I there is a connected component in $CG(I)$.

Figure 6.4 shows the connected component corresponding to net N_3 in the example in Figure 6.1. Given an instance I of the net segment selection problem, since all the hyperterminals in the same net are to be connected together for every net in I, it is necessary to find a spanning forest of $CG(I)$. Moreover, since the objective is to minimize the channel density, the density of the set of intervals associated with the edges in the spanning forest must be minimized.

Therefore, the net segment selection problem can be formulated as the minimum-density spanning forest problem (MDSFP). Given a weighted connection graph $CG = (V, E)$ and an integer D, determine a subset of edges $E' \subseteq E$ that forms a spanning forest of G, and the density of the interval set $\{w(e)|e \in E'\}$ is no more than D.

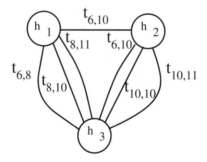

Fig. 6.4   The connected component induced by net.

In [CL90] it was shown that this problem is computationally hard.

Theorem 15   *The minimum-density spanning forest problem is NP-complete.*

In view of the NP-completeness of the MDSFP, an efficient heuristic algorithm has been developed for solving the net segment selection problem [CL90]. The heuristic algorithm is described as follows. Given an instance I of the net segment selection problem, a connection graph $CG = (V, E)$ is constructed. For each edge $e \in E$, the relative density of e, called $RD(e)$, is defined to be $d(e)/d(E)$, where $d(e)$ is the density of the set of intervals that intersect with the interval $w(e)$, and $d(E)$ is the density of the interval set $\{w(e)|e \in E\}$. The relative density of an edge measures the degree of congestion over the interval associated with the edge. The algorithm repeatedly removes edges from E until a spanning forest is obtained.

In summary, the MIS-based algorithm presented above contributed significantly to the understanding of the OTC routing problem. However, it does not
select nets to directly minimize channel density. Instead, it minimizes the total
number of hyperterminals and attempts to minimize the channel density in the second step. The algorithms, which will be described in subsequent sections, attempt
to rectify this problem.

6.3 Approximation Algorithms for the BTM

In this section, we review three approximation algorithms for BTM. The first two
algorithms deal with BTM-HDVC, while the third algorithm deals with BTM-
HCVC.

6.3.1 Improved OTC Routing Using Vacant Terminals

In this section, we present an overview of the algorithm called WISER for
OTC channel routing based on the BTM-HDVC. There are two key ideas: use of
vacant terminals to increase the number of nets that can be routed over the cells,
and selection of the "most suitable" nets for OTC routing. Consider the example
shown in Figure 6.5(a). OTC routing algorithms that do not use vacant terminals
can achieve at most a four-track solution (i.e., four tracks in the channel); however,
using the idea of vacant terminals, it is easy to find the two-track solution shown
in Figure 6.5(b). Furthermore, it is clear that the selection of nets which minimize
h_{max} is not sufficient to minimize the channel height. For example, the channel
height for the routing problem shown in Figure 6.5 is determined strictly by v_{max}
(that is, the longest path in the VCG). It is obvious that the nets that belong to
long paths in the VCG should be considered for routing over the cells, leading to
a better OTC routing solution.

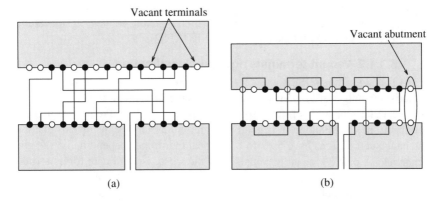

Fig. 6.5   (**a**) Routing without vacant terminals; (**b**) routing with vacant terminals.

6.3.1.1 Algorithm overview

We now give an informal description of each of the six steps of the algorithm WISER.

1. **Net Classification:** Each net is classified as one of three types, intuitively indicating the difficulty involved in routing this net over the cells.

2. **Vacant Terminal and Abutment Assignment:** Vacant terminals and abutments are assigned to each net depending on type and weight. The weight of a net intuitively indicates the improvement in channel congestion possible if this net can be routed over the cells.

3. **Net Selection:** Among all the nets that are suitable for routing over the cells, a maximum weighted subset is selected, which can be routed in a single layer.

4. **OTC Routing:** The selected nets are assigned exact geometric routes in the area over the cells.

5. **Channel Segment Assignment:** For multiterminal nets, it is possible that some net segments are not routed over the cells, and therefore must be routed in the channel. In this step, the "best possible" segments for routing of each net are selected to complete the net connection in the channel.

6. **Channel Routing:** The segments selected in the previous step are routed in the channel using a greedy channel router.

The most important steps in the algorithm WISER are net classification, vacant terminal and abutment assignment, and net selection. The algorithm for assigning the channel segments is similar to the one described in [CPL90], but in WISER's case, segments are iteratively added to the empty channel to achieve the minimum density. In [CPL90], segments are iteratively deleted from a channel containing all possible segments. When channel segment assignment is completed, a channel router is used to complete the connections within the channel. For this purpose, a greedy channel router is used, which typically achieves results at most one or two tracks beyond the channel density [RF82].

6.3.1.2 Vacant terminals and net classification

The algorithm WISER was developed to take advantage of the physical characteristics indigenous to cell-based designs. One such property is the abundance of vacant terminals. A terminal is said to be *vacant* if it is not required for any net connection. Examination of benchmarks and industrial designs reveals that most standard-cell designs have 50% to 80% vacant terminals depending on the given channel. A pair of vacant terminals with the same x-coordinate forms a *vacant abutment*. (See Figure 6.5.) In the average case, 30% to 70% of the columns in a given input channel are vacant abutments. The large number of vacant terminals and abutments in standard-cell designs is due to the fact that each logical terminal (inputs and outputs) is provided on both sides of a standard cell, but, in most cases, need only be connected on one side. For example, Table 6.1 shows the percentages of vacant terminals and abutments in the channels of the PRIMARY1 benchmark. The placement of PRIMARY1 was obtained by TimberWolfSC 5.1 [LS88] and the global routing was from [CSW89]. It should be noted that the actual number of vacant terminals and abutments and their locations cannot be obtained until global routing is completed.

Table 6.1:   Vacant Terminals and Abutments in PRIMARY1

Channel No.	% of Vacant Terminals	% of Vacant Abutments
1	85	70
2	74	56
3	64	42
4	60	38
5	64	43
6	52	30
7	58	35
8	50	27
9	53	30
10	60	38
11	63	44
12	64	46
13	62	39
14	64	43
15	63	44
16	67	48
17	66	45
18	91	82

To effectively use the vacant terminals and abutments available in a channel, the algorithm WISER categorizes nets according to the proximity of vacant terminals and abutments with respect to net terminals. Before classification, all m-terminal nets are decomposed into exactly $m - 1$ two-terminal nets at adjacent terminal locations. Every terminal t_i of an m-terminal net, except the left-most and right-most terminals, is decomposed into a two-terminal net t_{i1}, t_{i2}. For example, a four-terminal net $N = \{t_1, t_2, t_3, t_4\}$ is decomposed into three two-terminal nets: $N_1 = \{t_1, t_{21}\}$, $N_2 = \{t_{22}, t_{31}\}$, and $N_3 = \{t_{32}, t_4\}$. Clearly, $m - 1$ two-terminal nets are sufficient to preserve the connectivity of the original m-terminal net.

After decomposition, each net is classified as one of three basic types: Type I, Type II, or Type III net. The type of net intuitively indicates the difficulty involved in routing that net over the cell rows. In other words, Type III nets are most difficult to route, while Type I are relatively easy to route over the cell.

The following terminology is introduced in order to explain the net types. For any given t_i, the function $\text{OPP}(t_i)$ returns the terminal directly across the channel in the same column, the function $\text{ROW}(t_i)$ returns the row to which the terminal t_i belongs (either T or B), and $\text{COL}(t_i)$ returns the column of terminal t_i. Let $N = \{t_1, t_2\}$ be a net, where t_1 and t_2 are the left and right terminals of N, respectively.

A net $N = \{t_1, t_2\}$ is a Type I(a) net if $\text{ROW}(t_1) = \text{ROW}(t_2) = T$ and both $\text{OPP}(t_1)$ and $\text{OPP}(t_2)$ terminals are not vacant. Type I(a) may be routed only on the top OTC area. A net $N = \{t_1, t_2\}$ is a Type I(b) net if $\text{ROW}(t_1) = \text{ROW}(t_2) = B$ and both $\text{OPP}(t_1)$ and $\text{OPP}(t_2)$ terminals are not vacant. Type I(b) may be routed only on the bottom OTC area.

A net $N = \{t_1, t_2\}$ is a Type II net if $\text{OPP}(t_1)$ and $\text{OPP}(t_2)$ are both vacant. Type II nets may be routed over the cells on either side (top or bottom) of the channel.

A net N is a Type III net if $\text{ROW}(t_1) \neq \text{ROW}(t_2)$ and there exists a terminal t_i such that $\text{COL}(t_1) < \text{COL}(t_i) < \text{COL}(t_2)$ subject to the condition that t_i and $\text{OPP}(t_i)$ are vacant and $\text{OPP}(t_1)$ and $\text{OPP}(t_2)$ are not vacant. Type III nets are routed with the help of a vacant abutment, which converts a Type III net into a Type I(a) and a Type I(b) net.

Let $\mathcal{N}_{I(a)}$ refer to the Type I(a) nets. We define similar sets for all net types. The three basic net types are illustrated in Figure 6.6 along with possible OTC routes for each type.

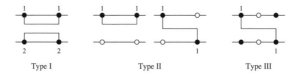

Fig. 6.6   Three basic net types.

A typical channel of a standard-cell design contains about 44% Type I nets, 41% Type II nets, and 10% Type III nets. Table 6.2 shows the breakdown of net types for the channels of PRIMARY1.

Observing that Type I and Type II nets constitute a majority of nets in the channel, one might suggest that it is sufficient to consider only these net types when routing over the cell rows. However, this is not the case, since removing Type III nets from the channel is critical in minimizing the length of the longest path ($v_{\max}$) in the VCG.

In the algorithm WISER, a net weighting function, $F_w : \mathcal{N} \to R^+$, which incorporates both the channel density and VCG path length criteria, is used to assess the suitability of a given net for OTC routing. The weight of a net $N = (t_1, t_2)$ is computed based on the relative density of the channel in the interval $[\text{COL}(t_1), \text{COL}(t_2)]$ and the ancestor and descendant weights of the net N. The

Table 6.2: Classification of Nets in PRIMARY1

Ch. No.	% of Type I Nets	% of Type II Nets	% of Type III Nets
1	13	87	1
2	35	54	3
3	44	40	8
4	50	32	13
5	55	31	10
6	60	23	13
7	49	33	12
8	57	20	19
9	54	28	13
10	45	34	15
11	54	29	10
12	51	27	13
13	42	43	9
14	38	41	10
15	43	37	11
16	46	39	6
17	42	48	5
18	9	84	0

relative density of net N can be computed by $r_d(N) = l_d(N)/h_{max}$, where $l_d(n)$ is the maximum of the local densities at each terminal location c, where $COL(t_1 \leq c \leq COL(t_2)$. The ancestor weight of a net N, denoted by ancw(N), is the length of the longest path from a vertex in the VCG with zero in-degree to the vertex representing N, and the descendant weight of N, denoted by dscw(N), is the length of the longest path from the vertex representing N to a vertex in VCG with zero out-degree. The general net weighting function is given below:

$$F_w(N) \quad = \quad k_1 \frac{r_d(N)}{v_{max}} + k_2 \frac{(\text{ancw}(N) + \text{dscw}(N)) - |(\text{ancw}(N) - \text{dscw}(N))|}{h_{max}}$$

where k_1 and k_2 are experimentally determined constants. Since the weight of a net N indicates the reduction possible in h_{max} and v_{max} if N is routed over the cell rows, the "best" set of nets to route over the cells is the one with maximum total weight.

6.3.1.3 Vacant terminal and abutment assignment

After classification and weighting, nets are allocated a subset of vacant terminals or vacant abutments, depending on their type, to help define their routing paths in the area over the cell rows. It should be noted that Type I nets, which have both of their terminals on the same boundary of the channel, can be routed in the area over the cells without using vacant terminals, as shown in Figure 6.7(a).

Therefore, the vacant terminal and abutment assignment problem is a matter of concern only for Type II and Type III nets. For Type II nets, the vacant terminals are "reserved" for a net; that is, only a particular net may use a particular vacant terminal, and as a result, the vacant terminal assignment for Type II nets is actually a net selection problem. On the other hand, a Type III net may use any abutment within its span, and therefore the vacant abutment assignment problem for Type III nets can be viewed as a matching problem.

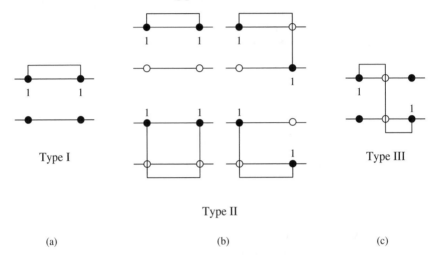

Type I

Type II

Type III

(a)   (b)   (c)

Fig. 6.7   Vacant terminal and abutment assignment for **(a)** Type I; **(b)** Type II; **(c)** Type III nets.

1. **Vacant Terminal Assignment for Type II Nets:** Recall that a Type II net has a vacant terminal located directly across from each net terminal. In Type II net routings, the horizontal segment of the net is "replaced" by either zero, one, or two direct vertical wire(s) in the channel. As a result, it does not contribute to channel height and it has effectively been "removed" from the channel.

   The vacant terminal assignment problem is indeed a net selection problem. In fact, since a Type II net is routable in either the top or bottom OTC area, the vacant terminal assignment problem for Type II nets can be simply modeled as a maximum bipartite subgraph (MBS) problem in an overlap graph, which is known to be NP-hard [SL89], establishing the following result.

Theorem 16   *The vacant terminal assignment problem for Type II nets is NP-complete.*

Using this theorem, it is easy to show that the problem of finding optimal routing using only k tracks in an OTC area is also NP-complete.

Corollary 1   *The vacant terminal assignment problem for Type II nets remains NP-complete when the number of tracks available over each cell row is restricted to k.*

Proof: Follows directly from the fact that the MBS problem for overlap graphs, even if restricted to k tracks. (See Chapter 2.)

$\square$

In the algorithm WISER, rather than solving the vacant terminal assignment problem for Type II nets directly, vacant terminals are assigned during the net selection phase. In this way, Type I and Type III nets may also have some effect on the vacant terminal assignment process for Type II nets, resulting in a better OTC routing solution.

1. **Vacant Abutment Assignment for Type III Nets:** It is easy to see that a Type III net may not be routed using only one vacant terminal, and, in fact, these nets require vacant abutments in their routings, as shown in Figure 6.7(c). A Type III net is decomposed into two Type I nets if the abutment is used. It is also clear that a net may only use those abutments that lie within its span. Given a net $N = (t_1, t_2)$, let us define a set $\mathcal{A}(N)$ of abutments that lie within the span of N. That is, $\mathcal{A}(N) = \{a | \text{COL}(t_1) < a < \text{COL}(t_2)\}$. If vacant abutment $a \in \mathcal{A}(N)$ is assigned to net N, then we define $\text{ASSIGN}(N) = a$, and the set of all abutments in the given input is referred to as $\mathcal{A}$.

   Given a set $\mathcal{N}_{III}$ of Type III nets, the vacant abutment assignment problem for Type III nets is to assign one abutment $a \in \mathcal{A}(N)$ to each net $N \in \mathcal{N}_{III}$, such that Type I(a) and Type I(b) nets thus formed from an independent set in the top and bottom OTC areas, respectively, and the total number of nets in both independents is maximum among all such assignments.

Theorem 17   *The vacant abutment assignment problem for Type III nets is NP-complete.*

Proof: See [HSS91b] for details.

$\square$

In view of the NP-completeness of the vacant abutment assignment problem for Type III nets, WISER uses a greedy heuristic based on certain necessary conditions for the routability of a pair of Type III nets. These necessary conditions are depicted in Figure 6.8.

The main idea of this algorithm is to assign vacant abutments to nets according to their weight. The "heaviest" nets are considered first.

Since the necessary conditions are satisfied for every pair of nets, and each net is assigned to a unique vacant abutment, the solution produced by the algorithm is always routable and the following lemma holds.

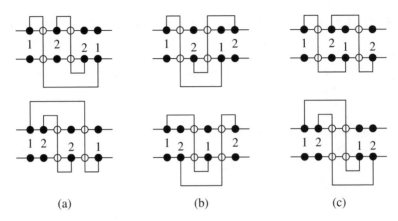

Fig. 6.8   Necessary conditions.

Lemma 5   *Algorithm Assign_Abutments produces a feasible solution in $O(dn^2)$.*

It should be noted that nonoptimality of the greedy approach used by the algorithm Assign_Abutments (see Figure 6.9) is not very significant because space available for OTC routing is limited and nets of Types I, II, and III must all compete for a position. Since the greedy algorithm allocates abutments to the maximum-weighted nets first, the most critical nets will have the greatest opportunity for abutment assignment.

6.3.1.4 Provably good algorithm for net selection

The net selection problem can be stated as follows. Given a set $\mathcal{N}$ of nets, select a maximum-weighted subset of nets $\mathcal{N}' \subseteq \mathcal{N}$, such that all the nets in $\mathcal{N}'$ can be routed in the area over the cell rows in planar fashion. Algorithm WISER uses a graph theoretic approach to net selection. An overlap graph G_O is defined for intervals of nets in set $\mathcal{N}$. It is easy to see that the net selection problem reduces to the problem of finding a maximum-weighted bipartite subgraph in the overlap graph G_O. However, the density of the nets in each partite set must be bounded by a constant k, which is the number of tracks available in the OTC region. The problem of computing an MBS in an overlap (circle) graph is known to be NP-complete [SL89], and, as a result, a 0.75 approximation algorithm is developed.

It is easy to see that there are several restrictions on the assignment of vertices to partite sets. For example, a vertex corresponding to a Type I(a) net may not be assigned to the partite set that is to be routed over the lower row of cells. On the other hand, a vertex corresponding to a Type I(b) net must be assigned to the partite set that is to be routed in the lower OTC area. As noted earlier, vertices representing Type II nets may be assigned to either partite set, since these nets can be routed over either the upper or lower cell row. As stated earlier, a Type III net is partitioned into two Type I nets at the location of its designated abutment.

Algorithm ASSIGN_ABUTMENTS()

Input: Set $\mathcal{N}_{III}$ of Type III nets and vacant abutment set $\mathcal{A}$.
Output: Vacant Abutment Assignment.

---

begin
/* Initialize array $ASSIGN[]$ to zero */
   For all $a \in \mathcal{A}$ do
   /* Nets are considered in the weight sorted order */
     For each net $N = (t_1, t_2)$ of $\mathcal{N}_{III}$ do
      If $(t_1 < a < t_2)$ Then $\mathcal{A}(N) = \mathcal{A}(N) \cup a$
   For each net $N \in \mathcal{N}_{III}$ do
   For all $a \in \mathcal{A}(N)$ do
     For all $N' \in \mathcal{N}_{III}$ such that $ASSIGN[N] \neq 0$ do
      If (Condition_Overlap($N', a, N, ASSIGN[N]$) = 1)
       Then { $ASSIGN[N] = a$; break; }
      If (Condition1_Containment($N', a, N, ASSIGN[N]$) = 1)
       Then { $ASSIGN[N] = a$; break; }
      If (Condition2_Containment($N', a, N, ASSIGN[N]$) = 1)
       Then { $ASSIGN[N] = a$; break; }
end.

Fig. 6.9   Algorithm ASSIGN_ABUTMENTS.

Let V_1 and V_2 be the sets of nets that can be routed in the top and bottom OTC areas, respectively, and let V_{12} represent the nets that are routable in either top or bottom area. Note that V_1 represents the nets in $\mathcal{N}_{I(a)} \cup \mathcal{N}_{III(a)}$, while V_2 represents the nets in $\mathcal{N}_{I(b)} \cup \mathcal{N}_{III(b)}$. The set V_{12} represents nets in $\mathcal{N}_{II}$. Let $V_1' \subseteq V_1 \cup V_{12}$ be the set of nets selected for routing in the top OTC area. Similarly, let $V_2' \subseteq V_2 \cup V_{12}$ be the set of nets selected for routing in the bottom OTC area. Note that $V_1' \cap V_2' = \phi$. Finally, let $\mathcal{S}' = (V_1', V_2')$ denote the solution.

The algorithm FIS (fixed independent set) for approximating a maximum-weighted bipartite subgraph in G_O ensures that the density of each partite set is less than or equal to the maximum number of tracks available over the cell rows; that is, k. Let the weight of a vertex v be denoted by $w(v)$, and the weight of a set of vertices is simply the sum of the weights of all the vertices in the set. The maximum-weighted k-density bipartite subgraph is approximated by finding two maximum-weighted independent sets in the graph G_O, one for V_1' and one for V_2'. Since some nets are not fixed to a specific partite set, the order in which the partite sets are computed is extremely important. In the FIS approximation, both possible orderings are considered.

Note that the procedure MIS_CIRCLE() finds the maximum-weighted k-DMIS in an overlap (circle) graph in $O(kn^2)$ time (see Chapter 2 for details).

To show that the algorithm FIS produces provably good results, a lower bound for the independent set approximation of a maximum-weighted bipartite subgraph is needed. Let us first consider the unweighted MBS problem without the restriction on the density of the partite sets. The weighted density-restricted version of the proof is a simple extension.

Theorem 18   *Let S^* denote the optimal maximum bipartite subgraph of an overlap graph. Let S' be the solution obtained by selecting two independent sets one at a time. Then*

$$\frac{S'}{S^*} \geq 0.75$$

is a lower bound on the approximation, which may be achieved in $O(n^2)$ time.
Proof: It is special case of MKIS for circle (overlap) graphs. See Chapter 2 for details.

□

In the algorithm FIS (see Figure 6.10) the interest is in solving the maximum-weighted bipartite subgraph problem where the density of the partite sets is restricted to a constant k. It is easy to see that the algorithm FIS does indeed satisfy the density requirement, since the maximum independent set algorithm used in net selection satisfies that requirement.

Algorithm FIS()

Input: Overlap Graph $G = (V, E)$, V_1, V_2, V_{12}.
Output: Set of vertices representing 2-independent sets.

> **begin**
> $V_1[1]' =$ MIS_CIRCLE($V_1 \cup V_{12}$);
> $V_2[1]' =$ MIS_CIRCLE($V_2 \cup V_{12} - V_1[1]'$);
> $V_2[2]' =$ MIS_CIRCLE($V_2 \cup V_{12}$);
> $V_1[2]' =$ MIS_CIRCLE($V_1 \cup V_{12} - V_2[2]'$);
> $S' = \emptyset$;
>   **for** $i = 1$ to 2 **do**
> $S' =$ Largest($S', V_1[i]' \cup V_2[i]'$);
>   **return** S';
> **end**.

Fig. 6.10   Algorithm FIS.

It follows from the above theorem that the approximation FIS is guaranteed to obtain solutions at least three-quarters of the optimal result, but in practice its performance is much better. FIS has been tested on randomly generated examples with the number of nets equal to 10, 15, and 20, where $2 \leq k \leq 4$. The results are listed in Table 6.3. Due to the exponential behavior of optimal algorithms for finding a k-density maximum-weighted bipartite subgraph in an overlap graph, it is not possible to test the performance of the FIS approximation on examples with more than 20 nets.

Table 6.3:   Experimental Results—Performance of FIS

10 Net Examples		15 Net Examples		20 Net Examples	
Example No.	% of Opt. Solution	Example No.	% of Opt. Solution	Example No.	% of Opt. Solution
1	100	1	100	1	100
2	100	2	100	2	100
3	100	3	91	3	100
4	100	4	92	4	100
5	100	5	100	5	92
6	100	6	100	6	100
7	100	7	100	7	100
8	100	8	100	8	93
9	100	9	100	9	100
10	100	10	91	10	93

As indicated in Table 6.3, FIS typically gives solutions that are at least 91% of the optimal result. In the average case, the performance of FIS is 98% optimal.

6.3.2 Improved Approximation Algorithm for Net Selection

The 0.75 lower bound developed for the algorithm FIS does not take into account the fact that some nets are only routable in upper or lower OTC areas. The bound developed for FIS depends on the cardinality of V_{12}. If V_{12} is null, the lower bound will indeed be less than 0.75. In order to distinguish between the problem addressed in Theorem 18, let us call the new problem the *maximum bipartite special subgraph problem* (MBSSP).

In this section, we present the proof that the lower bound of the algorithm FIS is 0.5 if all three types of nets are considered, and then will present an approximation algorithm that guarantees to provide a solution that is at least 60% of the optimal. Finally, we present examples showing that the bound is tight.

In order to get a lower bound on the algorithm FIS, the following terminology is introduced. All terms with star (*) as the superscript refer to the optimal solution, while the primed terms refer to the solution selected by the algorithm.

Let $S_1^* \subseteq V_1$, $S_2^* \subseteq V_2$, and $S_{12}^* \subseteq V_{12}$. In addition, let $S_{12-1}^* \subseteq V_{12}$, which is also a subset of V_1'. In other words, by the subscript 12-1 we wish to indicate that portion of the first partite set of the solution that has come from the set labeled V_{12}. Similarly, let the primed version of these subsets denote equivalent subsets derived by the algorithm FIS. For example, S_{12-2}' denotes that portion of the second partite set of the solution that has come from the set labeled 12.

Note that FIS basically provides a 0.75 approximation if all nets are in set V_{12}. For the case when $V_1 \neq \phi$ and $V_2 \neq \phi$, the following theorem is established.

Theorem 19   *Let S^* be the optimal maximum bipartite special subgraph of an overlap graph. Let S' be the solution obtained by the algorithm FIS. Then*

$$\frac{S'}{S^*} \geq 0.5$$

is a lower bound on the approximation, which may be achieved in $O(n^2)$ time.

However, FIS can be improved to provide an approximation of 0.6 as follows. The basic idea of the algorithm is to use three different strategies for selecting the sets and avoid missing the large subset of S_{12}^*, as is done in the algorithm FIS.

The formal algorithm IFIS (improved FIS) is shown in Figure 6.11.

In the above, the algorithm M2IS_CIRCLE stands for the MKIS_CIRCLE algorithm for $k = 2$.

Theorem 20   *Let S^* denote the optimal maximum bipartite special subgraph of an overlap graph. Let S' be the solution obtained by the algorithm FIS. Then*

$$\frac{S'}{S^*} \geq 0.60$$

is a lower bound on the approximation, which may be achieved in $O(n^2)$ time.
Proof: The M2IS_CIRCLE approximation algorithm as described in Chapter 2 produces a solution that is 75% of the optimal solution for $k = 2$. Therefore, line 1 of the algorithm IFIS produces a solution that is $0.75 \cdot \alpha \cdot |S^*|$.

Note the symmetric nature of steps 2 and 3 of the algorithm IFIS. We will now show that either step 2 or step 3 will produce a solution that is at least $0.5 \cdot \alpha \cdot |S^*|$.

Clearly, $|S_2'| \geq |S_2^*|$. Also, $|S_1' \cup S_{12-1}'| \geq |S_1^* \cup S_{12-1}^*|$.

Thus,

$$|MIS(V_1 \cup V_{12})| + |MIS(V_2)| = |S_1' \cup S_{12-1}'| + |S_2'|$$
$$\geq |S_1^* \cup S_{12-1}^*| + |S_2^*|$$
$$\text{(Since } S_1^* \cap S_{12-1}^* = \phi.\text{)}$$
$$= |S_1^*| + |S_{12-1}^*| + |S_2^*|$$
$$\geq (1 - \alpha)|S^*| + |S_{12-1}^*|$$
$$\text{either } |S_{12-1}^*| \geq 0.5 \cdot \alpha|S^*|$$
$$\text{or } |S_{12-2}^*| \geq 0.5 \cdot \alpha|S^*|$$
$$\text{So, } \geq (1 - \alpha)|S^*| + 0.5\alpha|S^*|$$
$$\text{Thus, } \geq (1 - 0.5\alpha)|S^*|$$

Algorithm IFIS()

Input: Overlap Graph $G = (V, E)$, V_1, V_2, V_{12}.
Output: Set of vertices representing 2-independent sets.

---

begin
    Step 1:
        $V_1[1]'$, $V_2[1]'$ =M2IS_CIRCLE(V_{12});
    Step 2:
        $V_1[2]'$ =MIS_CIRCLE($V_1 \cup V_{12}$);
        $V_2[2]'$ =MIS_CIRCLE(V_2);
    Step 3:
        $V_1[3]'$ =MIS_CIRCLE(V_1);
        $V_2[3]'$ =MIS_CIRCLE($V_2 \cup V_{12}$);
    $S' = \emptyset$;
    for $i = 1$ to 3 **do**
        $S' = \text{Largest}(S', V_1[i]' \cup V_2[i]')$;
    return S';
end.

Fig. 6.11   Algorithm improved FIS.

Therefore, $|S'| = \max\{0.75\alpha, 1 - 0.5\alpha\} \cdot |S^*|$. It is easy to see that the value of the max function is always greater than or equal to 0.60 with equality if and only if $\alpha = 0.8$.

☐

Corollary 2   *For a weighted case, if $w(S')$ represents the weight of the solution S' and $w(S^*)$ represents the weight of the solution S^*, then*

$$\frac{w(S')}{w\,S^*} \geq 0.60$$

is a lower bound on the approximation, which may be achieved in $O(n^2)$ time.

6.3.3 Approximation Algorithm for the BTM-HCVC Model

The cell rows in the BTM-HCVC consists of two columnwise-equipotential terminal rows; that is, $t_i = b_i$ for all $i = 1$ to L. As discussed in Chapter 5, the routing problem in the BTM-HCVC is a classical ETSPR problem. An interesting variation to this problem was considered by Cong, Preas, and Liu [CPL90], wherein the terminal rows are considered to be nonequipotential. The problem is now equivalent to the TSPR problem, discussed in Chapter 5. Figure 6.12 shows a valid solution to a TSPR problem. Note that the terminals in two different rows do not have to be connected, although they belong to the same net. This is because the two rows of terminals belong to two different channels, and connections for the same net in different channels have been accomplished by adding feedthroughs in the global routing phase [CP88]. Therefore, a routing solution S can be partitioned into the union of two planar routing solutions S^l and S^u for the lower and upper rows of terminals. The weight of S is defined to be the sum of the weights of S^l and S^u (i.e., $w(S) = w(S^l) + w(S^u)$). An *optimal solution* to the TSPR problem is a solution whose weight is maximum. For a solution S to a TSPR problem, $d(S)$ is used to denote the density of S. Clearly, if $d(S)$ is no more than the number of tracks (k) given in the problem, S is a valid solution to the TSPR problem. This is due to the fact that there are no vertical routing constraints in the TSPR problem, since the terminals in two different rows need not be connected.

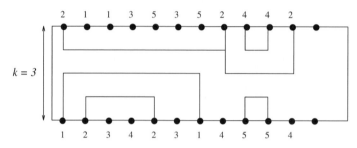

Fig. 6.12   A valid routing solution to TSPR.

The TSPR problem can be simplified if there is no track sharing between S^l and S^u. This is equivalent to routing that is similar to TSPR, except the number of tracks is not limited, as in the TSPR problem. The resulting problem is called the *two-row unlimited-height planar routing* (TSPR-1) problem (see bibliographic notes).

Clearly, the weight of any optimal solution to a TSPR problem is more than the weight of an optimal solution to the corresponding TSPR-1 problem. Figure 6.13 shows a valid solution to the corresponding TSPR-1 problem of the example in Figure 6.12 (each connected pair (net segment) is labeled for later reference).

However, solving the TSPR problem optimally is very difficult. So far, no polynomial algorithm has been obtained to solve the TSPR problem optimally, and the complexity of the TSPR problem is still unknown. Cong, Preas, and Liu [CPL90] have presented a two-step approach to obtain an approximation solution.

Let S^* denote an optimal solution and S the solution to be constructed by the algorithm. The algorithm obtained is as follows. In the first step, an optimal solution $\bar{S}$ is computed for the corresponding TSPR-1 problem (i.e., assuming that the height is unlimited). If $d(\bar{S}) \leq k$, then S is equal to $\bar{S}$. Clearly, in this case, S is an optimal solution to the given TSPR problem. If $d(\bar{S}) \geq k$, then S is chosen to

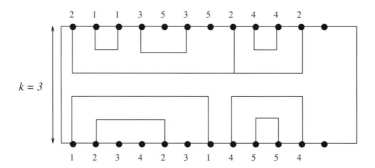

Fig. 6.13   A valid routing solution to the TSPR-1 problem of the example in Figure 6.12.

be a subset of connected pairs from $\bar{S}$ such that $d(S) = k$ and $w(S)$ is maximum. Clearly, S is a valid solution to the TSPR problem. (such an S can be computed in polynomial time, which is shown later). Moreover, it is shown that the weight of the solution S thus constructed will not be too small. In fact, it is shown that $w(S)/w(S^*) \geq k/d(\bar{S})$. Therefore, in either case,

$$\frac{w(S)}{w(S^*)} \geq \min(1, k/d(\bar{S})).$$

This bound ensures that the weight of the solution S is very close to the weight of an optimal solution to the original TSPR problem because of the following observation. In most cases, the density $d(\bar{S})$ of the optimal solution $\bar{S}$ to the corresponding TSPR-1 problem will not be too high. Although there is no limit on the number of tracks to be used in $\bar{S}$, due to the inherent restriction of planar routing, $\bar{S}$ cannot be very dense. Therefore, in many cases, $d(\bar{S}) \leq k$, so that the constructed solution S is optimal. In other cases, $d(\bar{S})$ usually exceeds k by a small constant so that the solution S is close to optimal, since $w(S)/w(S^*) \geq k/d(\bar{S})$. For example, for the 13 rows of cells in the PRIMARY1 circuit, the maximum of $d(\bar{S})$ is 13, and the average of $d(\bar{S})$ is 8.5. However, for the cell family used at Xerox PARC, the number of available tracks over the cells is 13 (i.e., $k = 13$). Thus, for all the rows in the PRIMARY1 circuit, the solutions to the TSPR problems constructed by this algorithm are optimal.

In the first step of the algorithm, an optimal solution $\bar{S}$ is computed for the corresponding TSPR-1 problem. $\bar{S}$ can be computed as follows. Maximum-weighted one-sided planar routing solutions, S^l and S^u, for the lower and upper terminals of R are computed independently, regardless of the availability of routing tracks. (A one-sided routing solution has the property that all the connections are on one side of a row of terminals. In our case, S^l is always above the lower terminals and S^u is always below the upper terminals.) Then the union of S^l and S^u is an optimal solution to the TSPR-1 problem. This is true because, in the TSPR-1 problem, track sharing is not considered in the routing solutions for the two rows, since an infinite number of tracks are available. In order to compute

S^l or S^u, a maximum-weighted one-sided planar routing solution must be found for a row of terminals. This problem is solved optimally using a weighted version of the algorithm for the SSPR problem in Section 6.2.2. Based on the results in Section 6.2.2, the following theorem is established.

Theorem 21   *For a TSPR problem, the corresponding TSPR-1 problem can be solved optimally in $O(k^2 n^2)$ time, where n is the number of terminals and k is the number of tracks. Thus, the first step of the approximation algorithm for the TSPR problem can be solved in $O(k^2 n^2)$ time.*

In the second step, which is performed only when $d(\bar{S}) > h$, the subset S of connected pairs is chosen from $\bar{S}$ such that $d(S)$ is no more than h and $w(S)$ is maximum. This problem can be solved by finding a maximum-weighted h-family in a partially ordered set. A *partially ordered set P* is a collection of elements together with a binary relation $\leftarrow$ defined on $P \times P$, which satisfies the following conditions:

1. *Reflexive*; i.e., $x \leftarrow x$ for all $x \in P$.
2. *Antisymmetric*; i.e., $x \leftarrow y$ and $y \leftarrow x$ implies $x \leftarrow y$ for all $x, y \in P$.
3. *Transitive*; i.e., $x \leftarrow y$ and $y \leftarrow z$ implies $x \leftarrow z$ for all $x, y, z \in P$.

We say that x and y are *related* if $x \leftarrow y$ or $y \leftarrow x$. A *chain* in P is a subset of elements in which every two of them are related. An *antichain* in P is a subset of elements in which no two are related. An h-family in P is a subset of elements that contains no chain of size $h + 1$. An integer weight $w(P)$ is associated with each element p in P. The weight of a subset Q of elements in P, denoted by $w(Q)$, is defined to be the sum of the weights of the elements in Q. A *maximum-weighted h-family* in P is an h-family whose weight is maximum. For each connected pair (net segment) A in S, it defines an interval $i(A) = [x, y]$, where x and y $(x \leq y)$ are the two column indexes of the two terminals in A. A partially ordered set $P(\bar{S})$ is constructed for the planar routing solution $\bar{S}$ computed in the first step as follows. Each element in $P(\bar{S})$ represents a net segment in $\bar{S}$. We say that a net segment A_1 *dominates* a net segment A_2 in $P(\bar{S})$ (or A_2 is *dominated* by A_1) if one of the following three conditions holds:

1. Both A_1 and A_2 are in the lower row, and $i(A_1)$ contains $i(A_2)$.
2. Both A_1 and A_2 are in the upper row and $i(A_2)$ contains $i(A_1)$.
3. A_1 is in the upper row and A_2 is in the lower row, and $i(A_1)$ intersects $i(A_2)$.

Let the dominance relation be the binary relation in $P(\bar{S})$. Then one can show that:

Lemma 6   *$P(\bar{S})$ thus constructed is partially ordered set.*

Since it is straightforward to verify that the dominance relation thus defined is reflexive, antisymmetric, and transitive, we leave the reader to complete the proof of Lemma 1. Intuitively, A_1 dominates A_2 if and only if the connection of A_1 must be above the connection of A_2. The purpose of the introduction of the notation of a partially ordered set is clear from the following result:

Lemma 7   *A subset S of net segments from $\bar{S}$ satisfies the condition $d(S) \leq h$ if and only if S is an h-family of $P(\bar{S})$.*

Proof: If $d(S) \leq h$, the connections of the pairs in S can be routed in at most h tracks (since there is no vertical constraint in this case). It is easy to verify that if the connections of two pairs share the same track, then these two pairs are not related in $P(\bar{S})$ under the dominance relation. Therefore, S can be partitioned into h antichains. So S is an h-family of $P(\bar{S})$.

On the other hand, if S is an h-family of $P(\bar{S})$, S can be partitioned into at most h antichains by recursively peeling off the maximal elements in S. It is easy to see that the density of the connections of the pairs in an antichain is 1. Thus, the density of the connections of the pairs in S is no more than h.

$\square$

According to Lemma 2, it is easy to see that the problem of finding a maximum-weighted subset of connected pairs s from $\bar{S}$, such that $d(S) \leq h$, is equivalent to the problem of finding a maximum-weighted h-family in $P(\bar{S})$. A maximum-weighted h-family in a partially ordered set can be computed in $O(h \cdot mn \log n^2/m)$ time, where n is the number of elements in the partially ordered set and m is the number of related pairs in the partially ordered set [CPL90]. Thus, S can be computed efficiently. Moreover, it can be shown that S is a good approximation of the optimal solution S^* to the TSPR problem. The following theorem states these results.

Theorem 22   *If the routing solution $\bar{S}$ computed in the first step is too dense (i.e., $d(\bar{S}) > h$), it is possible to choose a maximum-weighted subset of net segments S from $\bar{S}$ in $O(h \cdot mn \log n^2/m)$ time such that $d(S) \leq h$ and*

$$\frac{w(S)}{w(S^*)} \geq h/d(\bar{S})$$

Proof: According to Lemma 2, one can obtain S by computing a maximum-weighted h-family in $P(\bar{S})$, which can be carried in $O(h \cdot mn \log n^2/m)$. Now it can be shown that the weight of S satisfies the inequality stated above as follows.

Let g denote $d(\bar{S})$. According to Lemma 2, $\bar{S}$ is a g-family in $P(\bar{S})$. Thus, $\bar{S}$ can be decomposed into g antichains. Let $u_1, u_2, \ldots, u_g$ denote the weights of the g antichains sorted in nonincreasing order (the weight of an antichain in $P(\bar{S})$ is defined to be the sum of the weights of the elements in the antichain). Since S is a maximum-weighted h-family in $P(\bar{S})$,

$$w(S) \geq u_1 + u_2 + \cdots + u_h \geq h \cdot u_h$$

Moreover,

$$w(\bar{S}) = u_1 + u_2 + \cdots u_g \le w(S) + u_{h+1} + \cdots u_g$$

$$\le w(S) + (g - h) \cdot u_h \le w(S) + \frac{g - h}{h} w(S)$$

$$\frac{g}{h} w(S) = \frac{d(\bar{S})}{h} w(S)$$

Clearly, $w(S^*) \le w(\bar{S})$. Therefore,

$$\frac{w(S)}{w(S^*)} \ge \min(1, h/d(\bar{S}))$$

$\square$

According to this theorem, the time complexity of the second step (remember that it is carried out only when $d(\bar{S}) \ge h$) of the algorithm is $O(h \cdot mn \log n^2/m)$. Since the time complexity of the first step is $O(k^2 \cdot n^2)$, the overall complexity of the algorithm for the TSPR problem is $O(k^2 \cdot n^2 + h \cdot mn \log n^2/m)$, where n is the number of terminals, k is the maximum size of a net, and h is the number of available tracks. Since k is a constant for most circuits, the following corollary is true.

Corollary   *When the maximum size of a net is bounded by a constant, the approximation algorithm for the TSPR problem presented above produces a solution in S in $O(n^2 + h \cdot mn \log n^2/m)$ time such that*

$$\frac{w(S)}{w(S^*)} \ge \min(1, h/d(\bar{S}))$$

where S^* is an optimal solution to the TSPR routing problem.

Based on these results, the OTC routing algorithm works as follows. For each row of cells, the lower and upper terminals of each row are routed at the same time. First, an optimal solution $\bar{S}$ of the corresponding TSPR-1 problem is computed. If the density of $\bar{S}$ is no more than the number of tracks available, then the output is $\bar{S}$. Otherwise, construct the partially ordered set $P(S)$ associated with $\bar{S}$ and compute a maximum-weighted h-family.

6.4 Integer Linear Programming–Based Algorithms for the BTM

The research in OTC routing assumed that the more nets being routed over the cells, the greater the reduction in channel density. However, Lin, Perng, Hwang and Lin [LPHL91] showed that only the removal of critical nets contributes to the reduction in channel density. Since the height of the cell is fixed and the number of tracks in the OTC area of a cell row is limited, only those nets (called *critical nets*), when removed from the channel lead to reduction in channel density should be routed in OTC areas. The channel is divided into zones, and each zone has a *zone density* associated with it. Removal of a net from a zone reduces the zone density by 1. However, the channel density is minimized by routing nets from some critical zones in OTC areas. A bipartite graph is used to represent the relationship between the nets and zones. The problem is converted into a *constrained covering problem* and formulated as an ILP problem. The concept is first explained for a channel with only two-terminal nets and then extended to channels with multiterminal nets. The algorithm basically consists of three key steps: zone representation, bipartite graph modeling, and ILP formulation.

6.4.1 Algorithm Overview

The router optimally selects the nets that can be routed in OTC areas. The algorithm consists of the following four steps:

1. **Zone Representation:** The channel area is partitioned into zones. The zones are in fact a maximal clique in the interval graph, defined by the horizontal segments of the nets.

2. **Bipartite Graph Model:** The correlation between the nets in zones are modeled as a bipartite graph. The channel density reduction problem is now transformed into a constrained covering problem.

3. **ILP Formulation:** In this step, an integer linear program is formulated for choosing the nets that are to be routed in OTC areas.

4. **Multiterminal Nets:** The formulation developed in the previous step is extended for multiterminal nets. In order to extend the formulation, the multiterminal nets are decomposed into two-terminal nets and the terminals are assigned for routing in the channel.

6.4.2 Zone Representation

All the nets considered are assumed to be two-terminal nets. A net is considered for OTC routing if and only if both its terminals are located on the same side of the channel. The channel area is partitioned into zones by adopting the technique used in the YK algorithm, discussed in Chapter 2. Zones are in fact a maximal clique in the interval graph defined by the horizontal segments of the nets. The interval graph of the net list illustrated in the example channel in Figure 6.14(a) is shown in Figure 6.14(b). Let $S(i)$ be the set of nets whose horizontal segments (net intervals) intersect column i. Figure 6.14(c) shows $S(i)$ for all the columns and the zone allocation for the example shown in Figure 6.14(a). Let $\text{den}(z_i)$ be the density of an arbitrary zone i. The channel density D_{initial} is equal to $\max(\text{den}(z_i))$. It is clear that removal of any net from zone i will reduce $\text{den}(z_i)$ by 1. To reduce the density of every zone whose density equals D_{initial}, then the channel density is reduced by 1. Therefore, in order to reduce the channel density down to D_{final}, one must remove at least $\text{den}(z_i) - D_{\text{final}}$ nets from each zone i with density $\text{den}(z_i) > D_{\text{final}}$. A zone is critical if and only if its density is greater than D_{final}. Otherwise, it is noncritical.

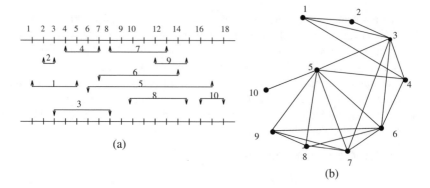

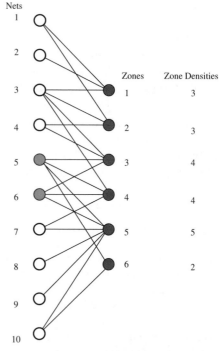

Column	S(i)	Zone
1	1	1
2	1 2	
3	1 2 3	
4	1 3 4	2
5	1 3 4	
6	3 4 5	3
7	3 4 5 6	
8	3 5 6 7	4
9	5 6 7	
10	5 6 7 8	5
11	5 6 7 8	
12	5 6 7 8 9	
13	5 6 7 8 9	
14	5 6 8 9	
15	5 8 9	
16	5 10	6
17	10	
18	10	

(c)

Nets 5 and 6 cannot be routed over the cells

(d)

Fig. 6.14 (a) An example channel; (b) interval graph; (c) Z assignment; (d) bipartite graph model.

6.4.3 Bipartite Graph Model

The correlation between nets in zones can be modeled as a bipartite graph $B = (V \cup U, E)$, where each vertex in $V(U)$ represents a net (zone) and $E = \{e_{i.j}v_j \in U$, and the net represented by v_i intersects the zone represented by $u_j\}$.

The degree of U_j, deg(u_j), equals den(z_j). The bipartite graph representation of the channel in Figure 6.14(a) is shown in Figure 6.14(d). The problem of reducing the channel density down to D_{final} now becomes a constrained covering problem. This problem is to find a subset of V that covers each $u \in U$ at least deg(u) $- D_{final}$ times, if deg(u) $> D_{final}$, subject to certain constraints. The constraints come from the case that a pair of nets may not be chosen simultaneously to be routed over the cells, because there is only one layer available. Such a pair of nets is termed incompatible. For example, in the channel of Figure 6.14(a), nets 1 and 4 are incompatible nets, because routing both the nets over the cells on the same layer will cause a short circuit.

To reduce the density to 3, the zones 1, 2, and 6 in Figure 6.14(a) can be ignored. The coverage relationship can be modeled as a reduced graph, shown in Figure 6.15(a).

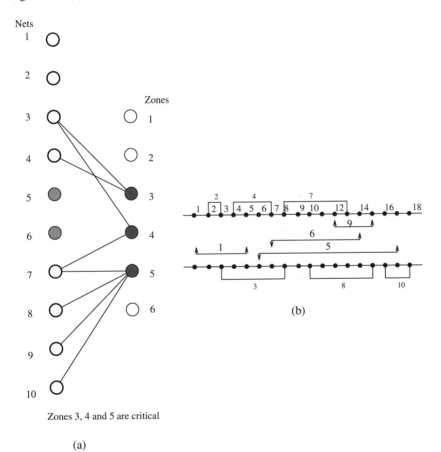

Fig. 6.15   (a) The reduced graph; (b) final solution.

6.4.4 ILP Formulation

Let D_{initial} be the initial channel density and D_{final} the final channel density. We define for each net i, x_i a 0-1 integer. The value of x_i is 1 if net i is routed over the cells, and 0 otherwise. Our goal can be formulated as follows:

Minimize D_{final} subject to $D_{\text{final}} \leq D_{\text{initial}}$ and

$$\sum_{e_{i \cdot j} \in E} x_i \geq \deg(z_i) - D_{\text{final}}$$

and $x_m + x_n \leq 1$ for each incompatible pair of nets m and n.

The first constraint ensures that D_{final} is no greater than D_{initial}. The second set of constraints indicate for each set of the critical zone the number of nets that must be routed over the cell in order to reduce the channel density to D_{final}. The final set of constraints make sure that all the nets selected are compatible. The ILP formulation for the channel in Figure 6.14(a) is shown in the following.

> Minimize D_{final}
> subject to
> /** Initial Density Constraint **/
> $D_{\text{final}} \leq 5.$
> /** Net-Incompatibility Constraints **/
> $x_1 + x_4 \leq 1$
> $x_7 + x_9 \leq 1$
>
> /** Zone Equations **/
> $D_{\text{final}} + x_1 + x_2 + x_3 \geq 3$
> $D_{\text{final}} + x_1 + x_3 + x_4 \geq 3$
> $D_{\text{final}} + x_3 + x_4 \geq 4$
> $D_{\text{final}} + x_4 + x_7 \geq 4$
> $D_{\text{final}} + x_7 + x_8 + x_9 \geq 5$
> $D_{\text{final}} + x_{10} \geq 2$
> end
>
> /** Variable Declaration **/
>
> integer $x_1, x_2, x_3, x_4, x_7, x_8, x_9, x_{10}$

Solving the above formulation, $x_2, x_3, x_4, x_7, x_8, x_{10}$ are set to 1. That is, nets 2, 3, 4, 7, 8, and 10 are chosen to be routed over the cells. D_{final} is minimized to 3. The result is shown in Figure 6.15(b).

6.4.5 Multiterminal Nets

In practice, a net may have more than two terminals. In order to extend the formulation, the multiterminal nets have to be decomposed into two-terminal subnets and are assigned terminals for routing within the channel. Since the terminals of a net may be on either side of the channel and the connectivity of a net has to be established at all times, the assumption of reduction in zone density on removal of a net from its zone is no longer valid. Without loss of generality, it is assumed that a multiterminal net does not have a pair of terminals located at the same column (one at the top and the other at the bottom) because if this is the case, the net can be treated as two independent nets separated by a column.

Let us consider an eight-terminal net, net 1, as shown in Figure 6.16(a). Net 1 has four terminals, t_1, t_3, t_5, and t_7, on the top terminal row and four terminals, b_2, b_4, b_8, and b_9, on the bottom terminal row. Our candidate subnets for routing over the cell are i_a, i_b, ..., i_f. These subnets divide the interval of net 1 into seven regions, R_1, R_2, ..., R_7, as shown in Figure 6.16(b). The regions can be classified into the following three categories:

1. **Singular Regions:** These regions are covered only by one subnet. In the example shown in Figure 6.16(b), R_1, R_6, and R_7 are singular regions. The density of the region is reduced by 1 if and only if both subnets are routed over the cells.

2. **Overlapping Regions:** These regions are covered by two subnets, such that neither of the subnets are dominated by the other. For example, i_e is completely dominated by i_d. The density in a dominating region will be reduced by 1 as long as the dominated subnet is routed over the cells.

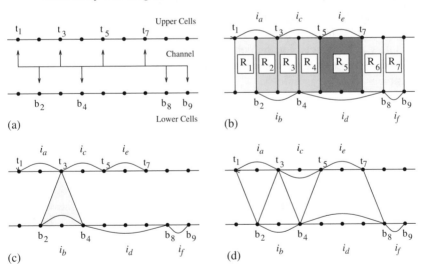

Fig. 6.16   (a) An eight-terminal net; (b) its decomposition; (c) a blue subnet is routed inside the channel; (d) all blue subnets are routed over the cells.

To model an overlapping region, a new variable $o_{i_{a \cdot b}} = x_{i_a} \, AND \, x_{i_b}$, where i_a and i_b are the overlapping subnets of net i. The ILP formulation is given below

$$x_{i_a} + x_{i_b} - o_{i_{a \cdot b}} \leq 1$$
$$x_{i_a} + x_{i_b} - 2o_{i_{a \cdot b}} \geq 0$$

$o_{i_{a \cdot b}}$ will be set to 1 if and only if both subnets (i_a and i_b) are routed in OTC areas. Adding $o_{i_{a \cdot b}}$ instead of x_{i_a} and x_{i_b} to the zone equation will precisely indicate the reduction in channel density. In the example shown in Figure 6.16(b), region R_2 should be formulated as

$$x_{1_a} + x_{1_b} - o_{1_{a \cdot b}} \leq 1$$
$$x_{1_a} + x_{1_b} - 2o_{1_{a \cdot b}} \geq 0$$

After some or all the nets are routed over the cell, the connectivity has to be preserved by connecting some terminals to be connected in the channel. According to our region classification, independently routing inside the channel, any subnets that are not routed over the cell will not increase the channel density. By doing so, we will have two-component partial routing (one connects all terminals on the top terminal row and another connects all terminals on the bottom terminal row). We need to make a connection from the top to the bottom. A subnet is *blue* if there is a terminal located inside its interval on the opposite bank, and *red* otherwise. In our example, subnets i_a, i_b, i_c, and i_d are blue, while i_e and i_f are red. If there is an unchosen blue subnet, the connectivity can be easily maintained by routing the subnet and the terminal on the opposite bank together (Figure 6.16(c)). However, if all the blue nets are routed in OTC areas, one of the overlapping regions should play the role of preserving connectivity, as shown in Figure 6.16(d). In our example, they are R_1, R_2, R_3, R_4, R_6, and R_7. Another variable s is introduced for the density calculations. $S_{k_{i \cdot j}}$ is 1 if the pair t_i and b_j are used to preserve the connectivity of net k, and 0 otherwise. Therefore, the example is modeled as follows:

if

$$x_{1_a} + x_{1_b} + x_{1_c} + x_{1_d} = 4$$

then

$$S_{1_{1 \cdot 2}} + S_{1_{3 \cdot 2}} + S_{1_{5 \cdot 4}} + S_{1_{7 \cdot 8}} = 1$$

else

$$S_{1_{1 \cdot 2}} + S_{1_{3 \cdot 2}} + S_{1_{5 \cdot 4}} + S_{1_{7 \cdot 8}} = 0$$

To formulate the if-then-else conditions, a new variable, c, is introduced as follows:

$$x_{1_a} + x_{1_b} + x_{1_c} + x_{1_d} - c \leq 3$$
$$x_{1_a} + x_{1_b} + x_{1_c} + x_{1_d} - 3c \geq 0$$
$$c - S_{1_{1 \cdot 2}} + S_{1_{3 \cdot 2}} + S_{1_{5 \cdot 4}} + S_{1_{7 \cdot 8}} = 0$$

The decomposition of the channel consisting of the multiterminal nets 1, 2, 3, 4, and 5 (Figure 6.17(a)) is shown in Figure 6.17(b). Net 3 can be treated as two independent nets because it has two terminals located at the same column. Nets 1 and 5 need o and s variables, while net 3 needs only o. The ILP formulation is shown in the following program on the next page:

Minimize D_{final}

subject to

/** Initial Density Constraints **/

$D_{\text{final}} \leq 4$

/** Net-Compatibility Constraints **/

$$
\begin{aligned}
x_{1_a} + x_2 &\leq 1 \\
x_2 + x_{5_a} &\leq 1 \\
x_4 + x_{3_b} &\leq 1 \\
x_{5_b} + x_6 &\leq 1
\end{aligned}
$$

/** Overlapping Region Density Calculation **/

$$
\begin{aligned}
x_{1_a} + x_{1_b} - o_{1_{a \cdot b}} &\leq 1 \\
x_{1_a} + x_{1_b} - 2o_{1_{a \cdot b}} &\geq 0 \\
x_{3_a} + x_{3_b} - o_{3_{a \cdot b}} &\leq 1 \\
x_{3_a} + x_{3_b} - 2o_{3_{a \cdot b}} &\geq 0 \\
x_{3_a} + x_{3_b} + x_{3_c} - o_{3_{a \cdot b \cdot c}} &\leq 1 \\
x_{3_a} + x_{3_b} + x_{3_c} - 3o_{3_{a \cdot b \cdot c}} &\geq 0 \\
x_{5_a} + x_{5_b} - o_{5_{a \cdot b}} &\leq 1 \\
x_{5_a} + x_{5_b} - 2o_{5_{a \cdot b}} &\geq 0 \\
x_{5_b} + x_{5_c} - o_{5_{b \cdot c}} &\leq 1 \\
x_{5_b} + x_{5_c} - 2o_{5_{b \cdot c}} &\geq 0 \\
x_{5_a} + x_{5_b} + x_{5_c} - o_{5_{a \cdot b \cdot c}} &\leq 1 \\
x_{5_a} + x_{5_b} + x_{5_c} - 3o_{5_{a \cdot b \cdot c}} &\geq 0
\end{aligned}
$$

/** Connectivity Preserving **/

$$
\begin{aligned}
c_1 - S_{1_{1 \cdot 2}} - S_{1_{3 \cdot 2}} - S_{1_{3 \cdot 4}} &= 0 \\
x_{1_a} + x_{1_b} - c_1 &\leq 1 \\
x_{1_a} + x_{1_b} - 2c_1 &\geq 0 \\
c_5 - S_{5_{4 \cdot 5}} - S_{5_{6 \cdot 5}} - S_{5_{6 \cdot 8}} - S_{5_{11 \cdot 8}} &= 0 \\
x_{5_a} + x_{5_b} + x_{5_c} - c_5 &\leq 2 \\
x_{5_a} + x_{5_b} + x_{5_c} - 3c_5 &\geq 0
\end{aligned}
$$

/** Zone Equations **/

$$
\begin{aligned}
D_{\text{final}} + o_{1_{a \cdot b}} - S_{1_{1 \cdot 2}} - S_{1_{3 \cdot 2}} &\geq 2 \\
D_{\text{final}} + x_2 + o_{1_{a \cdot b}} - S_{1_{3 \cdot 2}} - S_{1_{3 \cdot 4}} &\geq 3 \\
D_{\text{final}} + x_{1_b} + x_2 + x_{5_a} - S_{1_{3 \cdot 4}} - S_{5_{4 \cdot 5}} &\geq 4 \\
D_{\text{final}} + x_2 + o_{5_{a \cdot b}} - S_{5_{4 \cdot 5}} - S_{5_{6 \cdot 5}} &\geq 3 \\
D_{\text{final}} + x_{3_a} + o_{5_{a \cdot b \cdot c}} - S_{5_{6 \cdot 5}} - S_{5_{6 \cdot 8}} &\geq 3 \\
D_{\text{final}} + x_{3_a} + x_6 + o_{5_{b \cdot c}} - S_{5_{6 \cdot 8}} &\geq 4 \\
D_{\text{final}} + x_4 + x_6 + o_{3_{a \cdot b}} + o_{5_{b \cdot c}} - S_{5_{6 \cdot 8}} - S_{5_{11 \cdot 8}} &\geq 4 \\
D_{\text{final}} + x_{3_b} + x_4 x_{5_c} + x_6 - S_{5_{11 \cdot 8}} &\geq 4 \\
D_{\text{final}} + x_{5_c} + o_{3_{a \cdot b \cdot c}} - S_{5_{11 \cdot 8}} &\geq 2
\end{aligned}
$$

end

/** Variable Declaration **/

integer $x_{1_a}, x_{1_b}, x_2, x_{3_a}, x_{3_b}, x_{3_c}, x_4, x_{5_a}, x_{5_b}, x_{5_c}, x_6$.

integer $S_{1_{1 \cdot 2}}, S_{1_{3 \cdot 2}}, S_{1_{3 \cdot 4}}, S_{5_{4 \cdot 5}}, S_{5_{6 \cdot 5}}, S_{5_{6 \cdot 8}}, S_{5_{11 \cdot 8}}$.

integer $c_1, c_5, o_{1_{a \cdot b}}, o_{3_{a \cdot b}}, o_{3_{a \cdot b \cdot c}}, o_{5_{a \cdot b}}, o_{5_{b \cdot c}}, o_{5_{a \cdot b \cdot c}}$.

/** END **/

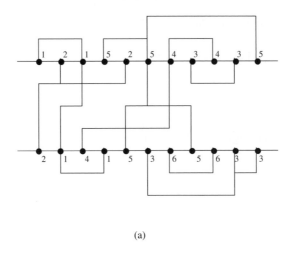

(a)

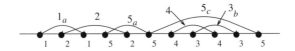

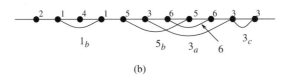

(b)

Fig. 6.17 (a) A channel with multiterminal nets; (b) decomposition in two-terminal nets.

6.5 Routing Algorithm for the CTM

In this section we present an overview of a two-layer OTC routing algorithm proposed by N. Holmes, M. Sarrafzadeh, N. Sherwani, and B. Wu [WHSS92]. The routing environment for the two-layer CTM is equivalent to the BTAP discussed in Chapter 5. OTC routing is required to be planar.

6.5.1 Algorithm Overview

The CTM router for the two-layer process consists of eight steps, as discussed below.

1. **Net Classification:** In this step, all the nets are classified into three types. *Type I nets* are the nets whose terminals are on the same terminal row.

Type II nets are two-terminal nets whose terminals are on different terminal rows (one on the top terminal row and one on the bottom terminal row). *Type III nets* are multiterminal nets whose terminals are on both top and bottom terminal rows.

2. **Net Weighting and Net Selection for M2 Layer:** In this step, all Type II nets are assigned weights. The weight of a net is based on several factors, such as criticality of the net and its contribution to channel congestion. Based on the weights assigned to the nets, an MIS of the nets is found. All the nets in this set are routed in M2 and they pass through the channel as a straight wire. These nets do not contribute to channel density; however, they partition M2 into several *regions*.

3. **Decomposition of Multiterminal Nets:** After obtaining the initial regions, the multiterminal nets, which have terminals on both top and bottom terminal rows (Type III nets), are decomposed. The objective of this operation is to increase the number of nets that can be routed straight in the channel on M2.

4. **Terminal Assignments of Critical Nets:** After the routing regions in M2 are topologically fixed, the geometric terminal positions are assigned to the critical nets to obtain the geometric definitions of the regions. This is done to maximize the routability of the nets in M2, as well as to satisfy the congestion requirement of each region. A dynamic programming approach is used to accomplish this.

5. **Terminal Assignments of Noncritical Nets:** Within each region, the terminal positions are assigned to the remaining nets to achieve three objectives: First, to eliminate vertical constraints; second, to minimize horizontal constraints; and third, to maximize the MBS in a special graph called the *overlap graph* defined by the intervals of the nets.

6. **M2 River Routing:** After all the terminal positions are fixed, a river router is used to route the nets in a planar fashion in the M2 OTC area.

7. **Channel Routing:** An HVH routing model can be used to route the remaining nets in the channel. It must be noted that since all vertical constraints are eliminated, a router based on LEA adapted for the three-layer routing environment can be used.

8. **Cleanup Routing:** Finally, the multiterminal nets are recomposed as much as possible. This step removes the redundant wiring and improves the layout.

In the subsequent sections, we discuss the details of steps 4 and 5. All other steps are self-explanatory.

6.5.2 Terminal Assignments of Critical Nets

In this section we present an algorithm for the problem of *terminal assignments of critical nets* (TAC) with capability constraints.

Given a set of nets $\mathcal{N} = \{N_1, N_2, \ldots, N_n\}$ needing to be routed in a given channel. Let $\mathcal{U} = \{u_1, u_2, \ldots, u_T\}$ denote the set of terminals on the top terminal row, and the set of terminals on the bottom terminal row is denoted by $\mathcal{L} = \{l_1, l_2, \ldots, l_B\}$, respectively. Let us consider the boundary of the cells as an imaginary row of terminals denoted by $M = \{m_1, m_2, \ldots, m_L\}$, where L is the number of terminal positions. Let $S = \{s_1, s_2, \ldots, s_K\}$ be the MIS of critical nets selected in step 2, where P is the size of the independent set. The entire M2 routing area is partitioned into $P+1$ regions, $R = \{r_1, r_2, \ldots, r_{P+1}\}$ by S, such that region r_i is defined by nets s_i and s_{i+1}. The region r_1 is defined by the left end of the channel and net s_1; similarly, r_{P+1} is defined by s_P and the right end of the channel. According to the region partitioning, the set of top terminals TOP is partitioned into $P+1$ subsets $\mathcal{U} = \{\mathcal{U}_1, \mathcal{U}_2, \ldots, \mathcal{U}_{P+1}\}$, where $\mathcal{U}_i = \{u_{i,1}, u_{i,2}, \ldots, u_{i,L_i}\}$. The set of bottom terminals $\mathcal{L}$ is partitioned into $P+1$ subsets $\mathcal{L} = \{\mathcal{L}_1, \mathcal{L}_2, \ldots, \mathcal{L}_{P+1}\}$, where $\mathcal{L}_i = \{l_{i,1}, l_{i,2}, \ldots, l_{i,L_i}\}$. The problem is to select a terminal $m_{s_i} \in M$ for each net N_i, such that the *separation constraints* required for elimination of vertical constraints and the *M2 OTC river routing capability constraints* are satisfied. Vertical constraints are eliminated if no two terminals from $\mathcal{U}_i \cup \mathcal{L}_i$ are assigned the same terminal between m_{s_i} and $m_{s_{i+1}}$. This requires m_{s_i} and $m_{s_{i+1}}$ be separated by $U_i + L_i$; that is, $(m_{s_i} - m_{s_{i-1}} \leq U_i + L_i)$ for all $i = 2, \ldots, P$, $m_{s_1} \geq U_1 + L_1$, and $m_P + 1 - m_{s_k} \geq U_{P+1} + L_{P+1}$. The river routing capability constraints require that $m_{s_i} \in (\text{range}\_u(s_i) \cap \text{range}\_l(s_i))$, where, $\text{range}\_u(s_i)$ and $\text{range}\_l(s_i)$ are the ranges of M allowed for s_i on the top and bottom side, respectively. These ranges can be obtained in $O(P)$ time, where P is the size of S. The objective function is to route the nets in S as straight as possible; that is, to minimize $\text{LEN}(S)$, where,

$$\text{LEN}(S) = \sum_{i=1}^{P+1} |u_{s_i} - m_{s_i}| + |l_{s_i} - m_{s_i}| - |u_{s_i} - l_{s_i}|$$

where u_{s_i} is the top terminal position of s_i, l_{s_i} is the bottom terminal position of s_i, m_{s_i} is the assigned terminal position of net s_i.

The formal statement of the TAC problem is as follows:

Instance: Let $\Phi = (\mathcal{N}, M, \mathcal{U}, \mathcal{L}, S, R, \text{LEN}(S))$ be a given instance of the TAC problem, where (1) $\mathcal{N}$ is the set of nets to be routed in the given channel; (2) $M = \{m_1, m_2, \ldots, m_L\}$, an imaginary row of terminals, where L is the number of terminal positions; (3) $\mathcal{U} = \{u_1, u_2, \ldots, u_T, T \leq L\}$, the set of terminals on the top terminal row; (4) $\mathcal{L} = \{l_1, l_2, \ldots, l_B, B \leq L\}$, the set of terminals on the bottom of the channel; (5) $S = \{s_1, s_2, \ldots, s_P\}$, the MIS of critical nets selected in the previous two steps, where P is the size of the independent set; (6) $R = \{r_1, r_2, \ldots, r_{P+1}\}$, the set of partitioned regions.

Problem: Define a function $f: S \to M$ (let us call $f(s_i)$ as m_{s_i} for short), such that (a) $(m_{s_i} - m_{s_{i-1}} \leq T_i + B_i)$ for all $i = 2, \ldots, P$, $m_{s_1} \geq T_1 + B_1$, and $m_L + 1 - m_{s_k} \geq T_{P+1} + B_{P+1}$; (b) $m_s \in (range\_u(s_i) \cap range\_l(s_i))$ for all $i = 1, \ldots, P$; (c) LEN(S) is minimum; (d) f is an injective function.

That is, a function f is an assignment of terminals in S to the imaginary row of terminals M, satisfying the two types of constraints, while minimizing the objective function. Intuitively, $f(s_i) = m_j$ means that both top and bottom terminals of s_i are assigned to the jth position (from the left) in the terminal imaginary row. Let us define the *region capability* cap(r_i) as follows:

$$cap(r_i) = T_i + B_i$$

where T_i is the number of top terminals in region r_i and B_i is the number of bottom terminals in region r_i.

Lemma 8 gives the condition for the existence of function f, such that both the separation and river routability constraints are satisfied at the same time.

Lemma 8   *Given any two nets s_i, s_j (assume $i < j$) and their terminal ranges $range\_u(s_i), range\_l(s_i)$, if s_i is assigned to $min(range\_u(s_i) \cap range\_l(s_i))$ and s_j is assigned to $max(range\_u(s_j) \cap range\_l(s_j))$, then a TAC problem is not feasible if*

$$\sum_{c=i+1}^{j-1} cap(r_c) > m_{s_j} - m_{s_i}$$

where $max(r)$ and $min(r)$ represent the maximum and minimum of the range r.
Proof: Since s_i is assigned to its left-most position $min(range\_u(s_i) \cap range\_l(s_i))$ and s_j is assigned to its right-most position $max(range\_u(s_j) \cap range\_l(s_j))$. Due to the river routing routability constraints, s_i cannot move left and s_j cannot move right. All the nets in $[s_i, s_j]$ must be assigned within the range

$$[min(range\_u(s_i) \cap range\_l(s_i)), max(range\_u(s_j) \cap range\_l(s_j))]$$

However, $\sum_{k=i+1}^{j-1} cap(r_k) > m_{s_j} - m_{s_i}$ means that the separation constraint cannot be satisfied within this range. Therefore, the TAC problem is not feasible.

□

Based on Lemma 8, the algorithm RANGE, shown in Figure 6.18, calculates the legal region range(s_i) of s_i, such that range(s_i) satisfies all the constraints. In the algorithm, range$L(s_i)$ denotes the left point of range(s_i), while range$R(s_i)$ denotes the right point of range(s_i). In case a net s_i does not satisfy Lemma 8, it is removed from S.

The function length(i, j) represents the cost of assigning the net s_i to the terminal position m_j, which is defined as follows:

$$length(i, j) = |u_{s_i} - m_j| + |l_{s_i} - m_j| - |u_{s_i} - l_{s_i}|$$

For an instance of Φ, a solution (i, j) (denoted by (i, j)−solution) represents the first i nets in S, $\{s_1, s_2, \ldots, s_i\}$ are assigned to the first j-terminal position in

Algorithm RANGE(st, end)
Input:
st : Start region number.
end : End region number.

---

begin
    /* Calculate left point of each range. */
    for $i = st$ to end **do**
        if $rangeL(s_{i-1}) + cap(r_i) < min(range\_u(s_i) \cap range\_l(s_i))$
            then $rangeL(s_i) = min(range\_u(s_i) \cap range\_l(s_i))$
        else if $rangeL(s_{i-1}) + cap(r_i) \in range\_u(s_i) \cap range\_l(s_i)$
            then $rangeL(s_i) = rangeL(s_{i-1}) + cap(r_i)$
        else if $rangeL(s_{i-1}) + cap(r_i) > max(range\_u(s_i) \cap range\_l(s_i))$
            then $rangeL(s_i) = -1$
    /* Calculate right point of each range. */
    $rangeR(s_{P+1}) = 0$
    for $i = end$ TO st DO
        if $rangeR(s_{i-1}) - cap(r_i) > max(range\_u(s_i) \cap range\_l(s_i))$
            then $rangeR(s_i) = max(range\_u(s_i) \cap range\_l(s_i))$
        else if $rangeR(s_{i-1}) - cap(r_i) \in range\_u(s_i) \cap range\_l(s_i)$
            then $rangeR(s_i) = rangeR(s_{i-1}) + cap(r_i)$
        else if $rangeR(s_{i-1}) - cap(r_i) > min(range\_u(s_i) \cap range\_l(s_i))$
            then $rangeR(s_i) = -1$
end.

Fig. 6.18   Algorithm RANGE.

$M, \{m_1, m_2, \ldots, m_j\}$, and len$(i, j)$ denotes the minimum cost of the solution. For a solution that does not satisfy the constraints, len$(i, j) = -1$.

We now present the algorithm for the TAC problem formally. The input of the algorithm is an instance $\Phi = (\mathcal{N}, M, \mathcal{U}, \mathcal{L}, S, R)$ of the TAC problem, and the output is a valid terminal assignment, or an indication that no such solution exists.

Theorem 23   *Algorithm ASSIGN_C solves the TAC problem in $O(LP^2)$ time and $O(LP^2)$ space, where L is the number of terminal positions and P is the size of the selected independent set.*

Proof: Let us first prove the correctness of the algorithm. The following proves by induction on j; that is, if Φ has a solution (i, j), then len(i, j) is equal to the minimum of all the (i, j) solution to Φ so far.

This is true for $j = 0$, where we chose the boundary values. Assume it is true for $j - 1$, $1 \leq j \leq P$. If Φ has an (i, j)−solution with minimum cost function, then it must be either an $(i, j - 1)-$solution or the cost of $(i - 1, k - 1)-$solution plus length(i, j). By the induction hypothesis and the fact that len(i, j) = min{len$(i, j - 1)$, length(i, j) + len$(i - 1, j - 1)$}, len(i, j) still holds the minimum cost $(i, k)-$solution of Φ.

Therefore, the algorithm ASSIGN_C, illustrated in Figure 6.19, solves the TAC problem Φ.

Algorithm ASSIGN_C

---

begin
  /* Initialization */
  for $i = 1$ to P **do**
   $len(i, 0) = -1$;
  for $j = 0$ to L **do**
   $len(0, k) = 0$;
  /* Computing objective function $len(P, L)$ using dynamic prog. */
  for $j = 1$ to L **do**
   for $i = 0$ to P **do**
     /* Based on the previous net assignments, */
     /* Compute legal ranges for the assigning net. */
     $RANGE(i, P)$
     if assignment is in legal range
     then $len(i, j) = min\{len(i, j - 1), length(i, j) + len(i - 1, j - 1)\}$
     else $len(i, j) = -1$
  if $len(P, L) = -1$
  then return "Φ is not feasible".
  else return $\pi = len(P, L)$
end.

Fig. 6.19   Algorithm ASSIGN_C.

The time complexity of the algorithm is clearly dominated by the computation of the function len(i, j). As the computation of len(i, j) can be finished in $O(LP)$ time and $O(LP^2)$ space, where L is the number of terminal positions and P is the size of the selected independent set.

□

6.5.3 Terminal Assignments of Noncritical Nets

In this section we describe the terminal permutation algorithm used by our router.

Given two sets where $T = \{T_1, T_2, \ldots, T_{k+1}\}$ and $B = \{B_1, B_2, \ldots, B_{k+1}\}$, each $T_i(B_i)$ consists of a nonpermutable list of terminals. Permutation of $P = \{P_1, P_2, \ldots, P_{k+1}\}$ is found, such that each P_i is a shuffle of T_i and B_i. The objective function of the shuffle operation is to minimize horizontal constraints and maximize the bipartite subgraph in the overlap graph.

We adopt the following heuristic. We start with the permutation $P_i = T_i B_i$ for all i, and then shuffle each pair $T_i B_i$. Inductively, we assume $T_1, \ldots, T_{i-1}$ and $B_1, \ldots, B_{i-1}$ have been pairwise shuffled. As for the basis, the first shuffle $P-1$ is $T_1 B_1$. We obtain P_i as follows. Consider $T_i = \{t_1, \ldots, t_s\}$ and $B_i = \{b_1, \ldots, b_s\}$. There are $s + 1$ positions to which b_i can be assigned: $b_1, \ldots, b_{i-1}$ are assigned to the left of t_1, and $b_{i+1}, \ldots, b_s$ are assigned to the right of t_s. For each position, we find the number of independent nets created.

Then a matching diagram is made. On one side it has the set of intervals $[-\infty, t_1], [t_1, t_2], \ldots, [t_s, +\infty]$, and on the other side it has $b_1, \ldots, b_s$. Each edge has the weight calculated before that represents the number of independent nets corresponding to that position. Now one can find a maximum-weight non-cross matching in $O(sxlogs)$ time. The matching dictates a shuffle. This finishes the inductive step; that is, finding a shuffle of T_i and B_i.

6.5.4 Formal Statement of the Algorithm

This section contains the formal details of the algorithm. It must be pointed out that once the set of nets to be routed over the cells is determined, it is rather easy to assign tracks to them so that they can be routed without violating any design rules. The sets of nets that are not routed in any of the previous steps is routed in the channel using an HV routing model.

6.6 OTC Routing Algorithm for the MTM

In case of MTM-V, routing in OTC areas in the layer over the cell must be planar. The OTC routing task in the MTM is comprised of two problems: first, planar routing between two equipotential terminal rows, which is a classical ETSPR problem; second, BTAP between the lower terminal row of the upper cell row and the upper terminal row of the lower cell row. This problem is equivalent to the BTAP discussed in Chapter 5.

6.6.1 Algorithm Overview

Based on the location of terminals, we classify the nets and assign them a weight. Higher weighted nets are critical nets and their removal results in a maximum reduction in channel height. The weight is computed based on vertical and horizontal constraints in the given channel [HSS91b]. Based on the weight, the MTM router first finds a maximum planar equipotential set (MPES). This is a set of intrarow and interrow nets that can be routed in the C area in a planar fashion.

Algorithm ICR()
Input: $\mathcal{N} = \{N_1, N_2, \ldots, N_n\}$ is a set of nets.
Output: Over-the-cell channel routing.

---

   begin
   step 1: Net Classification.
   step 2: Net Weighting and Finding the Maximum Independent Set
        in Permutation Graph.
      Construct_Permutation_Graph($G, \mathcal{N}$)
      Weights_Assignment_Graph($G, \mathcal{N}$)
      $\mathcal{N}1$=Max_Independent_SetS(G);
      $\mathcal{N}2 = \mathcal{N} - \mathcal{N}1$;
   step 3: TYPE III nets Partitioning.
   step 4: Terminal Assignments of Critical Nets and Region Partitioning.
   step 5: Terminal Assignments of Non-Critical Nets.
   step 6: M2 River Routing.
   step 7: Two Layer Channel Routing.
   step 8: Clean up Routing.
   end.

Fig. 6.20   Algorithm ICR.

The routing for each net in the C area can be accomplished in four different paths, as shown in Figure 6.21. In the next step, we do boundary terminal assignment in order to minimize the channel height. The cell boundary is essentially considered a row of virtual terminals. After a terminal assignment phase, we find two sets of independent nets to route in the M3 layer of the OTC area while minimizing the channel height. The remaining nets are routed in the channel and the cleanup routing phase is used to improve wiring lengths. The basic steps of the 2MTM router are given below:

1. **Net Classification, Decomposition, and Weighting:** In this step, all the nets are decomposed into two-terminal nets and classified into two types. A TYPE I net has terminals on the same cell row. A TYPE II net has terminals on different cell rows (one on the top terminal row and one on the bottom terminal row). After net decomposition and classification, all the nets are assigned weights. We use two weighting functions in the MTM-V router. The first weight function is based on criticality of nets. The weight of a net is directly proportional to its criticality. This enables the router to select most of the critical nets for routing in the M2 layer

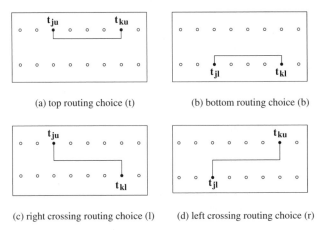

(a) top routing choice (t)   (b) bottom routing choice (b)

(c) right crossing routing choice (l)   (d) left crossing routing choice (r)

Fig. 6.21 Four routing choices of a TYPE I net.

of the C area. After the net selection in the C area, nets are reweighted using a second weighting function based on their contribution to channel congestion. The critical nets are routed with minimal length in the channel. Let the weight of a net n_i be represented as $w(n_i)$.

2. **MPES Selection for the M2 Layer in the C Area:** In this step, a subset of intrarow nets is selected for routing in the C area of a cell row R_i. Based on the weights assigned to the nets, we find a maximum planar independent subset of the intrarow nets, which can be routed in k_2 tracks available in the C area using one of the four different paths, as shown in Figure 6.21. We prove that this problem is NP-hard and, as a result, present an approximation algorithm to select such a subset. This step is executed for all the cell rows.

3. **Boundary Terminal Assignment and M2 River Routing:** The cell boundary is essentially considered a row of possible terminal locations. In this step, each net terminal is assigned a terminal location on the boundary considering river routing constraints while minimizing the channel density. A river router is used to complete the connections as specified by the terminal assignment in a planar fashion in the M2 layer of the T and B areas.

In the following, we discuss the second step of the algorithm. The first step is similar to algorithms for the BTM, and the third step is explained in Chapter 3.

6.6.2 MPES Selection for the M2 Layer in the C Area

In this step, a set of nets is selected for routing in the M2 layer in the C area of the cell row R_i. Since vias are not allowed over the cell, the nets in this set must be planar. Let $\mathcal{N}_1$ be the set of TYPE I nets for a given cell row. The terminal positions in the upper (lower) terminal row are numbered from $u_1(l_1)$ to $u_L(l_L)$. Let $N_i \in \mathcal{N}_1$ be represented as $N_i = (t_{i_j}, t_{i_k})$, such that $COL(t_{i_j}) < COL(t_{i_k})$. Since terminals u_i and l_i are equipotential, four different routing choices are available for routing N_i in the C area: (u_j, u_k) *top routing choice* (t), (l_j, l_k) *bottom routing choice* (b), (u_j, l_k) *right crossing routing choice* (r), and (l_j, u_k) *left crossing routing choice* (l) (see Figure 6.21).

Given, a positive integer α, a set $\beta \subseteq \{t, b, r, l\}$, and a set $(\mathcal{N}_1)$ of TYPE I nets, we say a set S is an EPS$(\alpha, \beta, \mathcal{N}_1)$ if $S \subseteq \mathcal{N}_1$, and all nets in S are routable using one of the routing choices in β in a planar fashion in α tracks. The maximum weighted EPS$(\alpha, \beta, \mathcal{N}_1)$ is referred to as MPES$(\alpha, \beta, \mathcal{N}_1)$. For short, we denote MPES$(\alpha, \beta, \mathcal{N}_1)$ as $S(\alpha, \beta, \mathcal{N}_1)$. The weight of a track in the routing of a net list $\mathcal{N}_1$ is defined as the summation of the weights of the nets routed on that track. Let $S^*(\alpha, \beta, \mathcal{N}_1)$ be a set of nets in the k_2 highest weighted tracks of the optimal no-dogleg planar routing of $S(\infty, \beta, \mathcal{N}_1)$.

Let the sets $S(k_2, \{t, b, r, l\}, \mathcal{N}_1)$, $S(\infty, \{t, b, r, l\}, \mathcal{N}_1)$, $S(\infty, \{b, r, l\}, \mathcal{N}_1)$, and $S^*(k_2, \{b, r, l\}, \mathcal{N}_1)$ be denoted by S_1, S_2, S_3, and S_4, respectively.

The objective of this step is to find $S(k_2, \{t, b, r, l\}, \mathcal{N}_1)$. We call the problem of finding $S(k_2, \{t, b, r, l\}, \mathcal{N}_1)$ the MES-1. We show that MES-1 is a computationally hard problem.

Theorem 24   *MES-1 is NP-hard.*

Proof: To prove that MES-1 is NP-hard, we reduce the instance of the problem of finding a maximum-weighted bipartite set (MBS) in a circle graph to an instance of MES-1. MBS is known to be NP-hard.

Let $\mathcal{I} = \{C\}$ be an instance of MBS, such that C is a set of k chords of a circle. Let the weight of a chord $c_i \in C$ be given as $w(c_i)$. Let the end points of these chords be numbered from 1 to $2k$. We construct the instance $\mathcal{I}'$ of MES-1 as follows. For each chord $c_i = (j, k) \in C$, we add a net $N_i = (t_{i_j}, t_{i_j})$ to $\mathcal{N}_1$ and assign a weight $w(c_i)$ to N_i. In addition, we add a net $N_0 = (t_0, t_{2k+1})$ to $\mathcal{N}_1$ and assign a weight $\sum_{i=1}^{k} w(c_i) + 1$ to N_0. Figure 6.22(b) shows an instance of MES-1 constructed from an instance of the MBS shown in Figure 6.22(a).

Let there be an algorithm A that optimally solves MES-1. Let η be the optimal solution to the instance I' of MES-1. η contains N_0, since N_0 has more weight than the summation of weights of all other nets in $\mathcal{N}_1$. Selecting N_0 does not allow the left and the right routing choices of the nets to be selected for the nets in $\mathcal{N}_1 - \{N_0\}$. Thus, a net in $\eta - \{N_0\}$ has either a top routing choice or a bottom routing choice. Let $X(Y)$ be the subset of $\eta - \{N_0\}$, such that each net in $X(Y)$ has top (bottom) routing choice.

We construct the solution to the instance $\mathcal{I}$ of MBS as follows. Let $C_1 = \{c_i | N_i \in X\}$ and $C_2 = \{c_i | N_i \in Y\}$. Then $C_1 \cup C_2$ is the maximum-weighted bipartite subset of C and the two partite sets are C_1 and C_2, respectively. Figure 6.22(c) shows the solution to the instance of MES-1 in Figure 6.22(b).

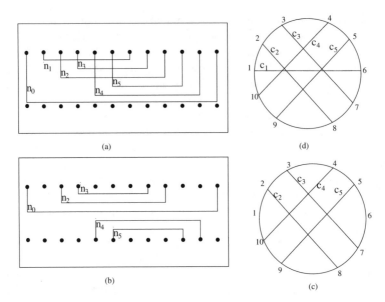

Fig. 6.22   (a) An instance of MBS; (b) MES-1 constructed for MBS; (c) solution to
MES; (d) solution to MBS.

Figure 6.22(d) shows the solution to the instance of MBS shown in Figure 6.22(a),
constructed from the solution of the corresponding instance of MES-1.

However, the MBS is known to be NP-hard; thus, MES-1 is NP-hard.

$\square$

In view of the NP-hardness of MES-1, we develop a $1/2 \min\{1, k_2/d\}$ approximation algorithm. Let the problem of finding S_2 be called MES-2. In order
to find S_2, we represent the terminal rows and the b, l, and r routing choices of the
nets in $\mathcal{N}_1$ using a circle diagram. The terminals u_1 through u_L are represented as
points p_{1u} through p_{Lu} on the circumference of a circle in the clockwise direction.
Similarly, terminals l_1 through l_L are represented as points p_{1l} through p_{Ll} on the
circumference of the circle in the counterclockwise direction. For each point p_{il},
let $p'_{il} = p_{il} + \epsilon$ and $p''_{il} = p_{il} - \epsilon$ be new points on the right and the left of
p_{il}, respectively. Let C represent the set of chords of the circle. For each net
$N_i = (t_j, t_k) \in \mathcal{N}_1$, C contains three chords $c_{ib} = (p_{jl}, p_{kl})$, $c_{ir} = (p_{ju}, p''_{kb})$,
and $c_{il} = (p'_{jl}, p_{ku})$ representing the top, right, and left routing choices of net N_i,
respectively (see Figure 6.24(b)). Each chord $c_{iz} \in C$ ($z \in \{t, r, l\}$) has a weight
$w(N_i)$ associated with it. Let C^* be the set of maximum-weighted independent
chords in C. Note that for a net N_i, C^* may contain at most one chord among
c_{ib}, c_{ir}, and c_{il} as each pair of chords in $\{c_{ib}, c_{ir}, c_{il}\}$ intersect each other. S_2 is
represented by $\{N_i | c_{iz} \in C^* \text{ and } z \in \{b, r, l\}\}$.

Based on the observation that $|C| = 3|\mathcal{N}_1|$ and that C^* can be found from C
in $O(|C|^2)$ time complexity, we state the following theorem.

Theorem 25   *MES-2 can be optimally solved in $O(|\mathcal{N}_1|^2)$ time complexity.*

Figure 6.24(c) shows the circle diagram of the nets in Figure 6.24(a). Figure 6.24(d) shows a maximum-weighted independent set of chords for the example in Figure 6.24(c).

Let S_2 be partitioned into two sets S_5 and S_6 such that $S_5 = \{N_i \mid N_i \in S_2$ and N_i has top routing choice$\}$, and $S_6 = S_2 - S_5$. Let $w(S)$ denote the summation of weights of nets in a set S.

The performance ratio of an approximation algorithm is defined as the ratio of the solution produced by the approximation to the optimal solution. That is,

$$\rho = \frac{w(S_4)}{w(S_1)}$$

where S_4 is the solution produced by the approximation algorithm and S_1 is the optimal solution.

Theorem 26   *The performance ratio of the algorithm for MES-1 is*

$$\rho \geq \frac{1}{2}(\min\{1, \frac{k_2}{d}\})$$

where d is the optimal number of tracks required for routing S_3 in a planar fashion without the use of doglegs.
Proof: First we prove that $w(S_3) \geq w(S_5)$ by contradiction. Let us assume $w(S_3) < w(S_5)$. Let S_7 be the set of all the nets in S_5, except that each of them is assigned the bottom routing choice. Since changing the routing choice of a net does not affect its weight, it is easy to note that $w(S_7) = w(S_5)$. Hence, $w(S_3) < w(S_7)$, which is a contradiction, since S_3 is $S(\infty, \{b, r, l\}, \mathcal{N}_1)$. Thus,

$$w(S_3) \geq w(S_5) \tag{6.1}$$

Similarly, it is easy to note that

$$w(S_3)) \geq w(S_6) \tag{6.2}$$

Next we prove that $w(S_3) \geq w(S_1)/2$. From equations (6.1) and (6.2), we have $2 \times w(S_3) \geq w(S_5) + w(S_6)$, since $w(S_2) = w(S_5) + w(S_6)$, $w(S_3) \geq \frac{1}{2}w(S_2)$ and since $w(S_2) \geq w(S_1)$, $w(S_3) \geq w(S_1)/2$. Therefore, if $k_2 \geq d$, then $w(S_4) = w(S_3)$. If $k_2 < d$, then the average track weight of the k_2 highest weighted tracks is greater than or equal to the average track weight of all d tracks; that is,

$$\frac{w(S_4)}{k_2} \geq \frac{w(S_3)}{d}$$

$$w(S_4) \geq \frac{w(S_1)}{2} \min\{1, \frac{k_2}{d}\}$$

$$\rho \geq \frac{1}{2} \min\{1, \frac{k_2}{d}\}$$

which concludes our proof.

Our experimental results have shown that $w(S_4)$ is 70% to 75% of $w(S_1)$ for most of the practical examples. Figure 6.23 shows the routing of S_4 for the set of nets shown in Figure 6.24(a). The utilization of the C area can be improved by routing a maximum-weighted planar subset of $\mathcal{N}_2 \cup S_4$ in the C area, where $\mathcal{N}_2 = \{N_i \mid N_i \in \mathcal{N}_1 \text{ and } N_i \notin S_4 \text{ and } N_i \text{ is assigned top routing choice}\}$. Such a set is represented by the maximum independent subset of the chords in the circle-chord representation of $\mathcal{N}_2 \cup S_4$. If this planar subset is not routable in the k_2 tracks, then the nets in k_2 highest weighted tracks are selected for routing in the C area.

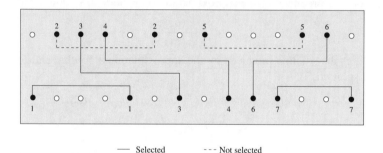

—— Selected   - - - Not selected

Fig. 6.23   Net in S_4 routed in the M2 layer of the C area ($k_2 = 3$).

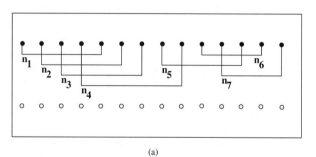

(a)

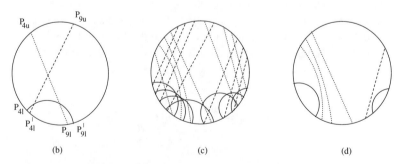

(b)   (c)   (d)

Fig. 6.24   **(a)** A set of TYPE I nets; **(b)** chords representing the b, l, and r routing choices of a net; **(c)** circle-chord representation; **(d)** maximum independent set of chords.

6.7 Experimental Evaluation of Two-Layer Routers

Several OTC routers have been discussed for the two-layer process in this chapter. In this section, we present a comparative performance evaluation of the OTC routers on some benchmark examples. Let us start with WISER.

6.7.1 Performance Evaluation of WISER

WISER considers the placement of PRIMARY1 from TimberWolfSC Version 5.1 [LS88], and the global routing is from [CSW89]. The PRIMARY1 example has 50% to 90% vacant terminals. To emphasize the effect of using vacant terminals, the algorithm WISER is compared with the implementation of the algorithm presented in [CL88], which is based on the same physical model as WISER, but does not use vacant terminals and abutments for routing. In the average case, the use of vacant terminals reduces the channel height by 15% to 20%.

Results for the PRIMARY1 example indicating the number of horizontal tracks used in the channel are given in Table 6.4. The performance of WISER is compared with results obtained by a greedy channel router and the implementation of the conventional OTC channel router (OTCR) from [CL88]. As can be seen in Table 6.4, WISER reduces the total height of PRIMARY1 by 17% as compared to the OTCR and 28% as compared to the greedy channel router [HSS91b]. On the average, for a given input channel, WISER performs 17% better than the conventional OTC router, and 29% better than conventional channel routers.

Furthermore, routings produced by the algorithm WISER have on the average 25% fewer vias than routings by a conventional OTC channel router and 32% fewer vias than routings by a greedy channel router. Experimental results indicating the number of vias per routing for PRIMARY1 channels are shown below in Table 6.5. It should be noted that for PRIMARY1 results, channels 1 and 18 are not included in any totals or average calculations, since they contain I/O pad connections and hence do not have OTC areas available on both sides of the channel. The average execution time of the algorithm WISER on the PRIMARY1 channel is 6.83 minutes. This is slightly slower than the average execution time for the conventional OTC router [CL88], which is 6.01 minutes.

WISER was also tested on Deutsch's difficult example, which has only 4% vacant abutments and 31% vacant terminals (including abutments). As indicated in Table 6.6, WISER produces a 15% reduction in channel height as compared to the greedy router, as well as a 17% reduction in vias. Due to the limited number of vacant terminals in this example, WISER was not able to further reduce the channel height as compared to the conventional OTC channel router, but does produce a solution with 6% fewer vias.

WISER reduces channel height by 46% as compared to the greedy channel router and 36% as compared to the conventional OTC router. Furthermore, the routing by WISER contains 47% fewer vias than the routing by the conventional OTC router and 64% fewer vias than the routing by the greedy channel router.

Table 6.4:   Experimental Results—Channel Height for PRIMARY1

Channel	% of Vacant	No. of Tracks Produced			% Imp. Over	
No.	Terminals	Greedy	OTCR	WISER	Greedy	OTCR
1	85	11	9	5	55	44
2	74	16	13	12	25	8
3	64	21	19	16	24	16
4	60	24	24	21	13	13
5	64	21	20	17	19	15
6	52	29	21	18	38	14
7	58	22	21	14	36	33
8	50	24	18	18	25	0
9	53	21	20	16	24	20
10	60	15	13	11	27	15
11	63	17	14	12	29	14
12	64	15	14	11	27	21
13	62	13	11	9	31	18
14	64	13	13	9	31	31
15	63	11	9	7	36	22
16	67	11	9	7	36	22
17	66	14	10	8	43	20
18	91	6	4	3	50	25
Total	-	287	249	206	28	17

Table 6.5:   Experimental Results—Number of Vias for PRIMARY1

Channel	% of Vacant	No. of Vias			% Imp. Over	
No.	Terminals	Greedy	OTCR	WISER	Greedy	OTCR
1	85	150	139	46	69	67
2	74	339	312	197	42	37
3	64	486	444	310	36	30
4	60	534	498	380	29	24
5	64	539	519	410	24	21
6	52	662	610	547	17	10
7	58	612	575	465	24	19
8	50	714	667	546	24	18
9	53	608	527	422	31	20
10	60	526	484	349	34	28
11	63	468	415	306	35	26
12	64	462	420	326	29	22
13	62	511	460	356	30	23
14	64	440	415	262	40	37
15	63	439	391	274	38	30
16	67	377	312	205	46	34
17	66	429	389	260	39	33
18	91	81	76	14	83	82
Total	-	8146	7438	5615	31	25

Table 6.6: Experimental Results—Deutsch's Difficult Example

Router	No. of Tracks	No. of Vias
Greedy	20	379
OTCR	17	336
WISER	17	315
% Imp. Over Greedy	15	17
% Imp. Over OTCR	0	6

6.7.2 Performance Evaluation of the ILP-Based Router

In [LPHL91], the ILP router was compared with the router presented in [CL90] by considering identical benchmarks. A program to automatically generate integer linear programming formulations for a channel was implemented in C [LPHL91] and a linear programming software, LINDO, was used to solve the formulation. Table 6.7 compares the results of the ILP router with the results presented in [CL90]. All the benchmarks except "De," Deutsch's Difficult Channel, are considered from [YK82]. The CPU consumption is not proportional to the channel size, but it depends on the difficulty of the formulation due to the branch-and-bound nature of the ILP. The ILP-based router outperformed the OTC router presented in [CL90] in all the examples [LPHL91]. In three out of the seven examples considered, greater channel density than was reported in [CL90] was achieved. Only in two (3c and 5) of these seven examples are more tracks used than in [CL90]. For the examples (1, 3a, 3b, De) in which the final density was found to be equal to that of [CL90], fewer tracks were used in OTC areas. However, since the authors of [LPHL91] have no experimental results for PRIMARY1, the ILP router is not compared with other routers for the benchmark example PRIMARY1.

Table 6.7: Experimental Results—ILP Router

No.	Org. Den.	Density Over Lower Cells		Density Over Upper Cells		Final Density		Time (seconds)	
		[CL90]	ILP	[CL90]	ILP	[CL90]	ILP	[CL90]	ILP
1	12	3	5	4	2	9	8	2.0	5.4
3a	15	6	5	3	3	12	11	2.9	12.6
3b	17	5	4	2	2	13	13	3.5	96.6
3c	18	4	4	3	4	14	12	4.5	58.7
4b	20	4	4	5	5	16	12	9.7	629.4
5	20	3	4	4	6	14	11	4.9	9.7
De	19	7	4	8	2	16	15	25.1	54.6

6.7.3 Performance Evaluation of the CTM Router

The CTM was basically developed for three-layer models [WHSS92], since in the two-layer environment OTC routing is required to be planar and the entire OTC area in M2 is required to route the terminals to the boundary. Unlike the BTM, the OTC area cannot be exclusively used for routing independent sets of nets in the OTC area. However, efficient river routing techniques can be used to route the terminals to the boundary, such that the vertical constraints can be minimized, thus minimizing the channel densities.

Experimentally, the performance of the CTM router for the two-metal-layer process, when compared with the results of routers on the BTM, is found to have deteriorated. As shown in Table 8, the CTM takes a much higher number of tracks when compared with the routers for other models.

6.7.4 Performance Evaluation of the MTM Router

The MTM router presented in [BPS93b] generates layouts with smaller heights than the corresponding layouts generated by routers for the BTM and CTM. This is because the MTM router uses the OTC area more efficiently. The nets are classified as interrow nets and intrarow nets. The C area is used for routing intrarow nets, while T and B areas are used for eliminating vertical constraints for interrow nets. When tested on PRIMARY1, the router generated a layout with a channel height of 205 tracks (see Table 6.8). The router took two minutes to complete the routing task for PRIMARY1, when tested on a SUN SPARC 1+ workstation.

Table 6.8: Experimental Results for Two-Layer Process: PRIMARY1 Benchmark

Model	Channel Height (in tracks)
BTM	214
CTM	298
MTM	205

6.7.5 Comparison of Two-Layer Routers

The routers for the BTM, CTM, and MTM are compared based on the performance on the benchmark example, PRIMARY1, for minimum channel height. BTM routers use the OTC area by routing large pairwise independent sets of nets in the OTC area. The CTM router uses the OTC area to river route the nets from the terminals to the boundary, so as to minimize vertical constraints. However, in most practical circuits, it can be seen that only five to six tracks are sufficient for river routing and only six to seven tracks are sufficient for routing the independent sets. This has been a motivating factor in the development of the MTM. For a cell with 24 OTC routing tracks, the MTM has six tracks each between the terminal rows and the boundaries, which may be sufficient to eliminate vertical constraints on both the top and bottom adjacent channels. Twelve tracks are provided between the terminal rows to route the independent sets for each terminal row [BPS93b]. It can be seen that the MTM generates layouts with smaller heights.

6.8 Summary

The two-layer process is currently the most widely used fabrication process. Obviously, vias are not used in OTC areas, since the routing is accomplished using only one metal layer. Hence, from the OTC routing point of view, only one layer is available for routing over the cells.

In this chapter, we have presented OTC routing algorithms for all applicable cell models. We discussed routing algorithms for the BTM, CTM, and MTM. The TBC cell model is not relevant, since it assumes M2 and M3 layers in OTC areas. First, we have classified the algorithms on the basis of the algorithmic techniques used. Second, we presented three different algorithms for BTM-based maximum independent sets in overlap graphs, vacant terminals, and integer linear programming. Third, we presented routing algorithms for the CTM and MTM. The BTM is a channel routing–oriented cell model, while the CTM and MTM are motivated by OTC routing.

PROBLEMS

6-1. In Section 6.3, an approximation algorithm was developed for the BTM. In the algorithm, only three types of nets were considered. However, more utilization of the OTC area is possible if we allow an additional net type (Type IV), as shown in Figure 6.25. Note that Type IV nets are not constrained to use abutments; however, they compete with Type II and Type III nets for the usage of vacant terminals. Modify WISER to use Type IV nets in addition to Type I, II, and III nets.

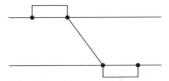

Fig. 6.25   A Type IV net.

6-2. In Section 6.6, an approximation algorithm was developed to solve the ETSPR problem for two-terminal nets in the C area of the MTM. Develop an approximation algorithm to solve an ETSPR problem consisting of all three-terminal nets.

6-3. In Section 6.4, an ILP formulation was presented for OTC routing in the BTM. Develop an ILP formulation for C area routing for the MTM. Extend the formulations for multiterminal nets.

6-4. In Section 3.3, an optimal solution for the BTAP was presented. In this problem, the terminal locations in the cell row are assumed to be fixed. However, in certain interesting variations in the cell model, the routers are given the flexibility to choose the terminals from a specific horizontal range of possible locations for each terminal. Develop river routing constraints for the BTAP when the terminals in a terminal row can be moved to the left or right by a specific number of columns.

6-5. In Section 6.6, an approximation algorithm was presented for routing in the C area of the MTM cell row. The terminal rows are assumed to be aligned. However, in some practical MTM cell designs, the terminal rows may be horizontally nonaligned, as shown in Figure 6.26(a). One possible solution is to generate aligned terminal rows by stretching the terminals using M2 vertical segments, as shown in Figure 6.26(b), and use the approximation algorithm given in Section 6.6. However, this results in inefficient use of the OTC area. Develop an approximation algorithm for routing in the C area of the MTM when the terminal rows are horizontally nonaligned.

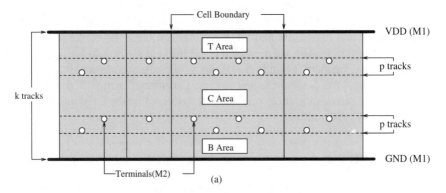

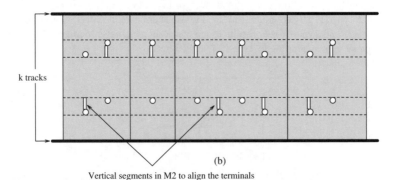

Fig. 6.26 (a) A cell row in the MTM with horizontally nonaligned terminal rows; (b) one possible routing solution.

BIBLIOGRAPHIC NOTES

An optimal algorithm for the BTM-HCVC has been recently developed [DMPS93]. The algorithm runs in $O(kn^2)$ time. However, the algorithm does not allow doglegs or bends in the nets. The complexity of the BTM-HCVC routing problem with an arbitrary number of doglegs per net is still open. BTM-HCVC algorithms presented in [CPL90] assume the terminals on either boundary are nonequipotential. However, when the terminals are considered to be equipotential, the routing problem will be equivalent to the C area routing problem in the MTM [BPS93b].

In [DMPS93], the authors also present a 0.5 approximation algorithm for the maximum planar equipotential problem, which is an improvement over the $0.5 \min\{1, k_2/d\}$ presented in this chapter.

LINDO is an excellent package for running the ILP formulations. The user manual by Linus Schrage provides all the necessary information on integer linear program execution.

A new design style called *Quickly Customized Logic* (QCL) has been introduced for quick turnaround times [NESY89]. The QCL routing problem is similar to the MTM routing problem. In [DSSL93], two channel routing algorithms have been presented for QCL designs. QCL routing problems are very similar to C area routing in the MTM. In fact, routes for QCL also need to route a set of nets from one set of terminals to another while maitaining planarity and satisfying river routing constraints.

7

Routing Algorithms for the Three-Layer Process

The three-metal-layer fabrication process allows two layers (M2 and M3) for OTC routing if only M1 is used for intracell routing. Obviously, when compared with the two-layer process, the three-metal-layer fabrication process has twice the OTC routing resources for a given layout. This results in reduced layout heights.

The key parameter that determines the OTC routing technique in the three-layer process is the *permissibility of vias in OTC areas.* The step-coverage and oxide undercut problems (discussed in Chapter 4) have restricted some fabrication process technologies from the usage of vias in OTC areas. A fabrication process that does not allow the usage of vias in OTC areas is referred to as a *via-less three-layer process*, or simply *three-layer process.* The routing environment for this process is similar to that of the two-layer process, except we have an additional layer in OTC areas for planar routing. A via is required from the M2 to the M3 layer in order to accomplish OTC routing in the M3 layer. However, due to the fabrication constraints, vias have to be located in the channels, adjacent to the cell boundaries. Therefore, OTC routing in the M3 layer is accomplished from the cell boundaries irrespective of the physical locations of the terminals. The basic concept is to first generate a two-layer OTC routing solution and then, from the remaining nets, find a planar subset of nets to be routed from the boundary in the M3 layer of the OTC areas.

In the BTM, after finding a two-layer solution (using the algorithms discussed in Chapter 6), the OTC routing problem in the M3 layer is equivalent to the EP routing problem. However, in the CTM and MTM, after OTC routing in the M2 layer, nonequipotential terminals are available on the top and bottom boundaries. Therefore, the M3 OTC routing problem is equivalent to the TP routing problem. TBC designs are not applicable in the three-layer process, since they use vias over active areas.

The critical concept in the current chip design trend is the performance optimization. The key factor in the development of routing algorithms for high-performance circuits is the minimization of wire lengths. Note that the lengths of the global wires are primarily dependent on placement and floorplanning. For a given placement, the wire lengths have to be reduced. Therefore, for high-performance circuits, the main objective of OTC routing algorithms is wire length minimization.

In this chapter, we first present approximation algorithms for three-layer OTC routing in the BTM, followed by a high-performance routing algorithm for BTM. We then present algorithms for the CTM and MTM. A comparison of the routers for the three models on the benchmark PRIMARY1 is presented at the end of the chapter. It can be seen that the MTM router generates layouts with minimum heights.

7.1 Routers Based on Approximation Algorithms for the BTM

Two different routers based the MKIS in an overlap graph have been developed. Since the MKIS problem for the overlap graph is NP-complete, approximation algorithms are developed.

7.1.1 Router Based on MKIS

In [HSS91a], a three-layer OTC router is proposed. The heart of the router is the algorithm ISA, which is an extension of the algorithm presented in [HSS91b]. The algorithm classifies nets into three different types, as in [HSS91b], and Type II nets are assigned vacant abutments. The basic idea is to choose four different independent sets in a similar fashion as is done in the case of the approximation algorithm for the MKIS in circle graphs. Let I_{t2}, I_{t3}, I_{b2}, and I_{b3} denote the independent sets selected for OTC routing, where I_{t2} and I_{t3} represent the sets of nets that will be routed over the upper row of cells, and I_{b2} and I_{b3} denote the sets of nets to be routed over the lower cell row. Since not all nets are fixed to a specific side of the channel, the order in which we select these independent sets is very important. In the algorithm, all six possible orderings are considered and the best one is selected.

Theorem 28   *The performance ratio of the algorithm ISA is $\rho \geq 0.68$.*

Proof: Follows directly from Theorem 7, Chapter 2, which states that for $k = 4$, the approximation ratio is lower bounded by 0.68.

□

Note that although the algorithm considers all types of nets, only Type II nets are considered in the proof above, since the proof assumes that all nets can be assigned to either top or bottom OTC areas.

Also notice that the OTC area is only blocked by power and ground lines in M2. As a result, another approach is possible. In this alternative approach, the routing solution only for the M2 layer of the OTC area is generated using two independent sets approach, discussed in Chapter 6. Now, the terminals for the nets that are not selected are located at the boundaries, and routing algorithm in the M3 layer is similar to the C area planar routing algorithm for the MTM.

7.1.2 Improved Router Based on Net Types

In [RS93], a new algorithm for three-layer OTC routing is presented, which concentrates on the development of an approximation algorithm in which both Type I and Type II nets are considered.

The algorithm uses similar terminology, as discussed in Chapter 6, Section 6.3.2. All terms with a star ($*$) as the superscript refer to the optimal solution, while the primed terms refer to the solution selected by the algorithm. Let $S_1^* \subseteq V_1$, $S_2^* \subseteq V_2$, and $S_{12}^* \subseteq V_{12}$. In addition, let $S_{12-1}^* \subseteq V_{12}$, which is also a subset of S_1'. In other words, the subscript $12 - 1$ indicates that portion of the first partite set of the solution that has come from the set labeled V_{12}. Similarly, S_{12-2}^* denotes the second partite set of the solution that was derived from the set labeled V_{12}.

The sets I_{t_2} and I_{t_3} refer to the independent sets to be routed above the OTC area in M_2 and M_3, respectively. Similarly, sets I_{b_2} and I_{b_3} refer to the independent sets below the OTC area in M_2 and M_3, respectively.

Theorem 29   *The performance ratio of the algorithm Improved_ISA( ) is given by $\rho \geq 0.483$.*

Proof: The algorithm Improved_ISA( ), shown in Figure 7.1, is based on three different strategies. The first strategy produces good results if the majority of nets in the optimal solution S^* are from set V_{12}. The second and third strategies are designed to produce good results if the majority of nets come from the sets V_1 and V_2.

By Theorem 7, strategy #1 generates

$$| I_{t_2}[1] \cup I_{t_3}[1] \cup i_{b_2}[1] \cup I_{b_3}[1] | \geq 0.68\alpha$$

Strategy #2 ensures that

$$| I_{t_2}[2] \cup I_{t_3}[2] | \geq 0.75 | (S_1^* \cup S_{12-1}^*) |$$
$$| I_{b_2}[2] \cup I_{b_3}[2] | \geq 0.75 | (S_2^* |$$

Since the maximum between strategy #2 and strategy #3 is chosen, the solution is

$$\geq 0.75 | (S_1^* \cup S_2^* \cup 0.5 S_{12}^*) |$$
$$\geq 0.75(1 - \alpha) + 0.375\alpha$$

Algorithm IMPROVED_ISA( )

Input: An Overlap Graph $G = (V, E)$, V_1, V_2 and V_{12}
Output: Sets I_{t_2}, I_{t_3}, I_{b_2}, I_{b_3} representing
      independent sets for top(bottom) M_2 and M_3 layers

Begin
  /* Strategy # 1 */
  $I_{t_2}[1]$, $I_{t_3}[1]$, $I_{b_2}[1]$, $I_{b_3}[1] = 4MIS(V_{12})$;
  /* Strategy # 2 */
  $I_{t_2}[2]$, $I_{t_3}[2] = 2MIS(V_1 \cup V_{12})$;
  $I_{b_2}[2]$, $I_{b_3}[2] = 2MIS(V_2)$;
  /* Strategy # 3 */
  $I_{t_2}[3]$, $I_{t_3}[3] = 2MIS(V_1)$;
  $I_{b_2}[3]$, $I_{b_3}[3] = 2MIS(V_{12} \cup V_2)$;
  /* Select Maximum */
  $I_{t_2} = I_{t_3} = I_{b_2} = I_{b_3} = \phi$;
  for $i = 1$ **to** 3
    if $\mid I_{t_2}[i] \cup I_{t_3}[i] \cup I_{b_2}[i] \cup I_{b_3}[i] \mid \geq$
    $\mid I_{t_2} \cup I_{t_3} \cup I_{b_2} \cup I_{b_3} \mid$ **then**
      $I_{t_2} = I_{t_2}[i]$;
      $I_{t_3} = I_{t_3}[i]$;
      $I_{b_2} = I_{b_2}[i]$;
      $I_{b_3} = I_{b_3}[i]$;
End;

Fig. 7.1 Pseudocode for algorithm Improved_ISA.

Finally, since the maximum is chosen between all three strategies, the following equation is true.

$$0.68\alpha = 0.75(1 - \alpha) + 0.375\alpha$$

$$\alpha = 0.7109$$

$$\rho = 0.483$$

□

Figure 7.2 shows the variation in performance ratio for different α values.

7.1.3 Router for High-Performance Circuits

The critical concept in the current chip design trend is performance optimization. With the increase in scale of integration and the increase in wafer size,

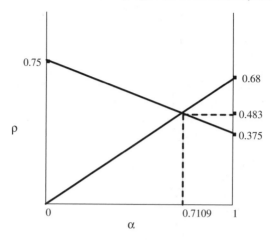

Fig. 7.2   Performance of algorithm Improved-ISA.

interconnections play an important role in determining the performance of VLSI circuits. An interconnection wire has an interconnect delay associated with it. The key factor in the development of routing algorithms for high-performance circuits is the minimization of wire lengths. Note that the lengths of the global wires are primarily dependent on placement and floorplanning. In this chapter, we present OTC routing algorithms for high-performance circuits based on wire length minimization. First we present the algorithms for the three-layer process, and then we present the algorithms for the advanced three-layer process.

Natarajan, Sherwani, Holmes, and Sarrafzadeh [NSHS92] presented a three-layer OTC channel routing algorithm, WILMA3 (wire length minimization algorithm), for high-speed circuits. The objectives of the router include minimization of the channel height by employing OTC routing techniques and net length minimization, while satisfying the bounds on the nets. The algorithm is novel in two aspects. First, the track assignment of each net is optimized with respect to delay. The track bound is computed for each net, which ensures that the net delay is no greater than the given bound (this bound can be obtained in a preprocessing step); a linear RC delay model is used. Using this track bound, nets are selected for OTC routing. For this purpose, an $O(dn)$ time algorithm is developed for finding an optimal subset of nets, where d is the density and n is the number of nets. Secondly, $45°$ segments are used to route the nets over the cells to further reduce the net length.

7.1.3.1 Algorithm overview

The algorithm WILMA3 consists of seven basic steps, which are described briefly as follows. Input to the algorithm is a global routing and a set of bounds on net length. These bounds either come from timing requirements or may be selected by a designer to reduce long nets in the circuit.

1. **Net Decomposition and Classification:** Multiterminal nets are decomposed into two-terminal nets and classified. This step is similar to the net classification step in [HSS91b].

2. **Net Weight Assignment:** The weight of a net intuitively indicates the improvement in channel congestion possible if this net can be routed over the cells. Selecting the nets with maximum weight is necessary to achieve maximal reduction in channel height.

3. **Estimation:** In this phase of the algorithm, a channel router is used to obtain the channel density (d_c) if routed in the channel. For each net N_i, the track used for routing N_i is recorded. An OTC router is used to obtain the channel density (d_o) for OTC routing. This information is used to compute track bounds.

4. **Track-Bound Computation:** For each net n_i, track bound k_i is computed, which ensures that if the net is routed OTC at a track less than or equal to k_i, it will have a wire length less than or equal to a given bound; this bound depends on the net's criticality and timing requirements. This is based on the estimated channel heights d_c and d_o.

5. **Net Selection with Track-Bound Constraint:** Among all the nets that are suitable for routing over the cells, four (two) maximum-weighted planar subsets are selected subject to the track-bound constraint for the three (two)-layer model. Once the nets are selected, a set of vacant terminals (vacant abutments) in the case of Type II (Type III) nets are assigned to each net N_i depending on its weight. These vacant terminal and abutment locations will later be used to determine an OTC routing for N_i.

6. **OTC Routing with 45° Segments:** Hybrid OTC routing is performed with 45° segments and rectilinear segments. In order to avoid design rule violations, any net n_i routed over the cell on track t_i must contain a vertical segment of length ρ_i before 45° segments can be used (a detailed discussion on ρ_i is given in subsequent sections).

7. **Channel Routing:** The net segments that have not been routed in the area over the cells are routed in the channel. In algorithms WILMA2 and WILMA3, channel routing algorithms from [RF82] and [CWL87] are used.

Note that if routability problems are encountered, then channel height is increased and the bounds are recomputed. The iterative process mentioned above takes place very rarely, since for most examples, the algorithm can complete the routing using no more tracks than d_o.

We discuss steps 4 and 5 in detail in the following sections. Step 6 will be discussed as a part of step 4, since understanding of angular routing is necessary to compute the track bound. The remaining steps are similar to the corresponding steps in the algorithms presented in Chapter 6.

7.1.3.2 Track-bound computation

This section first presents the strategy adopted in hybrid routing and then describes track-bound computing methodology. Track bound specifies the farthest track from the boundary that can be assigned to a net.

Hybrid routing for a net n_i with terminals l_i and r_i at track k_i will have (at most) five segments $s_1, s_2, \ldots, s_5$, as shown in Figure 7.3. The segments s_1 and s_5 are the *vertical segments*, s_2 and s_4 are the *angular segments*, and s_3 is the *horizontal segment* routed at track k_i. The vertical segment ρ_k^i for any track k_i is $\delta(k_i - 1)(1 - \sin\theta/\cos\theta)$, where δ is the minimum feature separation for the technology (which is 3λ for CMOS) and θ is the angle at which the angular segment is routed with respect to the horizontal. If $\text{len}(s_j)$ represents the length of a segment s_j, then the length of the net routed over the cell using hybrid routing is given by

$$\text{Length of net } n_i = \sum_{j=1}^{5} \text{len}(s_j^i)$$

where the lengths of the individual segments are given by

$$\text{len}(s_1^i) = \text{len}(s_5^i) = \rho_{k_i} = \delta(k_i - 1)\frac{(1 - \sin\theta)}{\cos\theta}$$

$$\text{len}(s_2^i) = \text{len}(s_4^i) = \frac{\delta k_i - s_1^i}{\sin\theta} = \frac{\delta(k\cos\theta + k\sin\theta - k + 1 - \sin\theta)}{\sin\theta\cos\theta}$$

$$\text{len}(s_3^i) = (r_i - l_i) - \frac{2(\delta k_i - s_1^i)}{\tan\theta}$$

$$= (r_i - l_i) - \frac{2\delta(k\cos\theta + k\sin\theta - k + 1 - \sin\theta)}{\sin\theta}$$

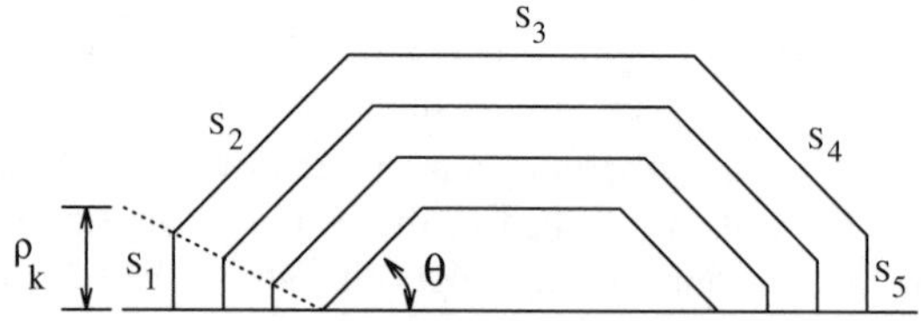

Fig. 7.3 OTC routing using 45° segments.

Let d_o and d_c be the channel densities as estimated in Phase 3. Let θ be the orientation of the angular segments used in routing with respect to the horizontal. Any given net n_i may be routed in a track in the channel (t_i^c), a track in the upper OTC area (t_i^u), or a track in the lower OTC area (t_i^l). Thus, it may be assigned to a track in one of the three possible ways. It is possible that the horizontal segment h_i of the net is split into two segments, one routed on the upper cell at track k_i^u and the other routed in the lower cell at track k_i^l, as in Type III nets. In order to compute the track bound, the following three lengths are required: the total length of the net when routed in the channel, denoted by L_i^c, the total length when routed over the cell, denoted by L_i^o, and the total length when routed over the cell using angular segments, denoted by L_i^θ.

The length L_i^c of a net n_i is equal to the sum of the length of the horizontal segment and the length of the vertical segment. The length of the horizontal segment is equal to $(r_i - l_i)$. The length of the vertical segment depends on whether both the terminals lie on the top or bottom of the channel, and is given by

$$\text{Length of the vertical segment of net} = \begin{cases} 2t_i^c & \text{if } (l_i, r_i) \text{ on } R_t \\ 2(d_c - t_i^c) & \text{if } (l_i, r_i) \text{ on } R_b \end{cases}$$

Without loss of generality, it is assumed that both terminals of n_i lie on R_t. Hence, the length of a net n_i routed in the channel is given by $L_i^c = (r_i - l_i) + 2t_i^c$. It should be noted that L_i^c is the same for all net types, whereas L_i^o and L_i^θ vary depending on the type of net. Since the length increase due to OTC routing varies with the type of net, for each class of net the track bound is identified in terms of the above-mentioned parameters.

1. **Type I Nets:** Recall that Type I nets are those nets that, due to the location of vacant terminals with respect to net terminals, may be routed over the cell rows on exactly one side of the channel. Type I(a) nets are those Type I nets with both terminals on a single row (top or bottom) of the channel, whereas Type I(b) nets have terminals on opposite channel boundaries. Each of these subtypes is separately considered.

   A net of Type I(a), with both its terminals on the same side of the cell, is routed over the cell on the side in which the terminals originate. The routings in the channel, over the cell, using hybrid routing for this net type are illustrated in Figure 7.4.

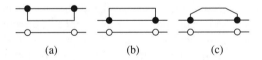

(a)   (b)   (c)

Fig. 7.4   Routing of Type I(a) net in **(a)** channel, **(b)** OTC, and **(c)** hybrid routing.

Lemma 9   *Given a net n_i of Type I(a), $L_i^\theta \le L_i^c$ if*

$$k_i \le \frac{t_i^c \sin\theta \cos\theta - \sin^2\theta + \cos\theta + 2\sin\theta - \sin\theta \cos\theta - 1}{2\sin\theta + 2\cos\theta - \sin\theta \cos\theta - 2}$$

Proof: Let $n_i = (l_i, r_i)$ be a Type I(a) net. Let t_i^c be the track on which net n_i is routed in an initial channel routing. The wire length L_i^c due to channel routing, when net n_i is routed in track t_i^c in the channel, is equal to the sum of the length of its vertical segments and horizontal segments.

Hence, the total length of the net when channel routing is employed is given by

$$L_i^c = r_i - l_i + 2\delta t_i^c$$

The total length of the net if routed over the cell with angular segments is given as

$$L_i^\theta = 2 \times \text{len}(s_1) + 2 \times \text{len}(s_2) + \text{len}(s_3)$$

Hence, L_i^θ is given by

$$L_i^\theta = 2\delta(k_i - 1)\frac{(1 - \sin\theta)}{\cos\theta}$$
$$+ \frac{2(k_i \cos\theta + k_i \sin\theta - k_i + 1 - \sin\theta)(1 - \cos\theta)}{\sin\theta \cos\theta} + (r_i - l_i)$$

(7.1)

Thus, the difference in routing in the channel and routing over the cell using angular segments is given by

$$\Delta L = L_i^\theta - L_i^c$$

In order to ensure that the net length does not increase if net n_i is routed over the cell using angular segments, ΔL must be zero. Hence, L_i^θ is equated to L_i^c to obtain the track bound as

$$k_i \leq \frac{t_i^c \sin\theta \cos\theta - \sin^2\theta + \cos\theta + 2\sin\theta - \sin\theta \cos\theta - 1}{2\sin\theta + 2\cos\theta - \sin\theta \cos\theta - 2}$$

$\square$

A net of Type I(b), with terminals on the same side of the cell, is routed on the other side of the cell. This type of OTC routing certainly uses more wire length than a Type I(a) net. However, in the case of availability of space over the cell on the opposite cell and no space on the side of the net, a potential reduction of one track in the channel is possible, which in turn may reduce the net length of some of the other nets. The three routing strategies for the net of Type I(b) are illustrated in Figure 7.5.

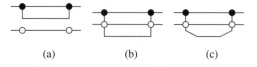

(a)   (b)   (c)

Fig. 7.5   Routing of Type I(b) net in **(a)** channel, **(b)** OTC, and **(c)** hybrid routing.

2. **Type II Nets:** For a net of Type II, with terminals on opposite sides of the channel, the net is routed over one of the cells subject to the availability of vacant terminals. Figure 7.6 illustrates routing for this net type for channel, OTC, and hybrid routing strategies.

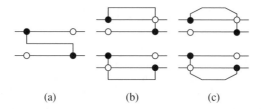

Fig. 7.6 Routing of Type II net in **(a)** channel, **(b)** OTC, and **(c)** hybrid routing.

3. **Type III Nets:** A Type III net, with terminals on opposite sides of the channel with occupied opposite terminals can be routed through a vacant abutment, such that the routed net spans the OTC areas both on the top and bottom cells. Figure 7.7 gives the net length difference due to different routing strategies employed for this net type.

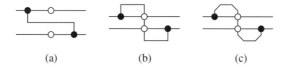

(a)   (b)   (c)

Fig. 7.7 Routing of Type III net in **(a)** channel, **(b)** OTC, and **(c)** hybrid routing.

In Table 7.1, the track bounds for all the above-mentioned net types are computed for $\theta = 45°$ and $\theta = 60°$. It has been reported that the net lengths are a minimum for angular routing for $\theta = 45°$. Hence, the algorithm uses 45° segments for routing over the cell. However, in asymmetric and low-density OTC routings, other orientations for segments may be considered.

Table 7.1: Comparison of Track Bounds for $\theta = 45°$ and $\theta = 60°$

Net Type	$\theta = 45°$	$\theta = 60°$
I(a)	$k_i = \dfrac{0.5t_i^c + 0.121}{0.328}$	$k_i = \dfrac{0.433t_i^c + 0.049}{0.299}$
I(b)	$k_i = \dfrac{0.5(t_i^c - d_o) + 0.121}{0.328}$	$k_i = \dfrac{0.433(t_i^c - d_o) + 0.049}{0.299}$
II	$k_i = \dfrac{0.5(\frac{d_c - d_o}{2}) + 0.121}{0.328}$	$k_i = \dfrac{0.433(\frac{d_c - d_o}{2}) + 0.049}{0.299}$
III	$k_i^l + k_i^u = \dfrac{0.5(\frac{d_c - d_o}{2}) + 0.242}{0.328}$	$k_i^l + k_i^u = \dfrac{0.433(\frac{d_c - d_o}{2}) + 0.098}{0.299}$

7.1.3.3 Net selection with track-bound constraint

In this section, an algorithm for finding an independent set in an unweighted graph with track-bound constraint is discussed. This algorithm runs in $O(dn)$ time, where d is the density of nets and n is the number of nets. For finding the independent set in a weighted graph, the $O(kn^2)$ time algorithm presented in [CPL90] is used.

Consider a set $\mathcal{I} = \{I_1, \ldots, I_n\}$ of intervals, where $I_i = (l_i, r_i)$ is specified by its two end points; that is, the *left point* l_i and the *right point* r_i, $l_i < r_i$. We assume all end points are distinct real numbers (the assumption simplifies our analysis and can be trivially removed). Two intervals $I_i = (l_i, r_i)$ and $I_j = (l_j, r_j)$ are *independent* if $r_i < l_j$ or $r_j < l_i$; otherwise, they are *dependent*. Two dependent intervals are *crossing* if $l_i < l_j < r_i < r_j$ or $l_j < l_i < r_j < r_i$. If I_i *contains* I_j, then $l_i < l_j < r_j < r_i$.

Assume track $i + 1$ is above track i, $1 \le i \le t - 1$. Given $\mathcal{I}$, the restricted track assignment problem (RTAP) is the problem of assigning intervals to tracks 1 to t such that:

p1) In each track, intervals are pairwise independent.

p2) If an interval I_i contains another interval I_j, then $\tau_i > \tau_j$, where τ_a is the track to which I_a is assigned.

In [SL90], Sarrafzadeh and Lou showed that RTAP is NP-hard to minimize the number of tracks, and proposed a two-approximation algorithm running in $O(n \log n)$ time, where n is the number of intervals. We will see that for our purpose an approximation algorithm is as good as an optimal algorithm (since the approximate bound will be a cost factor in the time complexity)!

Consider an overlap representation $\mathcal{I} = \{I_1, \ldots, I_n\}$ of an unweighted overlap graph, where each interval I_i is assigned a bound b_i being the maximum track to which it can be assigned. A track assignment satisfying the bounds is called a *bounded tracks assignment*. Among all bounded independent sets with density K, find the set with the maximum size. This problem is referred to as the *BMIS-K problem*.

For each interval $I_i = (l_j, r_j)$ in $\mathcal{I}$, calculate a size s_i being an MIS of the intervals (in the overlap model) with end points in the closed interval $[l_j, r_j]$ (i.e., it includes I_i) among all independent sets with density less than or equal to K. Also, for each MIS obtained, calculate its density k_i, $1 \le k_i \le K$. First, $\mathcal{I}$ is assigned to tracks 1 to t in a restricted manner, $d \le t \le 2d - 1$ (that is, a restricted track assignment is obtained). $\mathcal{I}_i$ denotes the set of intervals placed in track i, $1 \le i \le t$. $\mathcal{I}_{i+1} = \{I_{a_1}, \ldots, I_{a_m}\}$ denotes the set of intervals assigned to track $i + 1$. Certainly, intervals in $\mathcal{I}_{i+1}$ are pairwise independent. Let $\mathcal{J}_{a_j}$ be the set of intervals contained in I_{a_j}, $1 \le j \le m$. Clearly, $\mathcal{J}_{a_j} \cap \mathcal{J}_{a_k} = \phi$ for all j and k ($j \ne k$).

Inductively, assume that tracks 1 to i have been processed, and for each interval I_j in the set of processed tracks, the size s_j and its density k_j are known. The following approach is adopted. The basis is trivial—for each interval I_j in track 1, $s_j = 1$ and $k_j = 1$. Track $i + 1$ is processed as follows. For each $\mathcal{J}_{a_j}$ ($\mathcal{J}_{a_j}$ is the set of intervals contained in I_{a_j}, $1 \le j \le m$; see Figure 7.8), a maximum-weighted independent set—denoted by $\mathcal{S}_{a_j}$—in the interval graph representation (not the overlap representation) is found, where the weight of an interval is its size (and is known by the induction hypothesis). Let s'_{a_j} denote the weight of S_{a_j} and k'_{a_j} be the maximum of k_f for all I_f in S_{a_j}. There are two cases to be considered.

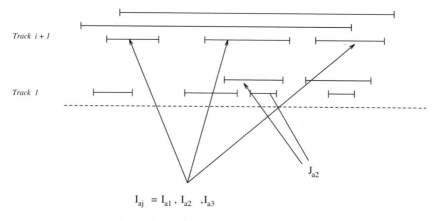

Fig. 7.8   A restricted track assignment of a set of intervals.

- **Case 1:** If $k'_{a_j} = K$ or if $k'_{a_j} > b_{a_j}$ then I_{a_j} will be deleted and will not be considered in subsequent steps. The reason is as follows. Consider an optimal solution that contains I_{a_j} (see Figure 7.9). Let S^* denote the set of intervals in this optimal solution that is contained in I_{a_j}. The density of S^* is less than K. Otherwise, the density of $S^* \cup \{I_{a_j}\}$ would be more than K. Certainly, $|S^*| \le |S_{a_j}| - 1$; otherwise, S^* should have been selected in the inductive step, since its density is less than k'_{a_j}. Now $S^* \cup \{I_{a_j}\}$ in the optimal solution can be replaced with S_{a_j} to obtain a solution of the same (or more) size.

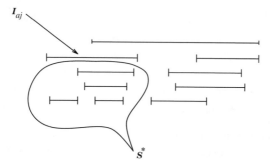

Fig. 7.9   Proof of Case 1.

- **Case 2:** If $k'_{a_j} < K$ and $k'_{a_j} \le b_{a_j}$, then we set the two parameters $k_{a_j} = k'_{a_j} + 1$ and $s_{a_j} = s'_{a_j} + 1$, and continue.

An interval $I_{n+1} = (l_{n+1}, r_{n+1})$ is introduced, where l_{n+1} is the minimum of l_i minus 1 and r_{n+1} is the maximum of r_i plus 1 (for $1 \leq i \leq n$). Repeating the process just described one more time results in an optimal solution. As before, it takes $O(nt) = O(nd^*)$ time to obtain the size s_i of all intervals.

Lemma 10 *An unweighted instance $\mathcal{I}$ of BMIS-K for overlap graphs can be solved in $O(dn + n \log n)$ time, where d is the density of $\mathcal{I}$ and $n = |\mathcal{I}|$ (note that $d \geq k$).*

Note that the proposed technique cannot be used for the weighted version of the problem, because even one interval in higher tracks may have a large weight and thus replaces (some or all) the previous intervals (the argument in Case 1 does not hold any more). The weighted version can be readily solved employing a straightforward modification of the algorithm proposed in [CPL90]. We conclude:

Lemma 11 *An arbitrary instance $\mathcal{I}$ of BMIS-K for overlap graphs can be solved in $O(Kn^2)$ time, where $n = |\mathcal{I}|$.*

Using this algorithm, four independent sets for the three-layer model to be routed on the top and bottom cell for the second metal layer (M2) and third metal layer (M3) are computed. In the two-layer model, only two sets are computed, since only one metal layer (M2) is available for OTC routing.

7.2 Routing Algorithm for the CTM

In this section, we present an overview of a three-layer OTC routing algorithm (3CTM-V) proposed by Wu, Holmes, Sarrafzadeh, and Sherwani [WHSS92].

The OTC routing problem in the three-layer CTM can be split into two subproblems. First, the terminals have to be routed to the boundary in a planar fashion in the M2 layer. This problem is exactly equivalent to the BT routing problem. Second, since the terminals are now available at the boundaries, the routing in the OTC areas is equivalent to the TP routing problem and the routing in the channels is a conventional channel (CC) routing problem. Therefore, the OTC routing problem for the three-layer process can be decomposed into BT, TP, and CC routing problems.

7.2.1 Algorithm Overview

The routing algorithm is similar to the algorithm for the two-layer CTM router presented in Chapter 6, except it uses the following additional steps:

1. **Net Selection for M3 Layer:** After completion of the routing in the M2 layer, among the remaining nets, a set of nets forming a maximum bipartite subgraph in a circle graph are selected. Since this problem is

known to be NP-hard, we discuss two approximation algorithms and an integer programming approach to solve it.

2. **M3 OTC Routing:** In this step, the nets selected in the previous step are routed in a planar fashion in M3 OTC areas on both sides of the channel. This step is similar to existing OTC routers.

3. **Channel Routing:** An HVH routing model can be used to route the remaining nets in the channel.

Since all vertical constraints are eliminated, a router based on the LEA adapted for the three-layer routing environment can be used. Hence, in the subsequent sections, we discuss the details of only the first two steps.

7.2.2 Net Selection for the M3 Layer

In this section, the procedure for M3 net selection is explained. The objective of this step is to minimize the channel height. Since all vertical constraints have already been eliminated, the channel height is decided by the size of the maximum clique in the HCG (H_{max}).

Since 3CTM-V does not allow vias in the OTC area, only two planar subsets of nets can be routed in the M3 OTC area. The objective of this step is to select these two sets of nets, such that H_{max} of remaining nets is minimized. Basically, this problem is a *maximum two independent sets* problem in a circle graph, such that the H_{max} of the remaining nets is minimized. Let us call such a set a 2HMIS. Finding optimal 2HMIS in a circle graph has already be shown to be NP-hard [HSS91a]. In [HSS91a], a two-step heuristic approach has been suggested. First, the nets are assigned weights, such that a net gets a higher weight if routing it in the OTC area may result in reduction of channel density. Then a *maximum-weighted two independent sets* 2MIS is found in a circle graph using a 0.75 approximation algorithm (see Chapter 2).

In [WHSS92], Sherwani et al. suggested an alternative approach. In this approach, *MIS*s are found one by one with the same objective as that of the original problem (i.e., minimization of the density of remaining nets). Let us call this problem the HMIS problem. This approach turns out to be an obvious 50% approximation solution to the optimal solution to the 2HMIS problem. The formal statement of the algorithm for finding the HMIS is in Figure 7.10. It uses the same terminology as that used for the MIS problem in Chapter 2.

Finally, since $G = G_{0,2N-1}$, the MIS is a maximum independent set of G.

The algorithm HMIS finds the MIS in a circle graph in $O(n^3)$ time, such that the size of the maximum clique in the HCG of remaining nets is minimum.

7.2.3 Integer Program for Optimal 2HMIS

Let the partite sets of 2HMIS be denoted as HMIS1 and HMIS2. Let x_i, y_i, and z_i be the three variables associated with each net $N_i \in L$, such that

Algorithm HMIS_CIRCLE(C)

Input: A set of chords C
Output: HMIS

```
begin
    for j = 0 to 2N − 1 do
        for i = 0 to j − 1 do
        Let Nₗ = {k, j} be a net.
            if i ≤ k ≤ j − 1
            then HMIS = HMIS(i, k − 1) ∪ {v_{k,j}} ∪ HMIS(k + 1, j − 1)
                if hmax(L − HMIS) < hmax(L − HMIS(i, j − 1))
                then HMIS(i, j) = HMIS
                else HMIS(i, j) = HMIS(i, j − 1)
            else HMIS(i, j) = HMIS(i, j − 1)
end.
```

Fig. 7.10   Algorithm HMIS_CIRCLE.

$$x_i = \begin{cases} 1 & \text{if } N_i \in \text{HMIS1} \\ 0 & \text{otherwise} \end{cases}$$

$$y_i = \begin{cases} 1 & \text{if } N_i \in L - \text{HMIS1-HMIS2} \\ 0 & \text{otherwise} \end{cases}$$

$$z_i = \begin{cases} 1 & \text{if } N_i \in \text{L-HMIS2} \\ 0 & \text{otherwise} \end{cases}$$

Let variable $D_{\max}$ indicate the maximum density of $L - \text{HMIS1} - \text{HMIS2}$. Thus, the objective of the integer program is to minimize $D_{\max}$, such that the following constraints are satisfied.

1. **Mutually Exclusive Constraints:** A net $N_i \in L$ can only be in one of the following three sets: HMIS1, HMIS2, and L−HMIS1−HMIS2.

$$x_i + y_i + z_i = 1 \quad \text{for all } i = 1 \text{ to } |L|$$

2. **Overlap Constraints:** The nets in set HMIS1 have to be pairwise independent; that is, if two nets N_i and N_j overlap, then either of them can be in HMIS1.

$$x_i + x_j = 1 \quad \text{if nets } N_i \text{ and } N_j \text{ overlap}$$

The same is the case for the nets in set HMIS2, Thus,

$$z_i + z_j = 1 \qquad \text{if nets } N_i \text{ and } N_j \text{ overlap}$$

3. **Density Constraints:** Since $D_{\max}$ indicate the maximum density of $L -$ HMIS1 $-$ HMIS2, the following density constraints are imposed.

$$\sum_{j \in \text{Int}(N_i)} y_i \qquad \text{for all } j = 1 \text{ to max\_col}$$

where, $\text{Int}(N_i)$ indicates the interval of net N_i.

7.3 Routing Algorithms for the MTM

The first subsection of this section contained an overview of the router, so the remaining subsections contain details of several steps of the router.

In the case of MTM-V, routing in OTC areas in the layer over the cell must be planar. Weights are assigned to nets such that higher weighted nets are critical nets and their removal results in maximum reduction in height. The weight is computed based on vertical and horizontal constraints in the given channel [HSS91a]. Based on the weight, the MTM router first finds an MPES. This is a set of intrarow and interrow nets, which can be routed in the C area in a planar fashion. The routing for each net in the C area can be accomplished in four different paths, as shown in Chapter 6. In the next step, we do boundary terminal assignment in order to minimize the channel height. The cell boundary is essentially considered a row of virtual terminals. After the terminal assignment phase, we find two sets of independent nets to route in the M3 layer of the OTC area while minimizing the channel height. The remaining nets are routed in the channel, and the cleanup routing phase is used to improve wiring lengths. For most of the examples, channel-less routing can be obtained, since the two layers in the OTC area are enough to complete the routing.

7.3.1 Algorithm Overview

The basic step of the 3MTM-V router is similar to 2MTM, but has the following additional steps:

1. **Additional Boundary Terminal Assignment:** In this step, the channel height is reduced by assigning the boundary terminal pairs to some of the nets in the channel. These terminal pairs are selected such that the resultant nets can be routed in the OTC areas.

2. **M3 OTC Routing:** In this step, the nets selected in the previous step are routed in a planar fashion in M3 OTC areas on both sides of the channel. This step is similar to that of existing OTC routers.

3. **Channel Routing:** Any channel router can be used to route the remaining nets in the channel.

7.3.2 Additional Boundary Terminal Assignment

In this step, the nets in the channel are assigned boundary terminal pairs so that they can be routed in the OTC area in order to reduce the channel density.

Nets in a channel are considered in a sequential manner according to their decreasing weights. For each net $n = (t_1, t_2)$ in the channel, with t_1 on cell row R_i and t_2 on cell row R_{i+1}, we do the following:

1. All terminal pairs (t_1', t_2') are found such that $t_1' \in U$, $t_2' \in L$, t_1' and t_2' are in the same column, net (t_1, t_1') is pairwise independent with the nets existing in the B area of R_i, and net (t_2, t_2') is pairwise independent with the nets existing in the T area of R_{i+1}.

2. If there does not exist any such terminal pair, then the net n is routed in the channel. Otherwise, among all such terminal pairs a pair (t_1'', t_2'') is chosen that would minimize the increase in the number of tracks in the OTC area. Ties are resolved arbitrarily.

3. Net (t_1, t_1'') is selected to route in the B area of R_i, net (t_1'', t_2'') is selected to route in the channel, and net (t_2'', t_2) is selected to route in the T area of R_{i+1}.

Figure 7.11(b) shows the effect of net selection for the M3 layer in the T and B areas on an example in Figure 7.11(a), whereas Figure 7.11(c) shows the effect of additional boundary terminal assignment on an example in Figure 7.11(b).

7.3.3 OTC Routing in the M3 Layer

After boundary terminal assignment, planar routing must be accomplished in the M3 layer in the OTC areas. The M3 OTC area is completely unblocked. The routing environment is as shown in Figure 7.12. The problem is similar to HCVC, but the objective function is slightly different. The objective here is to minimize the channel height between two cell rows. This is accomplished by using the technique discussed in Section 7.1.1.

7.4 Experimental Evaluation of Three-Layer Routers

Several OTC routers have been discussed for the three-layer process. In this section, we present a comparative performance evaluation of the OTC routers on some benchmark examples.

7.4.1 Performance Evaluation of the BTM Router

In this section, we present the evaluations of WILMA3 on the PRIMARY1 benchmark. The placement of PRIMARY1 was obtained from TimberWolfSC Version 5.1 [LS88], and the global routing is from [CSW89].

An RC model is used for computing the delays in the net. The per-unit resistance and capacitance of a net is 30 mΩ and 0.20 fF, respectively. This yields

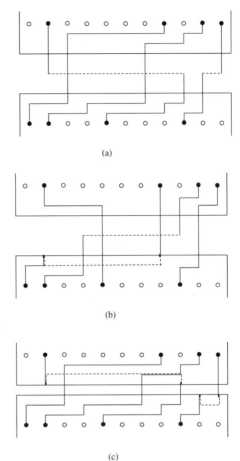

Fig. 7.11   **(a)** Example; **(b)** effect of net selection for M3 layer in T and B ar-
eas; **(c)** effect of additional boundary terminal assignment.

an effective per-unit delay of 6×10^{-9}ps. The effects of vias in the channel are
ignored while computing the net delay. This is because of the fact that vias are
process-dependent and the delay varies with the process used. Ignoring the delay
in vias gives the advantage to the channel routing—since it gives a conservative
delay bound for OTC routing.

Results for the PRIMARY1 indicating the number of horizontal tracks used in
the channel are given in Table 7.2 in comparison with results obtained by a greedy
channel router and a conventional channel router. As can be seen in Table 7.2,
WILMA3 reduces the channel height by an average of 52% as compared to the best
available three-layer greedy channel router. On the average, WILMA3 performs
73% better than a conventional two-layer channel router and 62% better than two-
layer OTC routers. Experimental results indicate significant reduction both in the
total wire length and in the length of the longest net.

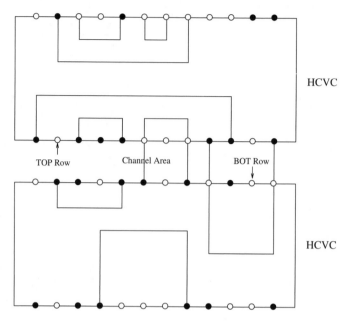

Fig. 7.12   OTC routing problem in the MTM.

Table 7.2:   Experimental Results: BTM Router on PRIMARY1

Chan.	No. of Tracks Produced					% Improvement by WILMA3		
	2CRP	2OTC	3CRP	3OTC	WILMA3	2CRP	2OTC	3CRP
1	11	5	6	1	2	82	60	66
2	16	12	9	5	5	69	58	44
3	21	16	11	6	6	71	62	45
4	24	21	15	6	9	62	57	40
5	21	17	11	6	7	67	58	36
6	23	18	12	7	7	69	61	41
7	22	14	14	4	5	77	64	64
8	24	18	16	7	7	71	61	56
9	21	16	13	6	6	71	62	53
10	15	11	8	4	5	67	54	37
11	17	12	12	3	5	70	58	58
12	15	11	8	3	4	73	64	50
13	13	9	8	3	3	77	67	62
14	13	9	7	3	4	69	55	42
15	11	7	6	2	2	81	71	66
16	11	7	6	2	2	81	71	66
17	14	8	7	3	3	78	62	57
18	6	3	3	0	1	83	66	66
Total	298	214	172	71	83	72	61	52

7.4.2 Performance Evaluation of the CTM Router

The performance of 3CTM-V on a benchmark, PRIMARY1 is compared with a conventional two-layer router [RF82], a two-layer OTC router [CPL90], and a conventional three-layer channel router [CWL87]. The placement of the PRIMARY1 example is obtained from TimberWolfSC Version 5.1 [SS85], and the global routing is from [CSW89].

The algorithm 3CTM-V, on the average, produces a 73% reduction in channel height as compared to a conventional two-layer channel (2CRP) router, a 58% reduction as compared to a two-layer OTC (2OTC) router, and a 47% reduction as compared to a conventional three-layer channel (3CRP) router. For PRIMARY1, it produces a routing which is 65%, 51%, and 40% better than 2CRP, 2OTC, and 3CRP routers, respectively. The percentage improvements of 3CTM-V over 2CRP, 2OTC, and 3CRP for the channels of PRIMARY1 are listed in Table 7.3.

Table 7.3:   Experimental Results: CTM Router on PRIMARY1 (channel height)

Chan.	% of Vac.	No. of Tracks Produced				% Improvement by ICR-3		
No.	Term.	2CRP	2OTC	3CRP	ICR-3	2CRP	2OTC	3CRP
1	85	11	5	6	3	73	40	50
2	74	16	12	9	6	63	50	33
3	64	21	16	11	8	62	50	27
4	60	24	21	15	10	58	62	33
5	64	21	17	11	8	62	53	27
6	52	23	18	12	10	57	44	17
7	58	22	14	14	7	68	50	50
8	50	24	18	16	9	63	50	44
9	53	21	16	13	7	67	56	46
10	60	15	11	8	7	53	36	13
11	63	17	12	12	6	65	50	50
12	64	15	11	8	5	67	55	38
13	62	13	9	8	4	69	56	50
14	64	13	9	7	4	69	56	43
15	63	11	7	6	3	73	57	50
16	67	11	7	6	3	73	57	50
17	66	14	8	7	3	79	63	57
18	91	6	3	3	1	83	67	67
Total	-	298	214	172	104	65	51	40

The algorithm 3CTM-V is also very successful in reducing the number of vias in a layout. Via minimization is an inherent feature of the 3CTM-V routing model due to the planarity requirement for nets in the OTC routing region. Our OTC routing algorithm typically reduces the number of vias per routing by 56% as compared to the 3CRP router. For the PRIMARY1 benchmark, 3CTM-V produces a solution that is 50%, 28%, and 56% better than 2CRP, 2OTC, and 3CRP, respectively. Via minimization details for each channel of the PRIMARY1 example are shown in Table 7.4

Table 7.4:   Experimental Results for PRIMARY1—Via Minimization

Channel	% of Vacant	No. of Vias			
No.	Terminals	2CRP	2OTC	3CRP	3CTM-V
1	85	150	46	159	103
2	74	339	197	382	183
3	64	486	310	557	248
4	60	534	380	600	273
5	64	539	410	601	233
6	52	662	547	731	339
7	58	612	465	670	265
8	50	714	546	780	342
9	53	608	422	660	329
10	60	526	349	520	255
11	63	468	306	507	213
12	64	462	326	507	202
13	62	511	356	566	213
14	64	440	262	483	201
15	63	439	274	498	199
16	67	377	205	399	180
17	66	429	260	478	216
18	91	81	14	82	45
Total	-	8,146	5,615	9,180	4,039

Algorithm 3CTM-V not only produces excellent routings, it is also very efficient. For the entire PRIMARY1 chip, 3CTM-V only needs 5.8 seconds of running time. The highest density channel (channel no. 8) is routed in 0.4 seconds.

7.4.3 Performance Evaluation of the MTM Router

The MTM router presented in [BPS93b] generates layouts with smaller heights than the corresponding layouts generated by routers for the BTM and CTM. This is because the MTM router uses the OTC area more effectively. When tested on PRIMARY1, the router generated a layout with a channel height of 65 tracks. The router takes two minutes to complete the routing task for PRIMARY1 when tested on a SUN SPARC 1+ workstation.

7.4.4 Comparison of Three-Layer Routers

When vias are not allowed in OTC areas, the routing in OTC areas is required to be planar. For planar OTC routing, the BTM is most advantageous for routing large pairwise independent sets of nets, thus reducing the horizontal constraints, while the CTM is useful for eliminating vertical constraints by river routing. The MTM has been developed with an intermediate approach. It eliminates vertical constraints and minimizes horizontal constraints [BPS93a]. However, in all three models, OTC routing in the M2 layer is the same as in the two-layer process. The OTC routing in the M3 layer in all three models is identical. In the BTM, the terminals are available at the boundary, while the nets in the CTM and MTM are first routed to the boundary. Therefore, the M3 OTC routing problem is equivalent to the BTM-HCVC routing problem, except that the M3 layer is used in OTC areas and M1, M2, and M3 layers are used in the channels.

The best available routers for the BTM generate a layout with a total height of 71 tracks [HSS91a], the CTM router requires 104 tracks [WHSS92], and the MTM router generates a layout with a total height of 63 tracks [BPS93b].

7.5 Summary

In the three-layer process, when vias are not allowed in OTC areas, the routing in these areas is required to be planar. The OTC routing in the M3 layer can only be accomplished when the terminals are available at the boundary. Therefore, the OTC routing technique in the three-layer process for all cell models consists of the two basic steps given below:

1. **Routing in the M2 layer:** This is equivalent to the two-layer routing algorithms presented in Chapter 6 for various models.

2. **Routing in the M3 layer:** This is a common step for routers in all cell models. The routing problem consists of two equipotential terminal rows located on either side of the boundary and a completely unblocked M3 layer. This problem is equivalent to the BTM-HCVC routing problem solved in Chapter 6.

In the BTM, for the two-layer process, efficient OTC routing is accomplished by routing large sets of independent nets, and for the three-layer process, an additional set of independent nets can be routed in the M3 layer of the OTC areas.

In the CTM and MTM, the nets that are to be routed in the M3 layer are first routed to the boundary. Now the OTC routing problem is equivalent to the BTM-HCVC routing problem with an objective of channel density minimization.

PROBLEMS

7-1. The three-layer OTC routing problem for the BTM-HCVC requires determination of two independent sets, one to be routed in the M2 layer and the other in the M3 layer.

Does the order of independent set selection make a difference in the final solution? Show with an example the worst-case results of this selection order.

7-2. Consider the following algorithm for BTM-HCVC.

For $i = 1$ to k,
$$S_i^1 = i - \text{MIS}(T);$$
$$S_i^2 = (k - i) - \text{MIS}(B);$$
$$S = \max_{i=1}^K S_i$$

Is the solution S better than the solution obtained by the algorithm presented for HCVC in Chapter 6? Provide a proof or a counterexample for your assertion.

7-3. If the above problem is weighted, does this algorithm provide a better solution than the solution for the previous problem.

7-4. Solve the BTA problem such that river routing constraints are satisfied and vertical constraints in the resulting channel are minimized.

7-5. Solve the BTA problem such that river routing constraints are satisfied and the longest path in the VCG is minimized.

7-6. Consider the following improvement to CTM and MTM routing problems, which improves both the wire length and channel density. Currently, in the M2 river routing phase, all the nets are routed toward the boundary. However, not all nets need to be routed to the boundary, especially if they have adjacent terminals. Find a maximum set of nets that can be routed in the OTC area without going to the boundary. (Hint: Use a stack-based algorithm.)

7-7. Develop an efficient approximation algorithm for three terminal nets in the C area of the MTM. Also, develop net types.

7-8. Consider an MTM cell row with three equipotential terminal rows. List all the routing choices for a net. Develop an efficient algorithm for C area routing.

7-9. Develop ILP formulations for the MTM using the three-metal-layer process.

BIBLIOGRAPHIC NOTES

Several efficient algorithms have been presented for three-layer channel routers and OTC routing. Rivest and Fiduccia [RF82] present a greedy channel router. Gudmundsson and Ntafos [GN87] present channel routing with superterminals. Cong, Wong, and Liu [CWL87] present a new approach to three-layer channel routing problems. An OTC gate-array channel router is presented in [Kro83]. An integer programming technique is adopted for selection of nets to be routed in OTC areas by Lin et al. [LPHL91]. Sarrafzadeh and Lou [SL90] present maximum-weighted K-coverings in transitive graphs. Chiang, Sarrafzadeh, and Wong [CSW89] present a global router based on Steiner min-max trees. The circle graphs and independent sets are discussed in detail in Chapter 2.

8

Routing Algorithms for the Advanced Three-Layer Process

The research in process technology in the past two decades has resulted in a vital breakthrough from the OTC routing point of view. It is the *permissibility of vias* in OTC areas. A three-layer process that permits the usage of vias in OTC areas is referred to as the *advanced three-layer process*. This process allows routers to treat OTC areas as two-layer routing regions, for which the routing algorithms are well advanced. As a result, this process has led to significant reductions in layout heights. For example, for the three-layer process discussed in Chapter 7, the CTM router needs 104 tracks and the MTM router needs 63 tracks in the channels to complete the routing of the PRIMARY1 benchmark. However, in the advanced three-layer process, both the routers generate channel-less layouts for PRIMARY1 [WHSS92, BPS93a]. As discussed in Chapter 1, the current research trend is to minimize the feature size while maximizing the chip size. This results in the increase in the transistor densities and in the number of interconnects. As the number of interconnections increases, the routers must be given the flexibility to switch layers in OTC areas so as to decrease the channel heights or to eliminate channels. With this flexibility, it is possible to achieve a zero-routing footprint (i.e., 100% silicon efficiency) for most of the existing circuits [BPS93a].

The routing techniques for the advanced three-layer process are radically different than the routing techniques discussed in Chapter 7. The planarity of the routing solution, which is a major concern in both two-layer and three-layer processes is of no concern here. The routing environment in the advanced three-layer process is similar to that of channel routing, except that we have two layers for routing in OTC areas and three layers in the channel. Several OTC routers have been presented for this process [SH93, BPS93a, WHSS92, FMMY92, KBPS94, TTNS91].

In the advanced three-layer process, since vias are allowed in OTC areas, the channel heights are significantly reduced, and for most of the current designs, channels are completely eliminated. Therefore, in order to further minimize the die size, the width of a layout has to be minimized. While the BTM was specifically designed for channel routing, TBCs are developed exclusively for the advanced three-layer process. The TBC designs provide the flexibility to the cell designers in designing cells that have minimum cell widths (see Chapter 4) and to the router to locate the terminals on a target. Thus, layout width is minimized due to reduction of cell widths, and layout height is minimized due to elimination of channels. Hence, TBC designs tend to produce minimum area layouts.

In this chapter, we present routers based on various cell models for the advanced three-layer process. First, we present an advanced three-layer router for the BTM and then the routers for the CTM-, MTM-, and TBC-based designs.

8.1 Routing Algorithm for the BTM

Strunk and Holmes [SH93] presented an advanced three-layer router called VICTOR for BTM-based designs. The algorithm uses the concept of vacant terminals and abutments discussed in Chapter 7. We start with an overview of the algorithm and then discuss the key steps in detail.

8.1.1 Algorithm Overview

The algorithm VICTOR is based on the idea of maximizing track utilization in the OTC routing region. Vacant terminals and abutments are used to increase the number of nets routed over the cell rows. The five basic steps of VICTOR are summarized as follows:

1. **Channel Decomposition and Net-Type Assignment:** The channel is decomposed into a series of blocks and free zones. The block/free zone model provides the means for efficient utilization of vacant abutment columns in the channel. After channel decomposition, each multiterminal net is decomposed into a series of two-terminal nets, and the resulting net list is analyzed. Each net is categorized as one of eight types, depending on the location of its terminals with respect to the blocks in the channel decomposition. Each net is assigned a weight, or criticality factor, which, intuitively, measures the reduction in channel height possible if this net is routed over the cell rows. Nets with high weights offer the largest potential for reduction in channel height; so it is desirable to maximize the total weight of the nets routed over the cells. In this phase, there are some steps that also determine if the given net list has any vertical constraints. If

there are no vertical constraints, a modified LEA is used to obtain optimal routing.

2. **Net Pairing:** Nets that may share OTC routing tracks are paired in a manner that maximizes track utilization in the area over the cell rows. The net pairing problem is modeled as a bipartite graph and solved using a minimum-weight matching algorithm. The net pairing is extended to form net families using a greedy algorithm to increase the utilization of the OTC tracks.

3. **Free Zone Assignment:** Each free zone in the channel decomposition is assigned a subset of net families (net pairs with some associated nets). The free zone assignment operation is also solved using a minimum-weight matching algorithm for bipartite graphs. Given a free zone assignment $\mathcal{A}$ in $\Theta(n)$, it can be determined if the assignment is routable in the region over the cell rows. If $\mathcal{A}$ is not routable, a maximum-weighted routable subset is determined in $O(k|\mathcal{F}|^2)$ time, where $\mathcal{F}$ is the set of net families in $\mathcal{A}$, and k is the number of tracks available for OTC routing.

4. **OTC Routing:** In this phase, net families are assigned tracks in OTC regions. Routing proceeds from left to right in a zonewise fashion. The track utilization is further increased by assigning additional nets if some track space is available in OTC regions.

5. **Channel Routing:** In some cases, not all nets may be routed over the cell rows. In this case, a channel router is used to complete the layout [CWL87].

Detailed descriptions of the steps in the algorithm VICTOR are provided in the following sections. The terminology used in VICTOR is the same as that in Chapter 6. In addition, we use the following terms. For a given interval I_i, LEFT(I_i) refers the left end point of the interval; similarly, RIGHT(I_i) refers to its right end point. LEN(I_i) is used to denote the length of the interval. If I_i is assigned to a region, then REGION(I_i) is used to refer to the region (either top or bottom) to which I_i is assigned.

8.1.2 Channel Decomposition and Net-Type Assignment

The algorithm VICTOR is based on decomposition of the channel using a *block/free zone model*. The channel is represented as a series of blocks and free zones defined by the location of vacant abutments in the channel. To start with, some definitions are presented.

A *block* $b_i = (c_p, c_q)$ is a set of continuous columns containing no vacant abutments; in other words, all columns c_i, $p \leq i \leq q$ have at least one terminal assigned to a net. For example, the channel in Figure 8.1 contains five blocks. The regions separating the blocks are called *free zones* and are composed strictly of vacant abutments. Therefore, a free zone $z_i = (c_r, c_s)$, such that every column

$c_j, r \leq j \leq s$ is an abutment. Note by definition that $s \geq r$. The capacity of a free zone $z_i = (c_r, c_s)$, denoted by $c(z_i)$, is equal to the number of vacant abutments within the zone; in other words, $c(z_i) = s - r + 1$. Furthermore, the function BLOCK(t_i) returns the block to which the terminal belongs.

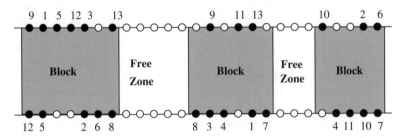

Fig. 8.1   Block and free zone decomposition.

After decomposing the channel, the nets of the net list are classified so that the "best" subset of nets is chosen to route over the cell rows. Net list classification is considered in two main phases: net categorization according to terminal location and net weighting. All m-terminal nets are decomposed into exactly $m - 1$ two-terminal nets at adjacent terminal locations. For example, a six-terminal net, $N_i = \{t_{i_1}, t_{i_2}, t_{i_3}, t_{i_4}, t_{i_5}, t_{i_6}\}$, is decomposed into five two-terminal nets: $N_{i1} = \{t_{i_1}, t_{i_2}\}$, $N_{i2} = \{t_{i_2}, t_{i_3}\}$, $N_{i3} = \{t_{i_3}, t_{i_4}\}$, $N_{i4} = \{t_{i_4}, t_{i_5}\}$, and $N_{i5} = \{t_{i_5}, t_{i_6}\}$. Clearly, $m - 1$ two-terminal nets are sufficient to preserve the connectivity of the original m-terminal net. Each net N_i of $\mathcal{N}$ can be broadly categorized as either an interblock net or an intrablock net. A net $N_j \in \mathcal{N}$ is called an *interblock net* if and only if BLOCK(t_{i_1}) $\neq$ BLOCK(t_{i_2}). On the other hand, a net $N_j \in \mathcal{N}$ is called an *intrablock net* if and only if BLOCK(t_{i_1}) $=$ BLOCK(t_{i_2}).

Both interblock and intrablock nets may be partitioned into five categories, $S^{tb}, S^{bt}, S^{tt}, S^{bb}, and S^{aa}$.

$$S^{tb} = \{N_i = (t_{i_1}, t_{i_2}) \mid \text{ROW}(t_{i_1}) = R_i \text{ and ROW}(t_{i_2}) = R_{i+1}\}$$
$$S^{bt} = \{N_i = (t_{i_1}, t_{i_2}) \mid \text{ROW}(t_{i_1}) = R_{i+1} \text{ and ROW}(t_{i_2}) = R_i\}$$
$$S^{tt} = \{N_i = (t_{i_1}, t_{i_2}) \mid \text{ROW}(t_{i_1}) = R_i \text{ and ROW}(t_{i_2}) = R_i\}$$
$$S^{bb} = \{N_i = (t_{i_1}, t_{i_2}) \mid \text{ROW}(t_{i_1}) = R_{i+1} \text{ and ROW}(t_{i_2}) = R_{i+1}\}$$
$$S^{aa} = \{N_i = (t_{i_1}, t_{i_2}) \mid \text{OPP}(t_{i_1}) \text{ and OPP}(t_{i_2}) \text{ are vacant}\}$$

To distinguish interblock sets from intrablock sets, a subscript is added to the notation. For example, S_E^{tb} denotes an interblock net set and S_I^{tb} represents an intrablock net set. Examples of these net types appear in Figure 8.2. For example, N_{13} is an S_E^{tt} net, N_3 is an S_E^{tb} net, and N_5 is an S_I^{bt} net.

In order to select nets for OTC routing, the effect of removing that net needs to be considered on the given channel routing problem. Weights are assigned to the nets in a similar fashion as that described in Chapter 6. The weights of nets will

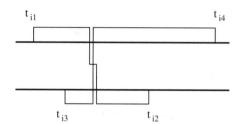

Fig. 8.2   An example of a net family.

be used in several phases in the algorithm VICTOR, including net pairing phase, net family formation phase, and the OTC routing cleanup phase.

8.1.3 Algorithms for Efficient Net Pairing

In order to optimally use the tracks in OTC regions, VICTOR pairs up compatible nets. This section discusses an algorithm for pairing nets. Net pairing is an important operation, since it is the key to maximization of track utilization.

Consider the routing options for each of the nets in sets S_I^{aa}, S_E^{aa}, S_I^{tt}, S_E^{tt}, S_I^{bb}, S_E^{bb}, S_I^{tb}, S_E^{tb}, S_I^{bt}, and S_E^{bt}. A net in set S_I^{aa} or S_E^{aa} can be routed in either the top or bottom OTC region, while a net in set S_I^{tt} or S_E^{tt} may only be routed in the top OTC region. Similarly, a net in set S_I^{bb} or S_E^{bb} may only be routed in the bottom OTC region. Routing of all these types of nets is rather straightforward. However, nets in sets S_E^{tb} and S_E^{bt} present some problems. First of all, these nets require a vertical abutment. Furthermore, they cause a blockage for other nets, since a track in the bottom OTC region and another track in the top OTC region are blocked. However, the effect of this blockage can be minimized if a net from set S_E^{tb} is paired up with a net in set S_E^{bt}. The basic idea is to route this pair in the same track in both OTC regions; as a result, both tracks are maximally utilized.

$N_{ij} = (N_i, N_j)$ is referred to as a *net pair*, where $N_i \in S_E^{tb}$, $N_j \in S_E^{bt}$, $N_i = (t_{i_1}, t_{i_2})$, and $N_j = (t_{i_3}, t_{i_4})$. The main considerations in net pairing is length of nets N_i and N_j, and the respective locations of the two nets. If $\text{COL}(t_{i_1}) = \text{COL}(t_{i_3})$ and $\text{COL}(t_{i_2}) = \text{COL}(t_{i_4})$, then N_{ij} is called a *perfect net pair*. It is easy to see that a set of perfect net pairs can be computed in $O(n)$, where n is the total number of nets in sets S_E^{tb} and S_E^{bt}. However, such a set is likely to be very short and does not serve the purpose of routing. If $\text{COL}(t_{i_1}) \neq \text{COL}(t_{i_3})$ and/or $\text{COL}(t_{i_2}) \neq \text{COL}(t_{i_4})$, an offset is introduced due to the nonaligned positions of N_i and N_j. This offset between N_i and N_j indicates the amount of unused track if N_i and N_j are selected to form a pair. Thus, the objective is to select net pairs with minimum offset.

Abstractly, a net can be considered as an interval bounded by its left and right terminals. Thus, $N_i = (t_{i_1}, t_{i_2})$ can be represented by an interval I_i bounded by $[\text{COL}(t_{i_1}), \text{COL}(t_{i_2})]$. By definition of a net pair N_{ij}, $I_i \cap I_j \neq 0$. Let I_{ij} be the union of intervals I_i and I_j. Then $I_{ij} - \{I_i \cap I_j\}$ is a set of two intervals I_{ij}^l and I_{ij}^r, where superscripts l and r refer to the relative side of the overlap interval. Given a netlist $\mathcal{N}$, the minimum offset net pairing (MONP) problem is the problem of assigning the nets to pairs such that:

p1) Each net N_i belongs to exactly one net pair.

p2) $\mathcal{O} = \sum_{\forall N_{ij}} O_{ij}$, where $O_{ij} = \max\{\text{LEN}(I_{ij}^l), \text{LEN}(I_{ij}^r)\}$ should be minimum.

Theorem 29   *MONP can be optimally solved in $O(n^{2.5})$ time.*

The basic idea of net pairing is to match the track utilization in a top and bottom track. In fact, if a net pair is perfect, then both the tracks are equally utilized. The net pairing obtained by the matching may not utilize both OTC tracks completely. It is easy to see that the potential unused track space (caused by mismatch of lengths of nets and their position in all the net pairs) is given by

$$S = \sum_{N_{ij} \in \mathcal{N}, s=(l,r)} \text{LEN}(I_{ij}^s)$$

In order to increase the utilization of the tracks, a few additional nets are assigned to each net pair. The net pair N_{ij} and its associated nets form a *net family* $F_{ij} = \{N_{ij}, N_1, N_2, \ldots, N_q\}$, where $N_q, 1 \le q \le m$ are the nets assigned to net pair N_{ij}. The problem of selecting additional nets for a net pair translates into an interval assignment problem (IAP) as described in the next section.

In order to properly assign additional nets to form net families, it should be noted that not all types of nets can be assigned to both top and bottom regions. Nets in S_I^{tt} and S_E^{tt} can only be assigned to the top OTC region, while nets in S_I^{bb} and S_E^{bb} can only be assigned to the bottom OTC region. Nets in S_I^{aa} and S_E^{aa} can be assigned to both top and bottom regions. In order to assign proper nets to I_{ij}^s, the region in which I_{ij}^s lies must be determined. Given a net pair $N_{ij} = (N_i, N_j)$, such that net $N_i \in S^{tb}$ and $N_j \in S^{bt}$, and $N_i = (t_{i_1}, t_{i_2})$ and $N_j = (t_{i_3}, t_{i_4})$,

$$\text{REGION}(I_{ij}^l) = \left\{ \begin{array}{ll} \text{top} & \text{if COL}(t_{i_3}) < \text{COL}(t_{i_1}) \\ \text{bottom} & \text{if COL}(t_{i_1}) < \text{COL}(t_{i_3}) \end{array} \right\}$$

$$\text{REGION}(I_{ij}^r) = \left\{ \begin{array}{ll} \text{top} & \text{if COL}(t_{i_2}) > \text{COL}(t_{i_4}) \\ \text{bottom} & \text{if COL}(t_4) > \text{COL}(t_2) \end{array} \right\}$$

Note that if $\text{COL}(t_{i_1}) = \text{COL}(t_{i_3})$, then $\text{LEN}(I_{ij}^l) = 0$. Similarly, if $\text{COL}(t_{i_2}) = \text{COL}(t_{i_4})$, then $\text{LEN}(I_{ij}^r) = 0$.

In order to formally define the IAP, the first two sets of intervals are defined. Let the set of the intervals formed due to net pairing be denoted by $\mathcal{I}$. More precisely,

$$\mathcal{I} = \{I_{ij}^s | N_{ij} \in \mathcal{N}, s = (l, r)\}$$

Also let the nets in sets S_I^{tt}, S_E^{tt}, S_I^{bb}, S_E^{bb}, S_I^{aa}, and S_E^{aa} each be assigned an interval J_i, and let that set of such intervals be called $\mathcal{J}$.

$$\mathcal{J} = \{J_i | J_i \text{ represents } n_i \in S_I^{tt}, S_E^{tt}, S_I^{bb}, S_E^{bb}, S_I^{aa}, \text{ or } S_E^{aa}\}$$

The objective of the IAP is to minimize unused space. More precisely, given $\mathcal{I}$ and $\mathcal{J}$, the IAP is the problem of assigning intervals from $\mathcal{J}$ to intervals in $\mathcal{I}$, such that

p1) Each interval $\mathcal{J}$ assigned to an interval in $\mathcal{I}$ is pairwise independent.

p2) An interval $J_j \in \mathcal{J}$ is assigned to I_{ij}^s if and only if REGION(I_{ij}^s) = top and $N_j \in S_I^{tt}$, S_E^{tt}, S_I^{aa}, S_E^{aa} or REGION(I_{ij}^s) = bottom, and $n_j \in S_I^{bb}$, S_E^{bb} S_I^{aa}, S_E^{aa}.

p3) The sum of the length of assigned intervals is maximum.

Unfortunately, it appears that the IAP may be a hard problem from a computational point of view, as stated in the theorem below.

Theorem 30   *The interval assignment problem is NP-complete.*

In view of the NP-completeness of the IAP, a greedy approach is used to solve the problem of assigning additional nets to net families. In this section, the approach and its complexity is discussed in detail.

The set $\mathcal{J}$ is used to form two sets of intervals, T and B. Sets T and B are subsets of $\mathcal{J}$, such that each interval in T and B can be assigned to a track in the top and bottom regions, respectively. Notice that set T and B are not disjoint. For an interval I_{ij}^s in $\mathcal{I}$, with REGION(I_{ij}^s) = top, a subset S' of T is constructed. The subset S' consists of only those intervals that lie completely within the interval [LEFT(I_{ij}^s), RIGHT(I_{ij}^s)]. The algorithm presented in [Gol80] is used to find an MWIS in S' and assign these intervals (nets) to I_{ij}^s (net pair N_{ij}). If REGION(I_{ij}^s) = bottom, then a subset of B is constructed. After the selection of intervals (nets) for I_{ij}^s, the selected intervals are deleted from both T and B, and this process of assignment is repeated for all intervals in $\mathcal{I}$. The formal algorithm is given in Figure 8.3.

The complexity of the algorithm MWIS is given by $O(|S'| \log |S'|)$. Since $|S'| = O(m)$, the complexity of the MWIS algorithm is $O(m \log m)$. Moreover, the MWIS algorithm is repeated twice for each net pair; thus, the overall complexity of the algorithm is $O(m^2 \log m)$. An example of a net family constructed by the algorithm given above is shown in Figure 8.4. The weight of a net pair $N_{ij} = (N_i, N_j)$ is given by $w(N_i) + w(N_j)$. Then the weight of net family $w(F_{ij})$ is given by

$$w(F_{ij}) = w(N_i) + w(N_j) + \sum_{N_q \in F_{ij}} w(N_q)$$

This weight will be used in the OTC routing phase of the algorithm for finding the best set of net families.

8.1.4 Free Zone Assignment and Routability

In order to optimize the use of abutments and OTC tracks, free zones need to be carefully allocated. Each net family F_{ij} needs at least two vacant abutments

Algorithm GREEDY_INTERVAL_ASSIGNMENT()

Input: $\mathcal{I}$ = set of intervals of net pairs

$\mathcal{J}$ = set on nets in sets S_I^{tt}, S_E^{tt}, S_I^{bb}, S_E^{bb}, S_I^{aa} or S_E^{aa}

Output: $\mathcal{J}'$ = subset of $\mathcal{J}$ assigned as per the Interval Assignment Problem

---

begin

/* Form T and B sets from $\mathcal{J}$ */

    for each $J_q \in \mathcal{J}$

        if (J_q) corresponds to a net in S_I^{aa}, S_E^{aa}

        then $T = T \cup J_q$

            $B = B \cup J_q$

        else if (J_q) corresponds to a net in S_I^{tt}, S_E^{tt}

            then $T = T \cup J_q$

        else if (J_q) corresponds to a net in S_I^{bb}, S_E^{bb}

            then $B = B \cup J_q$

/* Find a subset of intervals for each interval in $\mathcal{I}$ */

    for each $I_{ij}^s \in \mathcal{I}$

        $S' = \phi$

        if $REGION(I_{ij}^s) = top$ **then**

            for each interval $T_p \in T$

                if $LEFT(T_p) \geq LEFT(I_{ij}^s)$ and $RIGHT(T_P) \leq RIGHT(I_{ij}^s)$

                then $S' = S' \cup T_p$

        else

            for each interval $B_p \in B$

                if $LEFT(B_p) \geq LEFT(I_{ij}^s)$ and $RIGHT(B_P) \leq RIGHT(I_{ij}^s)$

                then $S' = S' \cup B_p$

        M = MAX_WEIGHTED_IND_SET(S')

        $F_{ij} = F_{ij} \cup M$

        $T = T - M$

        $B = B - M$

end.

Fig. 8.3 Algorithm GR-INTERVAL_ASSIGNMENT.

in a zone that lies within I_{ij}. A net family spanning several zones has a choice of several free zones. On the other hand, net families with relatively short intervals may compete for a particular free zone. Thus, an optimal assignment of net families to free zone ensures optimal use of free zones and maximization of the OTC track utilization. This section describes the algorithm used for assigning net families to free zones.

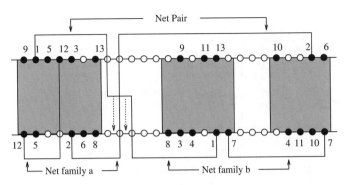

Fig. 8.4 Net family.

There are three factors that need to be considered in free zone assignment. First, the net families must be assigned in such a way that the capacity of free zone $c(z_i)$ is not exceeded. Second, the vacant terminals that may be available in the column adjacent to the free zone must be fully utilized. Third, the track blockages both in columns and horizontal tracks must be minimized. Track blockages are caused by the routing of a net family if routing on the boundary of the channel is not possible, as explained below. An upper bound is also presented on the number of net families that can be assigned to a free zone.

Let the maximum number of net families that can be assigned to a free zone z_i be denoted by $f(z_i)$. Also, let $vt(z_i)$ denote the number of vacant terminals in the columns adjacent to z_i. It is clear that $vt(z_i) \in \{0, 1, 2\}$ for all z_i. Consider the routing of a net family as shown in Figure 8.4. It can been seen that three columns are needed to complete this routing. However, if $c(z_i) = 2$, then routing is still possible if one of the terminals in a column adjacent to z_i is vacant; that is, $vt(z_i) \geq 1$. This can be done by routing a short segment on the boundary using the vacant terminal as shown in Figure 8.5(a). If $vt(z_i) = 0$, then routing is possible by creating a track blockage in an OTC track. Figure 8.5(b) shows an example of track blockage. Let $tb(z_i)$ be the number of track blockages allowed for a free zone z_i.

Theorem 31 *The maximum number of net families that can be assigned to a free zone z_i is given by*

$$
f(z_i) = \begin{cases} \lfloor \frac{c(z_i) - 2vt(z_i)}{3} \rfloor + vt(z_i) & \text{if } tb(z_i) = 0 \\[2mm] \lfloor \frac{c(z_i) - 2\{vt(z_i) + tb(z_i)\}}{3} \rfloor + vt(z_i) + tb(z_i) & \text{otherwise} \end{cases}
$$

Proof: The proof follows directly from the observation that only two columns are needed to route a net pair if an adjacent vacant terminal is used. Similarly, for each track blockage allowed, another net pair can be routed using only two columns. All the remaining net pairs need three columns each to complete the routing.   □

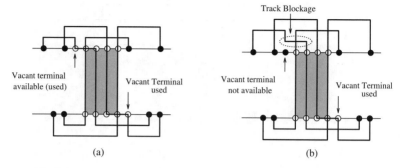

Fig. 8.5   Effect of vacant terminals and track blockage on zone assignment.

Given $\mathcal{F}$ and $\mathcal{Z}$, the free zone asssignment problem (FZAP) is the problem of assigning net family $\mathcal{F}$ to a free zone $\mathcal{Z}$ such that:

p1) In each net family $F_{ij} \in \mathcal{F}$ is assigned to exactly one free zone $z_i \in \mathcal{Z}$.

p2) For each free zone $z_i \in \mathcal{Z}$, the number of net families assigned to the free zone is less than or equal to $f(z_i)$.

Theorem 32   *The FZAP can be optimally solved in $O(n^{2.5})$ time.*

Proof: The relationship between net families and free zones is represented by a bipartite graph $G_{fz} = (V_f \cup V_z, E_{fz})$. A vertex v_{ij} of V_f represents a net family F_{ij}. The set $V_z = V_{z_1} \cup V_{z_2} \ldots \cup V_{z_y}$, where y is the total number of free zones in the given channel. Each V_{z_i} consists of $f(z_i)$ vertices. Let I_x^z be the interval defined by the right-most and left-most abutments in zone z_x. The edge set E_{fz} is defined as follows. An edge is drawn between $v_{ij} \in V_p$ and $v_x \in V_{z_i}$ if and only if the interval I_{ij} corresponding to the net family F_{ij} properly contains I_x^z. All edges incident on $v_{ij} \in V_f$ are assigned the weight of the net family F_{ij} corresponding to the vertex v_{ij}. An optimal free zone assignment can be determined by finding a maximum-weighted matching in the graph G_{fz}. Note that at most $f(z_i)$ net families can be assigned to a free zone. Thus, no free zone capacities are violated. Similarly, notice that each family is assigned to exactly one free zone. The maximum-weighted matching problem can be solved optimally in $O(n^{2.5})$ time [HK73]. Therefore, it can be concluded that the FZAP can also be solved in $O(n^{2.5})$ time.

$\square$

It should be noted that $f(z_i) = 0$ for free zones, except z_1. The $c(z_1) = 5$ and $vt(z_1) = 1$; therefore, $F(z_1) = \frac{5-2}{3} + 1 = 2$. Many fast and efficient heuristics for matching are available as well [GGS89, GMG86, Tar83] and may be used if a faster but suboptimal solution for the FZAP is acceptable.

Given a free zone assignment, it is necessary to determine whether the assignment is routable in the region over the cell rows. Recall that each OTC routing region has k available tracks. Since many net families are competing for a fixed number of tracks in OTC regions, a free zone assignment may not always be routable. If free zone assignment is not routable, a maximum-weighted routable subset of the free zone assignment is computed. In other words, given $\mathcal{F}$ and $\mathcal{Z}$, the maximum-weighted routable subset problem (MWRSP) is the problem of selection of a maximum-weighted routable subset of $\mathcal{F}$.

Theorem 33   *Given the set $\mathcal{F}$ of net families, MWRSP can be solved in $O(k|\mathcal{F}|^2)$ time, where k is the number of tracks available for OTC routing.*

Proof: A vertex weighted interval graph $G_I = (V_I, E_I)$ is constructed, where each vertex $v_{ij} \in V_I$ represents the net family F_{ij}, and an edge is defined between two vertices v_{ij} and v_{xy} if two intervals I_{ij} and I_{xy} intersect. Each vertex v_{ij} is assigned the weight $w(F_{ij})$. It is easy to see that optimal solution to the MWRSP is equivalent to choosing a maximum-weighted k-density independent set in G_I, which can be computed in $O(k|\mathcal{F}|^2)$ time [SL90]. ☐

It should be noted that the theorem above in fact produces a track assignment. The actual routing of the net families will be discussed in the next section.

8.1.5 Over-the-Cell Routing

This section describes the OTC routing algorithms. The routing proceeds in a left-to-right fashion by completing OTC track routing for each zone z_i, $i = 1$ to $maxzone$. The track assignment is produced by the algorithm presented in the previous section. The actual routing of a zone z_i depends on the value of $vt(z_i)$ as explained below.

The routing of each net family requires two tracks, one in the upper OTC region and one in the lower OTC region. However, due to the symmetry, the net family is assigned to one region and its mirror image is used for track assignment in the other region. After track assignment, three basic templates (shown in Figure 8.6) are used to complete the routing in the free zones, depending on the value of $vt(z_i)$. It should be noted that routing may be done on the channel boundaries in the free zones, since no terminal in a free zone is required for interconnection. Boundary routing is used to ensure that each net family $\mathcal{F}_{ij}$ may be routed over the cells using exactly one track in the top OTC routing region and one track in the bottom OTC routing region. The templates for each free zone are combined to obtain the completed OTC routing in linear time.

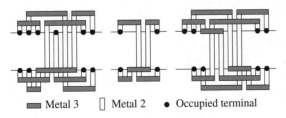

━━ Metal 3   ▯ Metal 2   ● Occupied terminal

Fig. 8.6   Templates for routing.

After the selected net families have been routed according to free zone assignment, the remaining nets are selected and routed over the cells in a cleanup phase. During this phase, nets are selected for OTC routing in a greedy fashion, and vacant terminals are used to increase the number of nets routed in OTC regions.

After OTC routing is completed, the remaining net segments, along with any "direct" vertical segments constructed using vacant terminals and abutments for OTC routing, are routed in the channel using a conventional three-layer channel router [CWL87]. A formal statement of the algorithm VICTOR is given in Figures 8.7 and 8.8.

Algorithm VICTOR()

Input: $\mathcal{N} = \{N_1, N_2, \ldots, N_n\}$ is a set of nets where each net

$N_j = \{t_i \mid ROW(t_i) \in \{R_i, R_{i+1}\}, 1 \leq i \leq$ num_of_columns$\}$

/* Each t_i is a terminal of N_j lying either in R_i or R_{i+1}. */

Output: Over-the-cell routing.

---

begin

PHASE 1(a): Channel Decomposition

/* Each zone/block is defined by its left and right column numbers */

PHASE 1(b): Net Decomposition

PHASE 1(c): Net Type Assignment($\mathcal{N}$, S_I^{aa}, S_E^{aa}, S_I^{tb}, S_E^{tb}, S_I^{bt},

S_E^{bt}, S_I^{tt}, S_E^{tt}, S_I^{tt}, S_E^{bb})

PHASE 1(d): Formation of Constraint Graphs and Net Weight Assignment

G_v = FORM_VC_GRAPH($\mathcal{N}$)

G_h = FORM_HC_GRAPH($\mathcal{N}$)

if $|E_v| = 0$ **then goto** Phase 5(b)

/* Assign weights to the nets */

for each $N_i \in \mathcal{N}$ **do**

$w(N_i) = \frac{h_{max}}{v_{max}} r_d(N_i) + \frac{v_{max}}{h_{max}} r_p(N_i)$

PHASE 2: Net Pairing and formation of Net Families

G_p = FORM_GP_GRAPH($\mathcal{N}$)

$\mathcal{N}_{\Updownarrow}$ = MATCHING(G_p)

FORM $\mathcal{I}$ and $\mathcal{J}$

$\mathcal{F}$ = GREEDY_INTERVAL_ASSIGNMENT ($\mathcal{I}$, $\mathcal{J}$)

PHASE 3: Free Zone Assignment for Net Families

/* $\mathcal{A}$ contains the free zone assignment and track assignment */

G_{fz} = FORM_GFZ_GRAPH($\mathcal{F}$, $\mathcal{Z}$)

$\mathcal{A}$ = MATCHING(G_{fz})

if NOT_ROUTABLE ($\mathcal{A}$)

then $\mathcal{A}'$ = MAX_WT_ROUTABLE_SUBSET($\mathcal{A}$, k)

Fig. 8.7   Algorithm VICTOR.

Algorithm VICTOR() (Contd.)

PHASE 4(a): Over-The-Cell Routing Using Templates
    for $i = 1$ *to maxzone*
        $vt(z_i) = \text{COMPUTE\_ADJ\_VAC\_TERM}(z_i)$
    for $i = 1$ to *maxzone*
        if $(vt(z_i) = 0)$
        then APPLY_TEMPLATE$(0, z_i, \mathcal{A})$
        else if $(vt(z_i) = 1)$
            then APPLY_TEMPLATE$(1, z_i, \mathcal{A})$
            else APPLY_TEMPLATE$(2, z_i, \mathcal{A})$
PHASE 4(b): Greedy Clean up Over-The-Cell Routing
    /* Let $\mathcal{N}'$ be the the set of net segments not routed in PHASE 4(a) */
    for each $N_j \in \mathcal{N}'$ do
        for each $t = 1$ *to* 26 do /* Try top/bottom OTC tracks */
            if EMPTY_TRACK(t, I_j)
            then ASSIGN_GREEDY(t, N_j); $\mathcal{N}' = \mathcal{N}' - N_j$
PHASE 5(a): Channel Segment Assignment and Channel Routing
    $\mathcal{CA} = \text{ASSIGN\_CHANNEL\_SEGMENTS}(\mathcal{N}')$
    CHANNEL_ROUTE$(\mathcal{CA})$
    goto end
PHASE 5(b): OTC and Channel routing with no vertical constraints
    MOD_LEFT_EDGE_ALGORITHM$(\mathcal{N})$
end.

Fig. 8.8   Algorithm VICTOR (contd.)

The complexity of the algorithm is determined by the complexity of the matching algorithms; thus, the algorithm VICTOR runs in $O(n^{2.5})$ time. As pointed out earlier, suboptimal heuristic algorithms for matching can be used to improve the time complexity of the algorithm.

8.2 Routing Algorithms for the CTM

The OTC routing problem in the CTM for the advanced three-layer process is a classical NLCR problem. Since vias are allowed in 3CTM+V, unlike 3CTM-V, it is not necessary to select pairwise independent nets in the OTC area. In this section, we present two algorithms: a conventional CTM routing algorithm, in which the terminals are assumed to be aligned, and an algorithm for CTM designs, which have nonaligned terminals located around a horizontal center line in the cell.

8.2.1 Conventional CTM Routing Algorithm

As a matter of fact, the problem of OTC routing in 3CTM+V is much simpler and need not be considered independent of channel routing. The area between terminal rows of two adjacent cell rows is the routing area. The routing area consists of two fixed-height OTC regions, one between the upper terminal row and the top boundary (T area) and another between the lower terminal row and the bottom boundary (B area), and a channel between them. (See Figure 8.9.) The following changes can be made in a greedy routing algorithm (see Chapter 3) to complete routing in such a routing environment:

1. Use the HV routing model in OTC areas, but use the HVH model in channel areas.

2. Allocate a track in the channel areas area only when no track is available in T or B areas.

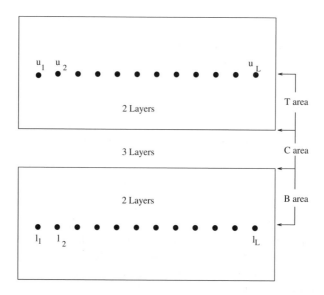

Fig. 8.9 Routing regions in 3CTM+V.

8.2.2 OCTM Routing Algorithm

Fujii, Mima, Matsuda, and Yoshimura [FMMY92] presented a new style of OTC routing. The terminals are assumed to be located around a horizontal center line in the cell. The region between the central terminals in the upper cell row and the ones in the lower cell row is an *expanded channel*. A new channel router handles that region and where connections to terminals are achieved by avoiding the obstacles. It is assumed that four layers, poly, M1, M2, and M3, are available for routing. Poly and M2 layers are used for vertical routing segments, while M1 and M3 are used for horizontal routing. The cell structure is shown in Figure 8.10. External obstacles may exist just outside the upper and lower cell boundaries. The shape is regarded as a rectilinear polygon. The terminals are located not only on the boundary of the external obstacles, but also around the horizontal center line. Equipotential terminals are only allowed to be on the same vertical column. The terminals around the central line are referred to as *the central terminals*. The terminals in the poly layer exist only on the boundary. The central terminals are in the M2 layer.

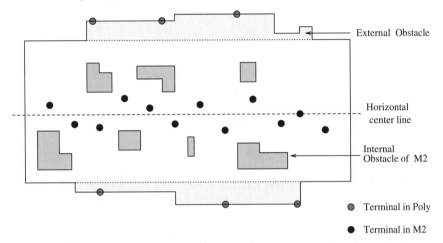

Fig. 8.10   New cell structure presented in [FMMY92].

On the basis of the cell model, a new routing region is considered, as shown in Figure 8.11. This region is an expanded channel. It is defined to be a maximal rectilinear polygon surrounded by the central terminals and horizontal center lines in the upper and lower cell rows. It is assumed that in the OTC region, only the wires of M2 and M3 can be used. The routing problem for the expanded channel is different from conventional channel routing problems due to the following:

1. There may exist obstacles in the routing region.
2. The shape of the expanded channel is not rectangular, since the central terminals are not aligned horizontally.
3. M1 and M3 may be used by horizontal wires in the conventional channel; however, only M3 can be used in the OTC region.
4. Poly terminals are located inside the expanded channel.
5. The grid interval of M3 need not be equal to 1 in the conventional channel.

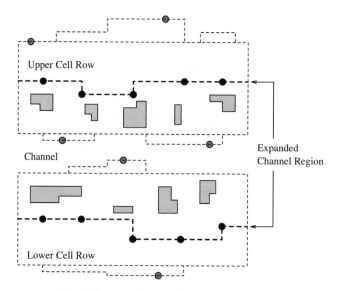

Fig. 8.11 Expanded channel of OCTM.

8.2.2.1 Algorithm overview

The presence of obstacles in OTC regions makes the connections to central terminals difficult. For this problem, the router presented in [FMMY92] finds points where wires from central terminals can virtually pass to the outlines of the cells, avoiding all obstacles. The router has the following four steps:

1. **Specification of the Channel:** The expanded channel is defined based on the cell data, such as the locations of the obstacles and terminals.

2. **Routing from Central Terminals:** In this step, the connections to central terminals are performed avoiding obstacles in cells. In order to completely perform the routing from each central terminal, the following two processes are required. The cell column on which a central terminal t is located is denoted by $COL(t)$.

    (a) **Selection of Drawing Direction:** If $COL(t)$ intersects an obstacle P for a central terminal t, then we must determine the direction (left or right) to which the routing from t detours, and avoids the obstacle P.

    (b) **Selection of Cell Track:** We must determine a cell track on which the detour runs in order to avoid the obstacle, taking account of the shape of the obstacle.

    For example, consider the central terminals u_1, u_2, and u_3 in Figure 8.12. The detour for terminals u_1, u_2, and u_3 will run along cell tracks T_2, T_1, T_1, respectively. The direction of routing from u_1 and u_2 is left while the direction of routing from u_3 is right.

(c) **Changing Poly Terminals to Terminals in M2:** The nets in the poly layer connecting to the poly terminals are routed to the M2 layer for as many poly terminals as possible.

(d) **Track Assignment:** According to vertical constraints, horizontal wires of subnets are assigned to tracks M1 and M3. The algorithm for track assignment based on [Yos84] is modified to take into account 3.5-layer routing.

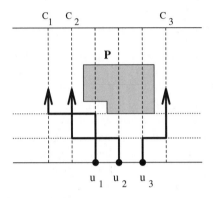

Fig. 8.12   An example of routing from central terminals.

Consider the routing example shown in Figure 8.13(a). The cell track T_1 is broken by central terminal u_1. Thus, the detours from central terminals u_2 and u_3 to the left side of P cannot be achieved by using T_1. For u_2, two cell tracks T_2 and T_3 are listed up. Four cell columns c_1 to c_4 intersect T_2 in the left side of P. Since

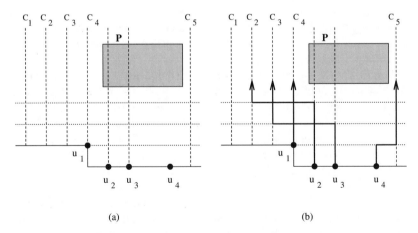

Fig. 8.13   Process of routing from central terminals; **(a)** an example; **(b)** a routing result.

$COL(u_1)$ is c_4, three cell columns c_1 to c_3 are considered as candidates for the cell columns used to detouring from u_2. For u_2, c_3 is chosen, since c_3 is nearest to P. In a similar way, for u_3, T_3 and c_2 are chosen. As shown in Figure 8.13(b), the routing from u_4 is achieved by using T_1 and c_5 on the right side of P. The experimental results have proven that the OCTM routing algorithm is effective in reducing the die size.

In the next section, we present routing algorithms for MTM-based designs.

8.3 Routing Algorithms for the MTM

In this section, we discuss the MTM+V router presented in [BPS93a]. Since vias are allowed over the cell in MTM+V, the overall problem becomes similar to the channel routing problem with two layers available over the cell and three layers available in the channel area. The presence of two terminal rows in each cell row gives rise to a problem of terminal row selection in order to minimize the overall channel height. Figure 8.14(a) shows an example routed by using a two-layer conventional channel router. An overall channel height of 168 tracks was obtained for this example. Figure 8.14(b) shows that an overall height of 8 tracks is obtained when terminal row assignment is done in a sequential manner from top to bottom, as in Figure 8.14(a). Employing the *high-density first* strategy for terminal row assignment to the same example results in an overall channel height of 10, as shown in Figure 8.14(c). However, Figure 8.14(d) shows an optimal terminal row assignment giving an overall channel height of 6. In Figures 8.14(a), (b), (c), and (d), the numbers between the two adjacent cell rows indicate the channel height in number of tracks, whereas the number at the bottom indicates the overall channel height.

The remaining part of this section addresses the terminal row assignment problem in detail. First, the required terminology is developed, and then the problem is transformed to a well-known shortest path problem related to weighted, directed, multistage graphs.

For 3MTM+V, the best routing can be accomplished by using only one terminal row ($\mathcal{U}$ or $\mathcal{L}$) in each cell row. This transforms the MTM cell layout into a CTM layout with an offset in the terminal row locations. Let $f: \mathcal{R} \rightarrow \{\mathcal{U}, \mathcal{L}\}$ be a function determining the assignment of the terminal row to the cell rows in $\mathcal{R}$. For two adjacent cell rows R_i and R_{i+1} in $\mathcal{R}$, there are four choices for selection of the terminal rows. These are (1) $f(R_i) = f(R_{i+1}) = \mathcal{U}$, (2) $f(R_i) = f(R_{i+1}) = \mathcal{L}$, (3) $f(R_i) = \mathcal{L}$, $f(R_{i+1}) = \mathcal{U}$, and (4) $f(R_i) = \mathcal{U}$, $f(R_{i+1}) = \mathcal{L}$.

Let $g: \{1, 2, \ldots, K-1\} \rightarrow I$ be another function that gives the number of tracks available in the OTC area between the selected terminal rows of R_i and R_{i+1} of cell rows. Function g is defined below (p. 258) and illustrated in Figure 8.15.

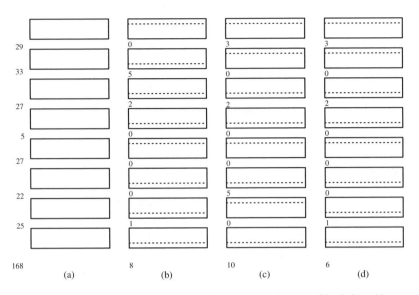

Fig. 8.14   An example of terminal row assignment: **(a)** an instance; **(b)** solution with sequentional manner; **(c)** solution with high-density first strategy; **(d)** optimal solution.

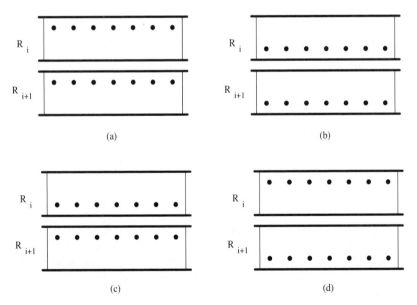

Fig. 8.15   Choices of terminal row selection for two adjacent cell rows: **(a)** $f(R_i) = f(R_{i+1}) = \mathcal{U}$; **(b)** $f(R_i) = f(R_{i+1}) = \mathcal{L}$; **(c)** $f(R_i) = \mathcal{L}$, $f(R_{i+1}) = \mathcal{U}$; **(d)** $f(R_i) = \mathcal{U}$, $f(R_{i+1}) = \mathcal{L}$.

$$g(i) = \begin{cases} k_2 + k_3 + k_1 & \text{if } f(R_i) = f(R_{i+1}) = \mathcal{U} \\ k_2 + k_3 + k_1 & \text{if } f(R_i) = f(R_{i+1}) = \mathcal{L} \\ k_3 + k_1 & \text{if } f(R_i) = \mathcal{L} \text{ and } f(R_{i+1}) = \mathcal{U} \\ k_3 + 2k_2 + k_1 & \text{if } f(R_i) = \mathcal{U} \text{ and } f(R_{i+1}) = \mathcal{L} \end{cases}$$

Let the net density between the cell rows R_i and R_{i+1} using conventional two-layer channel routing be given by d_i. In MTM+V, if $g(i)$ is more than or equal to d_i, then there will not be a channel between R_i and R_{i+1}. If $g(i)$ is less that d_i then in addition to the tracks available in the OTC area, a channel of height $\lceil g(i) - d_i/2 \rceil$ tracks should be used for completing the routing. Let $t: \{1, 2, \ldots, K-1\} \times \{\mathcal{U}, \mathcal{L}\} \times \{\mathcal{U}, \mathcal{L}\} \to I$ be a function such that $t(i, p, q)$ gives the height of the ith channel when a terminal row p is selected for R_i and q for R_{i+1}.

The objective of this step is to find if there is any terminal row assignment to the cell rows such that a channel-less layout can be obtained. The formal statement of this problem (TRSP-1) is as follows:

Instance: Given:

1. $\mathcal{R} = \{R_1, R_2, \ldots, R_K\}$, a set of cell rows, where a cell row $R_i \in \mathcal{R}$ can have one of the two possible terminal rows $\mathcal{L}$ or $\mathcal{U}$.
2. $D = \{d_1, d_2, \ldots, d_{K-1}\}$, a set of densities, where d_i indicates the density of nets between cell row i and $i + 1$.

Question: Does there exist a function $f: \mathcal{R} \to \{\mathcal{L}, \mathcal{U}\}$, such that

$$\sum_{i=1}^{K-1} t(i, f(i), f(i + 1)) = 0$$

In order to answer this question, we model the instance of TRSP-1 using a weighted, directed, multistage graph $G = (V, E)$, as follows.

Corresponding to two terminal row positions ($\mathcal{L}$ and $\mathcal{U}$) in R_i, G has two vertices, $v_{i\mathcal{L}}$ and $v_{i\mathcal{U}}$, respectively. In addition, it has a source vertex s and a destination vertex t; that is,

$$V = \{v_{ij} | 1 \le i \le K, j \in \{\mathcal{L}, \mathcal{U}\}\} \cup \{s, t\}$$

G has directed edges between the vertices corresponding to the adjacent cell rows. In addition, G has the edges $(s, v_{1\mathcal{L}})$, $(s, v_{1\mathcal{U}})$, $(v_{K\mathcal{L}}, t)$, and $(v_{K\mathcal{U}}, t)$. The direction of an edge between v_{ij} and $v_{(i+1)k}$ is from v_{ij} to $v_{(i+1)k}$. Edges $(s, v_{1\mathcal{L}})$ and $(s, v_{1\mathcal{U}})$ are directed away from s, whereas $(v_{K\mathcal{L}}, t)$ and $(v_{K\mathcal{U}}, t)$ are directed toward t.

$$E = \{(v_{ij}, v_{(i+1)k}) | 1 \le i \le K - 1, j \in \{\mathcal{L}, \cup U\}, k \in \{\mathcal{L}, \mathcal{U}\}\}$$
$$\cup \{(s, v_{1\mathcal{L}}), (s, v_{1\mathcal{U}}), (v_{K\mathcal{L}}, t), (v_{K\mathcal{L}}, t)\}$$

The weight of an edge $(v_{ij}, v_{(i+1)k})$ indicates the number of tracks introduced in the ith channel (channel between ith and $(i+1)$th cell row) when terminal rows j and k are selected on the ith and $(i+1)$th cell row, respectively.

The weight of an edge $(u, v) \in E$ is given by the following function:

$$
w(u, v) = \begin{cases}
0 & \text{if } u = s \text{ or } v = t \\
\max(\lceil \frac{d[i]-(k_2+k_3+k_1)}{2}\rceil, 0) & \text{if } u = v_{i\mathcal{U}} \text{ and } v = v_{(i+1)\mathcal{U}} \\
\max(\lceil \frac{d[i]-(k_2+k_3+k_1+k_2)}{2}\rceil, 0) & u = v_{i\mathcal{U}} \text{ and } v = v_{(i+1)l} \\
\max(\lceil \frac{d[i]-(k_3+k_1)}{2}\rceil, 0) & u = v_{i\mathcal{L}} \text{ and } v = v_{(i+1)\mathcal{U}} \\
\max(\lceil \frac{d[i]-(k_3+k_1+k_2)}{2}\rceil, 0) & u = v_{i\mathcal{L}} \text{ and } v = v_{(i+1)\mathcal{L}}
\end{cases}
$$

Figure 8.16 illustrates the construction of such a graph.

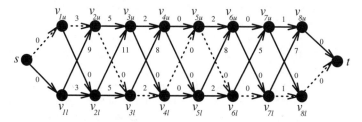

Fig. 8.16   Modeling TRSP using weighted, directed, multistage graph.

A (v_i, v_j) path is defined as a path from vertex v_i to vertex v_j. The cost of a (v_i, v_j) path is defined as the sum of the weights of the edges along the (v_i, v_j) path.

Theorem 34   *A channel-less layout can be obtained for an instance of TRSP-1 if and only if their exists a zero-cost (s, t) path in the corresponding G.*
Proof: Let there be a zero-cost s to t in the weighted, directed, multistage graph G obtained as explained above. Let S be the set of vertices in the shortest path from s and t and let $S' = S - \{s, t\}$. Note that S' contains exactly one vertex corresponding to each cell row. If $v_{ij} \in S'$, then for the ith cell row, the terminal at position j can be selected; that is, $f(R_i) = j$. A channel-less layout can be obtained with the terminal row assignment given by function f.

□

When a channel-less layout is not possible, the objective of this step should be to minimize the overall height of the channel. The formal statement of this problem (TRSP-2) is as follows:

Instance: Given:

1. $\mathcal{R} = \{R_1, R_2, \ldots, R_K\}$, a set of cell rows, where a cell row $R_i \in \mathcal{R}$ can have one of the two possible terminal rows L or U.

2. $D = \{d_1, d_2, \ldots, d_{K-1}\}$, a set of densities, where d_i indicates the density of nets between cell row i and $i + 1$.

Question: Does there exist a function $f : \mathcal{R} \rightarrow \{\mathcal{L}, \mathcal{U}\}$, such that

$$\sum_{i=1}^{K-1} t(i, f(i), f(i+1))$$

is minimized.

TRSP-2 can be solved by finding a minimum-cost s to t path in G.

Theorem 35   *TRSP-2 can be optimally solved in $\Theta(K)$ time.*

Proof: TRSP-2 can also be solved by modeling its instance using a weighted, directed, multistage graph and finding a minimum cost shortest path (MCSP) from s to t as described above. Let S be the set of vertices in the shortest path from s and t. Let $S' = S - \{s, t\}$. Note that S' contains exactly one vertex corresponding to each cell row. If $v_{ij} \in S'$, then for the ith cell row, the terminal at position j can be selected; that is, $f(R_i) = j$.

The time complexity of finding an MCSP is $O(n + e)$ in a weighted, directed, multistage graph, where n and e indicate the number of vertices and edges in the graph. Since the graph G has $2K + 2$ vertices and $4K$ edges, respectively, the complexity of solving TRSP-2 is $O(K)$. Note that the time complexity of solving TRSP-2 cannot be reduced further, since any other algorithm to solve TRSP-2 would have K assignments, one each for a cell row. Hence, we solve the TRSP-2 in $\Theta(K)$ time complexity.

$\square$

Let the total number of tracks introduced in the channels of an OTC routing problem P, using the cell models M, be given by $height(P, M)$. The following theorem proves that for an example P, a smaller or same layout height can be obtained by using the MTM+V model than that obtained by using the CTM+V model.

Theorem 36   *Given an OTC routing problem P, height(P,3MTM+V) $\leq$ height(P,3CTM+V).*

Proof: Let the number of tracks in the OTC area between two terminal rows using 3CTM+V model be k. Then by selecting the upper terminal row in all the cell rows, the same number of tracks can be obtained between any two adjacent cell rows in 3MTM+V. If the channel height between two adjacent cell rows using 3CTM+V is p, then the same channel can be routed by using a channel height of $k + 2p$ tracks using a two-layer conventional channel router. The same example can, therefore, be routed by using the 3MTM+V model with a channel height of p tracks. Thus, in this case, height(P,3MTM+V) = height(P,3CTM+V).

MTMs give flexibility in selecting one of the two terminal rows. This flex- ibility can be used for reducing the channel height. Figure 8.17(a) shows an example routed by using a two-layer conventional channel router. An overall channel height of 54 tracks is obtained for this example. Figure 8.17(b) shows that an overall height of three tracks is obtained using 3CTM+V for the example in Figure 8.17(a). However, Figure 8.17(c) shows that by using 3MTM+V with an optimal terminal row selection, an overall height of one track is obtained for the same example in Figure 8.17(a).

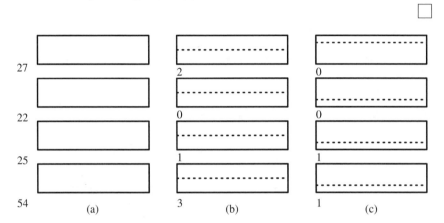

Fig. 8.17   3MTM+V gives less overall channel height when compared with 3CTM+V: **(a)** an instance; **(b)** solution using 3CTM+V; **(c)** solution using 3MTM+V with optimal row selection.

8.4 Routing Algorithms for the ATM

Terai, Takahashi, Nakajima, and Sato [TTNS91] presented a new technique for OTC routing of ATM models with three layers. The model consists of two channels and the routing area over a cell row between them. The two-channel model is shown in Figure 8.18. The routing model consists of a cell row R_i and two channels R_{i-1} and R_i. The M1 layer is used for intracell routing. Vertical net segments are routed in the M2 layer and horizontal segments are routed in the M3 layer. The terminals are located in the M2 layer inside the cell area. The terminals may be nonaligned, and at most one terminal exists in each column. A *net number* is assigned to each terminal. Terminals with the same number i must be connected by net i, while number 0 is assigned to the vacant terminals. Each net in a channel is assigned trunks as shown in Figure 8.18. Each trunk t is represented with the net number denoted by $N(t)$ and the right and left column numbers denoted by $L(t)$ and $R(t)$, respectively, together with the arrows that indicate the directions of connection to the terminals.

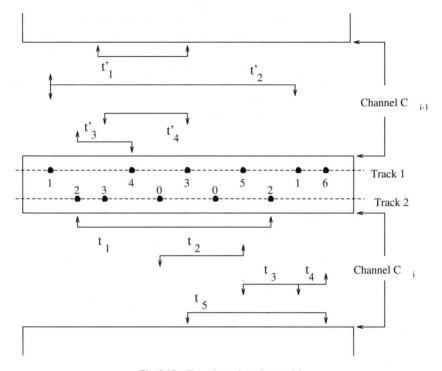

Fig. 8.18 Two-channel routing model.

8.4.1 Algorithm Overview

Three and two layers are available for routing in the channels and OTC areas, respectively. The problem is decomposed into two phases as given below:

1. **OTC Routing:** OTC routing is formulated in the same way as channel routing, with the additional constraints of assignment of trunks to a track in OTC areas. Based on this formulation, an efficient heuristic OTC routing algorithm is developed.

2. **Channel Routing:** To solve the channel routing problem, Yoshimura's channel router [Yos84] is extended to the case of three layers.

Since channel routing is an extension of the well-known Yoshimura channel router, the OTC routing phase is explained in detail. First, the constraints imposed on the assignment of each trunk to a track in the OTC area of cell row R_i are discussed.

Consider the OTC routing area over cell row R_i (henceforth in this section, for ease of presentation, we refer to the OTC area over a cell row R_i as OTC routing region R), shown in Figure 8.18.

8.5 Routing Algorithms for the TBC

In this section, we discuss the TBC router presented in [KBPS94]. The TBC router includes two key features: an optimal $O(KL)$ algorithm (where K and L are the number of cell rows and the layout width, respectively) for assigning the OTC area to each channel, so as to minimize total layout height, and an irregular boundary channel HV-HVH-HV router for OTC and channel areas between terminal rows, which optimally utilizes the OTC area. The key difference between the TBC router and the existing OTC routers for other cell models is the approach adopted in minimizing the total layout height. The TBC router minimizes the overall layout height by minimizing the total channel height in each column, while the OTC routers for the MTM, CTM, and BTM minimize the layout height by minimizing maximum heights in individual channels.

8.5.1 Algorithm Overview

The TBC router assumes placement and global routing stages have been completed. As a result, we have the net list, which specifies the interconnection between different targets (terminals) for each channel. The following are the various phases of the TBC router.

1. **Density Computation:** In this step, columnwise net densities are computed for each channel in the layout. The density calculation is simple, since global routing is already completed and the net list is available for each channel.

2. **Net Migration:** Many global routing algorithms do not consider local densities in the channel while assigning global routes to nets. As a result, it is quite possible that we may have very uneven column densities in the original net list which may lead to a very irregular terminal boundary allocation and difficult routing. In this step, the irregular boundaries are minimized by migrating the nets from one routing region to the adjacent regions. This step may be viewed as a modification of global routing.

3. **Terminal Position Assignment and Boundary Generation:** After computing columnwise net densities for all the channels, the terminal positions are assigned for each cell row in a top-down approach. The terminal positions are selected based on the columnwise net densities between two adjacent cell rows. Assigning terminal rows generates the routing boundaries in the OTC areas. An $O(KL)$ algorithm is presented for computing terminal locations for each terminal row, such that total layout height is minimized. The basic idea of the algorithm is to compute the terminal locations in a left-to-right scan, starting with the top-most row.

4. **Boundary Refinement:** The terminal position assignment may generate highly irregular boundaries. This is due to the abutment of dense columns

in adjacent terminal rows. It is a classical case of a crest meeting a crest in two adjacent terminal rows. Step 2 minimizes irregularity in the boundary to a great extent. However, we may have critical regions in the boundary, which constitute dead areas that cannot be used for rectilinear routing.

In order to ease the task of routing, the boundaries must be made as straight as possible. This is accomplished by first eliminating insignificant boundary jogs by considering the routability constraints. Then further net migration techniques are adopted to reduce or eliminate irregularity in the boundary.

5. **Irregular Boundary Hybrid Routing:** A hybrid greedy-based router is developed to route in two layers in OTC areas with irregular boundaries and the route using an HVH reserved-layer model in the channel areas. The router scans from left to right and allocates a track in the channel only after all OTC areas have been used.

The steps 2, 3, 4, and 5 are the key steps in the TBC router and are described in detail in the following sections.

8.5.2 Net Migration

As discussed earlier, irregular boundaries of a routing area lead to routability problems. In most cases, irregular boundaries have dead areas, which cannot be used for routing, and therefore they result in an increase in channel width. Thus, it is necessary to reduce the irregularity of the boundaries of routing regions. Since the shape of a routing region is determined at the terminal selection phase, based on the net densities, the irregularity of a routing region can be reduced by migrating some nets from its neighboring routing region, such that the net density is equally distributed over all of its columns.

Let us start with some required terminology. The notion of vacant terminals was first introduced and exploited in [HSS91a]. Notice that in the TBC, a vacant terminal location is simply an empty area in M2. For any given t_i, the function $\text{OPP}(t_i)$ returns the terminal directly across in the adjacent terminal row in the same column. The function $\text{ROW}(t_i)$ returns the row to which the terminal t_i belongs (either R_i or R_{i+1}), and $\text{COL}(t_i)$ returns the column of terminal t_i.

Since net migration requires an empty area in M2, nets are classified according to the location of net terminals with respect to vacant terminals and abutments in the channel. In a typical standard-cell design, approximately 50% to 80% of the possible terminal space is vacant.

To ease the routing task, we decompose each m-terminal net into exactly $m - 1$ two-terminal nets at adjacent terminal locations. Clearly, $m - 1$ two-terminal nets are sufficient to preserve the connectivity of the original m-terminal net. After decomposition, we classify all the nets as one of three basic types. Let $N_j = \{t_{i_1}, t_{i_2}\}$ be a net where t_{i_1} and t_{i_2} are the left-most and right-most terminals, respectively, of N_j. Let us assume that the net is in channel i. A net N_j is a Type I net if $\text{ROW}(t_{i_1}) = \text{ROW}(t_{i_2})$; that is, N_j is an intrarow net. Notice that a Type I net may be routed in channel i or channel $i - 1$. However, if some additional conditions are satisfied, then we may be able to move the net to even channels $i - 2$ and $i + 1$.

Net N_j is a Type II net if $\text{ROW}(t_{i_1}) = \text{ROW}(t_{i_2})$, and $\text{OPP}(t_{i_1})$ and $\text{OPP}(t_{i_2})$ are both vacant. In this case, net N_j may be routed in channels $i + 1$, i, $i - 1$, and $i - 2$. Type III is a special case of Type II nets if one or both terminals on one side are not vacant. For example, if $\text{OPP}(t_{i_1})$ is not vacant and all three terminals are vacant, then the net may be routed in channels $i + 1$, i, and $i - 1$.

Let us consider the net migration problem for Type I nets first. Let $\mathcal{N}_i$ and $\mathcal{N}_{i-1}$ be two sets of nets in the channels i and $i - 1$, respectively. The net migration problem (NMP-I) is to assign Type I nets in the ith row of terminals to channels i and $i - 1$, such that the maximum difference in density in either channel is minimized.

The formal statement of the net migration problem (NMP-I) is as follows:

Instance: Given two sets of nets $\mathcal{N}_i$ and $\mathcal{N}_{i-1}$ for channels i and $i - 1$, and a set $\mathcal{N}'$ of Type I nets.

Problem: Partition $\mathcal{N}'$ into two subsets $\mathcal{N}'_1$ $\mathcal{N}'_2$, such that

$$\text{max\_density}(\mathcal{N}_i \cup \mathcal{N}'_1) - \text{min\_density}(\mathcal{N}_i \cup \mathcal{N}'_1) \text{ is minimized}$$

$$\text{max\_density}(\mathcal{N}_{i-1} \cup \mathcal{N}'_2) - \text{min\_density}(\mathcal{N}_{i-1} \cup \mathcal{N}'_2) \text{ is minimized}$$

Theorem 37   *The net migration problem for Type I nets is NP-hard.*

In view of the NP-hardness of net migration problems, the following greedy heuristic is proposed. Each net in $\mathcal{N}'$ is assigned a weight equal to the length of the net. Nets are sorted on their weights and considered one at a time. A net N_j is assigned to channel i if it leads to a minimum increase in channel density in that channel; otherwise it is assigned to channel $i - 1$. This process is repeated until nets are assigned to channels. The heuristic for Type II and Type III nets is similar, except more channels are considered for assignment.

8.5.3 Terminal Position Assignment and Boundary Generation

In this section, the algorithm is first developed for the terminal position assignment for a *theoretical-TBC cell design* and then is extended to general TBC designs. The theoretical-TBC designs have uniformly placed targets with identical heights. A top-down approach is considered for terminal position assignment. The upper-most positions are selected for each column of target areas, depending on the net densities. Let us start with a formal statement of the terminal position selection problem (TPSP-t). A greedy algorithm called GR-TPSP is then proposed to solve TPSP-t.

Instance: Let $\mathcal{R} = \{R_1, R_2, \ldots, R_K\}$ be the set of cell rows. Let L be the length of the channel. Let the terminal locations in each column of a cell row be numbered from 1 to k_2. Let $\mathcal{D} = \{D_1, D_2, \ldots, D_{K-1}\}$, such that $D_i = \{d_{i1}, d_{i2}, \ldots, d_{iL}\}$, where d_{ij} is the density of nets in the jth column of the ith channel.

Problem: Select a terminal position in the jth column of R_i, that is, p_{ij} from $\{1, 2, \ldots, k_2\}$ for $i = 1$ to K and $j = 1$ to L, such that

$$H = \max_{j=1}^{L}\{\sum_{i=1}^{K-1} h(p_{ij}, p_{(i+1)j}, d_{ij})\}$$

is minimized. Let h_{ij} indicate the part of density not accommodated in the OTC area at the jth column. Therefore, for the base case $h_{1j} = d_{1j} - 2(k_1 + k_2)/2$. Let $h_{\max}^1 = \max(h_{ij})$ for $j = 1, \ldots, L$. A boundary is considered optimal if it achieves minimum height at all columns. Let us first state the algorithm formally, as shown in Figure 8.19.

Theorem 38   *The algorithm GR-TPSP generates an optimal solution for TPSP-t in $O(KL)$.*

Proof: The proof is based by induction on K. First, we consider the base case, determining minimum heights for all columns by considering the first channel. By construction, this can be optimally obtained. Then, in the inductive step, we assume that the minimum column heights have been obtained up to $(i-1)$ channels, and hence we prove that optimal column heights are obtained for the ith channel and, therefore, minimum H.

For the basis, consider the case of two cell rows, and hence only one channel. In this case, we select $p_{1j} = 1$ for all $j = 1$ to L. This is natural, since we adopt a top-down approach and use the entire OTC area in R_1 for routing channel 1 nets. Hence, in the base case, the minimum $h_{\max}^1$ is obtained.

For the inductive step, we assume all channels up to $(i-1)$th have been routed and we assume that $h_{\max}^1 + h_{\max}^2 + \cdots + h_{\max}^{i-1}$ is minimized, and hence the allocation of the terminal boundary to the ith boundary is also optimal. Hence, by induction, the optimal terminal location is selected for each column in all the rows. The time complexity follows directly from the algorithm.

$\square$

The selection of terminal positions for the TBC is similar to the theoretical TBC. The difference is in the range of available terminal positions to make the selection. The available terminal positions in the TBC is different in each column, and they originate and terminate at different positions. Therefore, the TPSP for the TBC has a limited range of locations for selecting the terminal positions. Let m and n be the top-most and bottom-most locations, respectively, in the available position range in a column. If $m = 1$ and $n = k_2$, then the TPSP is transformed to an instance in TPSP-t.

Algorithm GR-TPSP($\mathcal{R}, \mathcal{D}, L$)

Input: $\mathcal{R}, \mathcal{D}, L$

Output: Terminal position on each target.

---

begin Algorithm

    /* Assign top most terminals for the first row */

    for $j = 1$ **to** L **do**

        If $(d_{1j} \geq 2(k_1 + k_2))$

            then $h_{1j} = \frac{d_{1j} - 2(k_1 + k_2)}{2}$;

        $p_{2j} = k_2$.

        if $((2k_1 + k_2) \leq d_{1j} \leq 2(k_1 + k_2))$ **then**

            if $h_{max}^1 \leq k_2$, $p_{2j} = d_{1j} - (h_{max}^1 + 2k_1 + k_2)$

            if $h_{max}^1 > k_2$, $p_{2j} = 1$.

        if $d_{1j} \leq (2k_1 + k_2)$, then $p_{2j} = 1$.

        $p_{(i-1)j} = k'_{ij}$; $x_{ij} = k_2 - k'_{ij}$.

    /* for remaining rows, select the terminal

    in a greedy manner */

    for $i = 2$ **to** K **do**

        for $j = 1$ **to** L **do**

        if $(d_{ij} \geq 2k_1 + x_{ij} + k_2))$

            then $h_{ij} = \frac{d_{ij} - (2k_1 + x_{ij} + k_2)}{2}$;

        $p_{ij} = k_2$.

        if $(2k_1 + x_{ij}) \leq d_{ij} \leq (2k_1 + k_2 + x_{ij})$

        then

            if $(h_{max}^i \leq k_2)$ **then**

                $p_{ij} = d_{ij} - (h_{max}^i + 2k_1 + x_{ij}))$

        if $(h_{max}^i > k_2)$ **then** $p_{ij} = 1$.

        if $(d_{ij} \leq (2k_1 + x_{ij}))$ **then** $p_{ij} = 1$.

    end Algorithm

Fig. 8.19   Algorithm for terminal position assignment in TBC.

Instance: Let $\mathcal{R} = \{R_1, R_2, \ldots, R_K\}$ be the set of cell rows. Let L be the length of the channel. Let the terminal locations in each column of a cell row be numbered from 1 to k_2. Let the range of possible terminal locations in each column be U_i to B_i, where m and n are the top-most and bottom-most locations in the available position range in a column. Let $\mathcal{D} = \{D_1, D_2, \ldots, D_{K-1}\}$, such that $D_i = \{d_{i1}, d_{i2}, \ldots, d_{iL}\}$, where d_{ij} is the density of nets in the jth column of the ith channel.

Problem: Select a terminal position in the jth column of R_i, that is, p_{ij} from $\{U_i, U_i + 1, \ldots, B_i\}$ for $i = 1$ to K and $j = 1$ to L, such that

$$H = \max_{j=1}^{L}\{\sum_{i=1}^{K-1} h(p_{ij}, p_{(i+1)j}, d_{ij})\}$$

is minimized.

The solution to the TPSP can be obtained optimally by modifying the algorithm GR-TPSP, such that terminals in the allowable range are selected. The proof that such an algorithm achieves the minimum total height is similar to the theorem presented.

Theorem 39   *The algorithm GR-TPSP generates an optimal solution for TPSP in $O(KL)$.*

8.5.4 Boundary Refinement

The boundaries generated by the terminal position assignment may be highly irregular in nature. (See Figure 8.20(a).) In order to ease the task of routing, the boundaries must be made as straight as possible. This is accomplished by eliminating insignificant boundary jogs, by considering the routability constraints, and by using M2 vertical segments to smooth out the boundary. The horizontal flow of the nets is along the boundaries, and when a boundary has jogs, so do the nets close to that boundary. The area between two terminal boundaries (A_i) is partitioned into routing regions (r_j) depending on the boundary irregularities, as shown in Figure 8.20(b). The following constraints must be considered in checking the routability of each region:

1. **Critical Region:** This is the region preceding a step up in the boundary, as shown in Figure 8.20(c). For a given routing area A_i, a critical region r_{jc} is said to satisfy the routability constraints if it satisfies the vertical distances required to accommodate d_{ij}, and also the horizontal distances such that all d_{ij} can be routed. Given a column j in a routing region r of a routing area A_i, the region satisfies vertical and horizontal constraints if height$(r, i) \geq d_{ij}$ and $w \geq h + a$, respectively, where a is the number of terminals starting in the critical region.

2. **Valley Region:** This is the region following a step down, as shown in Figure 8.20(b). The vertical region is said to satisfy the routability constraints if it satisfies the vertical distances required to accommodate d_{ij} and horizontal distances, such that all d_{ij} can be routed. The width of the vertical region, w_v, must be greater than or equal to $(h + a)$.

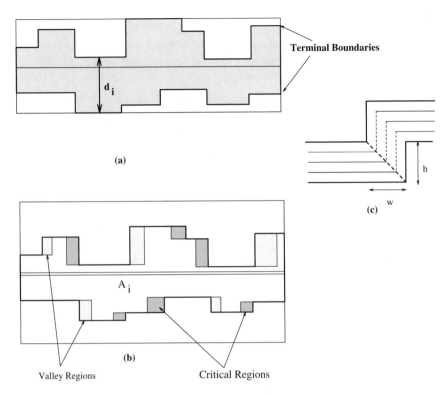

Fig. 8.20   Irregular boundaries generated by terminal position assignment: **(a)** irregular channel; **(b)** two types of routing regions; **(c)** critical region routing.

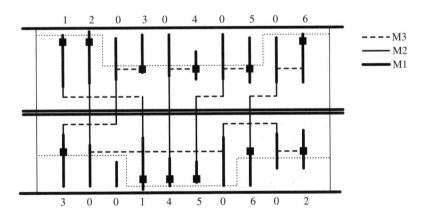

Fig. 8.21   A routing example using a TBC router.

Algorithm: TBC Router
Input: $Upper\_boundary, Lower\_boundary, Netlist$
Output: A hybrid routing

---

begin Algorithm
For $i = 1$ *to number_of_columns_in_channel* **do**
   begin
         $T_1 = GET\_EMPTY\_TRACK;$
           If $T_1 = 0$ then
/* Add two tracks in channel. */
             $ADD\_TRACK;$
         $T_2 = GET\_EMPTY\_TRACK;$
           If $T_2 = 0$ then
/* Add two tracks in HVH Channel Layer. */
             $ADD\_TRACK;$
/* Connect the top terminal to the track */
           $CONNECT(T_i, T_1);$
/* Connect terminal to bottom boundary */
           CONNECT$(B_i, T_2);$
         if CRITICAL(upper_boundary, i) then
           SMOOTHER (upper_boundary, c_u_h, i);
           UPDATE (c_u_h);
         if CRITICAL(lower_boundary, i) then
           SMOOTHER (lower_boundary, c_l_h, i);
           UPDATE (c_l_h);
/* Join split nets */
           JOIN ;
/* Move tracks towards boundary */
           $MOVE\_TRACKS;$
     end; /* columns */
end /* Algorithm ends */

c_u_h:Current upper height; c_l_h:Current lower height

Fig. 8.22   An algorithm for irregular boundary channel.

8.5.5 Irregular Boundary Channel Routing

The TBC router developed to route channels consists of two-layer routing in OTC areas with irregular boundaries and three-layer HVH routing in the channels (see Figure 8.23). The router is greedy in nature and is an extension of the greedy router presented in [Luk85]. The router scans the channel area column by column starting from the left-most column. The first available unused track is assigned to the nets starting from each column. If an unrouted net is already assigned a track and has a terminal in the top boundary, then the net jogs toward the top boundary; otherwise the net jogs toward the bottom boundary. When all the tracks in the OTC areas are occupied, two new tracks are added in the channel area. The formal algorithm is given below.

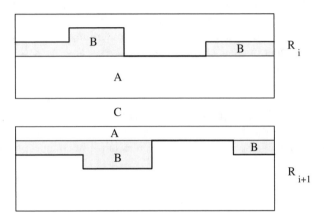

A = HV Routing With Straight Boundaries

B = HV Routing With Irregular Boundaries

C = HVH Channel Routing

Fig. 8.23   Routing model for TBC router.

At any column i, the function CRITICAL scans the next $2 \times K \times L$ columns to check if any critical region is starting in that range. The procedure SMOOTHER brings up the nets that are below the height of the next boundary obstacle, and the current boundary height is updated by the UPDATE procedure.

In channel routing, we are interested in finding a routing solution that uses as few tracks as possible. A common lower bound for this solution is the channel density. The sparseness of a channel can be measured from the density distribution d of the channel. The heuristic is based on the following reasonings: Optimality can be achieved if we can guarantee, for each column, that there is only one horizontal track for each net. The objective of the routing procedure is to minimize the number of horizontal tracks per column per net. The method is to scan columns and try to join the split horizontal tracks (if there are any) that belong to the same net as much as possible. Apart from that, an objective of minimizing the overall layout height is also met. The factors considered by the router are the following:

1. The net migration algorithm has smoothed out the valley regions, having a width less than that of the oncoming critical region in that boundary.

2. The given channel has almost 50% of null terminals. These are uniformly scattered in the channel.

3. The terminal positions are fixed and extra tracks are obtained by expanding the channel.

The width, cw, of a critical region is defined as the minimum distance before getting into deadlock, as shown in the figure. It is observed that this width is bounded by $2 \times K \times N$, where K is the number of terminal locations over the cell and N is the number of channels. The router always scans $2 \times K \times N$ terminals ahead in the top and bottom boundaries, since this is the only region where there can be a start of valley or critical region.

All the nets lying below the critical region height are brought to the new boundary height. These nets would be treated just like the regular terminals. There are some deadlock possibilities shown, and the way they are dealt with is shown in Figure 8.24. See Figure 8.21 for an example of complete routing using the TBC router.

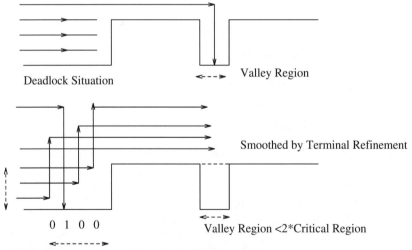

Fig. 8.24 Deadlocks in greedy routing.

8.6 Experimental Evaluation of Routers

The routers in the three models CTM, BTM, and MTM generate layouts with smaller channel heights than the corresponding layouts generated for the three-layer process, discussed in Chapter 7. This is due to the fact that the routers are given the flexibility to switch layers in OTC areas. When tested on the benchmark example, PRIMARY1, the BTM router requires 19 tracks [SH93] to complete the routing, while the other three models, the CTM [WHSS92], MTM [BPS93a], and TBC [KBPS94], generate channel-less layouts for PRIMARY1. Hence, in order to evaluate the performance of these three routers, new and dense benchmarks that specify the locations of the target areas must be used.

Since the cell designer has complete freedom in placing terminals (targets), the cells in TBC designs have the smallest cell widths and hence minimum cell area (see Chapter 4). The minimum area for each cell leads to the minimum width of the layout, while the optimal terminal assignment algorithm minimizes the height. The combined effect leads to minimum chip areas for the TBC-based designs.

8.7 Summary

The effect on the layout height when vias are allowed in OTC areas is evident from the experimental results. While the BTM router in the three-layer process needs 71 tracks to route the benchmark PRIMARY1, the router in the advanced three-layer process needs only 19 tracks. The CTM and MTM routers require 63 and 104 tracks, respectively, in the three-layer process to route PRIMARY1. Channel-less layouts are generated for the advanced three-layer processes. The routing technique in the advanced three-layer process is completely different from the technique used in the three-layer process. Since vias are allowed in OTC areas, a typical routing strategy uses an HV reserved-layer channel routing model to route in OTC areas, and an HVH reserved-layer model for routing in the channel.

In this chapter, we have discussed various routers for all four models. Though the CTM and MTM routers also generate channel-less layouts for PRIMARY1, note that layouts generated by TBC designs have less area, since the cells in these designs have smaller widths. Also, for denser circuits, TBC routers generate better results than CTM and MTM routers due to the flexibility in locating the terminals on a target.

BTM-based layouts have larger channel heights when compared with the corresponding layouts in other cell models. The BTM is primarily developed for channel routing–based designs and does not provide the flexibility provided by other models for OTC routing, and experimental results clearly show the difference in results. On the other hand, TBCs are developed exclusively for the advanced three-layer process.

PROBLEMS

8-1. Prove Theorem 30 stated in Section 8.1.3.

8-2. Consider two channels on either side of a BTM-HCVD-based cell row. Let d_1 and d_2 be the densities of upper and lower channels. Select a set of nets S containing nets from both the channels, such that $\max(d'_1, d'_2)$ is minimized, where d'_1 and d'_2 are the densities of nets in the upper and lower channels, respectively, that are not selected in S.

8-3. The terminal row selection problem in the MTM is based on the minimization of overall channel height. However, in high-performance circuits, the wire lengths have to be minimized. Develop a terminal row selection technique such that the length of the longest wire is minimized. (Assume a channel-less layout for all choices of terminal rows.)

8-4. In the cell structure presented in Section 8.4, assume that the terminals are located in the center and do not have any external obstacles. Develop an efficient BTA algorithm for this cell structure. (Notice that this cell model is essentially a CTM with obstacles.)

8-5. In high-performance designs, some specific OTC areas called *forbidden regions* are restricted from the use of vias. The routing algorithms presented in this chapter cannot be directly adopted for such designs. Develop an efficient routing algorithm for such designs. Assume a BTM-based design.

8-6. What is the effect of forbidden regions on terminal row selection in the MTM? Develop an efficient algorithm for terminal row selection in the presence of forbidden regions.

8-7. Generalize the terminal row selection technique presented in Section 8.3 to L terminal rows in each cell row. Assume k_i tracks between terminal rows T_i and T_{i+1} for $i = 1$ to L.

8-8. Prove Theorem 37.

BIBLIOGRAPHIC NOTES

Chen and Liu [CL84] present a three-layer channel routing algorithm. A greedy three-layer channel routing is presented by [BS86]. Edahiro and Yoshimura [EY90] develop placement and global routing algorithms for standard-cell layouts. An integrated three-layer gridless channel router and compactor is presented in [GS90].

9

Routing Algorithms for Advanced VLSI and Thin-Film MCMs

The rapid decrease in feature size and a corresponding increase in chip size has led to a significant increase in net densities. The increase in net densities has necessitated the use of more layers for routing. A chip using four- to eight-layer processes is called an *advanced VLSI chip*. The advanced VLSI chips have chip-wide routing. The MCMs with four to eight layers for routing are generally called thin-film MCMs. Due to the availability of large OTC routing resources, routability is not a primary issue in establishing the interconnections.

The cell models, which were considered to be the key factors in efficient utilization of OTC areas for routing, become insignificant when more layers are available for routing. Therefore, cells should be designed with the lowest possible areas so that the die size is minimized. The objective of the chip designer should be to minimize areas. At the chip level, the cell areas are minimized and at the MCM level, the chip areas are minimized. In this chapter, we use the following sets of words interchangeably: when we discuss MCMs, we use *chips* or *substrate*, and we use the word *cell* or *chip* in the context of VLSI chips. However, note that the routing problem remains the same for both advanced VLSI and thin-film MCMs.

In the two-metal-layer process, OTC routing is accomplished in the M2 layer in a planar fashion. Traditionally, the terminals were conveniently located at the boundary, and a planar subset of nets could be routed in the OTC area, while the remaining nets were routed in the channel. Several routers have been presented for two-layer processes [SS87, CPL90]. Though OTC routing minimized the layout heights, channel-less layouts were not possible, even for low-density designs. With the introduction of the three-metal-layer process, two layers were made available for OTC routing, thus allowing more nets to be routed in OTC areas. When the fabrication process does not permit the usage of vias in OTC areas, the OTC routing in M2 and M3 layers is planar. When vias are allowed in OTC areas, the router gets additional flexibility in switching the layers in OTC areas, which leads to a significant reduction in the layout heights. Several three-layer OTC routers have been developed [BPS93a, CPL90, WHSS92].

The effectiveness of OTC routing in two- and three-layer processes [BPS93c, CPL90, SS87, WHSS92] depends on the locations of the terminals and the power and ground lines, which are specified by the cell model. Therefore, the cell model plays a significant role in minimizing the layout heights for two- and three-layer processes. Several cell models and the corresponding routers were developed for the minimization of the layout heights [CPL90, HSS91a, WHSS92]. All the existing OTC routers assume that certain restrictions have been placed on terminal locations. Usually it is assumed that all the terminals are aligned and located in one or two rows on the boundary, in the center or at a specific offset from the boundary.

In [CPL90], it is assumed that the terminals are on the boundary (BTM), and due to the placement of feed-throughs and power and ground lines, three cell layout styles, namely HDVC, HCVC, and HCVD, are used for OTC routing. In [WHSS92], the terminals are assumed to be in the center (CTM), while in [BPS93a ,BPS93b], it is assumed that the terminals are located at a specific offset from the boundary (MTM), the primary objective being the minimization of the channel heights or the elimination of the channels, thus minimizing the layout height.

With the advent of multilayer processes, more OTC area is now available for routing, and hence further reduction in the layout height can be accomplished. The total height in a standard-cell layout is given by

$$H = K \times h_{\text{cell}} + \sum_{i=1}^{K-1} h_i$$

where K is the total number of cell rows, h_{cell} is the height of the cells, and h_i is the channel height in the ith channel. The minimum height of the layout is achieved when all the channels are eliminated. The total layout area for a channel-less layout is given by $A_T = W \times K \times h_{\text{cell}}$, where W is the width of the layout. The layout area can be further minimized only by minimizing the layout width W. This can be accomplished by designing cells with minimum cell widths. For a given functionality, the cell width basically depends on the terminal alignment and intracell routing.

In [BKPS94], a new cell model, the TBC model, was presented, which allows flexibility in the terminal locations. In this cell model, long vertical columns called *targets* are provided in the M1 layer, instead of terminals located at fixed positions. Cells designed using this methodology have smaller widths compared to other cell models.

9.1 Minimum-Width Cell Layouts

The total area of a standard-cell layout is dependent on the layout height and the layout width. While the effective utilization of OTC area leads to smaller layout heights, the layout width is determined by the cell widths. Since current process technologies lead to channel-less layouts, the further die size minimization can only be achieved by minimizing the layout widths. The layout width depends on the cell widths. Therefore, it is important to minimize cell widths. In this section, we focus on the development of minimum-width cell layouts.

Conventional cell layouts require the terminals to be located at the cell boundaries. This was required to ease the task of channel routing. Since all the interconnections within a cell are accomplished in poly and M1 layers, the terminals have to be brought to the boundaries either in poly or in M1 layers. However, for circuits with dense intracell routing, a predefined position for the terminals and bringing the terminals to the boundaries lead to an increase in cell widths. Therefore, imposing a rigidity on the locations of terminals leads to an increase in cell width.

In two- and three-layer processes, the total available OTC area is limited. Therefore, in order to efficiently use the limited OTC area for routing, the terminals were required to be aligned. But with the advent of the multilayer process technology, the OTC areas are not limited anymore and channel-less designs can be achieved, irrespective of where the terminals are located. Therefore, terminal alignment is not necessary and the cell designer is given the flexibility to locate the terminals at the most convenient place with a single criterion of cell width minimization. A cell designer should now be given the following task: "Do whatever you want to, but give us minimum cell width." The intracell routing is accomplished in poly and M1 layers, while the power and ground lines are located in the M1 layer. In order to compute the reduction in cell widths, several leaf cells were laid out in both the conventional and the new minimum-cell-width style.

A three-input NAND gate and a DELAY cell gate laid out using MAGIC Version 6.3 are shown in Figures 9.1 and 9.2, respectively. As shown in Figure 9.1(a), the gate poly is routed from the top cell boundary to the bottom cell boundary, forming p-type and n-type transistors. These poly routes can be directly extended beyond the power and ground lines to form the input terminals. However, the cell outputs are usually available in the M1 layer and have to be routed to the boundary in the poly layer, outside the diffusion areas, which may tend to increase the cell width in conventional cells. On the other hand, minimum-width designs have flexibility in the location of the terminals. As shown in Figure 9.3, the DELAY cell has two terminals, T1 and T2. In the conventional layout, since the terminals are located at the boundary in the poly layer, the routing to the boundary is accomplished outside the diffusion areas.

For analysis, a cell library comprising over 20 leaf cells has been generated in conventional and minimum-width designs. Table 9.1 gives a comparison between the conventional cell layouts and the cells designed with minimum cell widths.

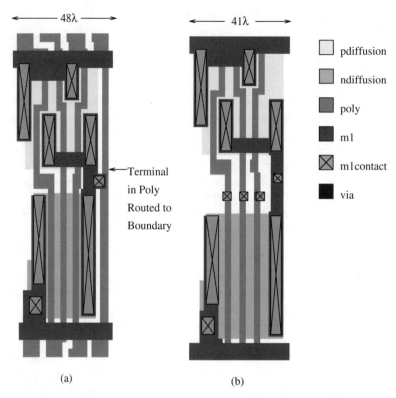

Terminal
in Poly
Routed to
Boundary

pdiffusion

ndiffusion

poly

m1

m1contact

via

(a)   (b)

Fig. 9.1 Layout of a three-input NAND gate: **(a)** conventional; **(b)** minimum cell-width.

The variation in cell widths between the cell layout styles can be attributed to intracell routing and terminal alignment. In the conventional design style, the intracell routing over the diffusion areas is accomplished in the M1 layer, and the routing between the diffusion areas is accomplished in both poly and M1 layers. For circuits with dense intracell routing, a predefined location for the terminals may lead to an increase in the cell width. The minimum-cell-width designs have an advantage over the conventional designs, since there is no fixed location for terminals. Hence, the cell widths in minimum-width designs are always smaller than the corresponding widths in the other cell models. The conventional cell style required the terminals to be brought to the boundary, while the terminals in the minimum-cell-width designs can be conveniently located. This flexibility in the terminal locations led to a 10.71% decrease in cell width for the DELAY gate and a 14.58% reduction in cell width for the three-input NAND gate. Our experimental results on the 23 randomly selected leaf cells laid out in both layout styles show that this flexibility leads to a significant decrease in cell width when compared with the corresponding cells in the conventional cell layouts. As can be seen from the table, only three cells have similar widths, while the rest are wider

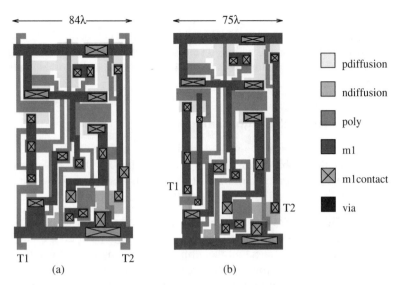

Fig. 9.2 Layout of a DELAY gate: **(a)** conventional; **(b)** minimum cell-width.

Table 9.1: Cell Width Comparison

Sl. No.	Cell Description	Cell Width in λ		% Reduction in
		Conventional	Proposed	Cell Width
1	Two-input NOR gate	36	36	0.00
2	Three-input NOR gate	48	42	12.50
3	Four-input NOR gate	73	67	8.22
4	Two-input NAND gate	36	35	2.78
5	Three-input NAND gate	48	41	14.58
6	Inverter	36	31	13.89
7	Noninverter	36	32	11.11
8	Transmission gate	72	72	0.00
9	Nor-latch	60	60	0.00
10	Pull-up	36	18	50.00
11	Pull-down	36	18	50.00
12	High-impedance buffer	72	67	6.94
13	Delay gate	84	75	10.71
14	D flip-flop	156	153	1.92
15	Two-input NAND and AND gate	49	47	4.08
16	Three-input NAND and AND gate	60	56	6.67
17	Four-input NAND and AND gate	72	66	8.33
18	Four-input OR gate	60	54	10.00
19	Two-input OR and NOR gate	48	45	6.25
20	Three-input OR and NOR gate	60	56	6.67
21	Two-bit full adder	168	156	7.14
22	Exclusive-OR gate	74	68	8.11
23	Exclusive-NOR gate	72	66	8.33

in the conventional layout style. The minimization in cell widths ranges up to 25% when compared with the BTM, with the *pull-up* cell being exceptional, with a reduction of 50% in cell width. On an average, it can be seen that the cell widths in the new minimum width layouts are 9.43% smaller than the corresponding cells in the conventional layout style.

9.2 Overview of the Algorithm

In this section, we present a unified routing approach to multilayer OTC routing. We assume that global routing is completed and the feed-through assignment has been done such that multiterminal nets (nets with more than two terminals) span contiguous cell rows. The routing algorithm has the following steps:

1. **Multirow Net Connection Assignment:** In this phase, the multirow nets are decomposed into same-row nets and adjacent-row nets. Same-row nets are used to connect all the terminals belonging to the same net in a cell row. Adjacent-row nets are used to connect two same-row nets belonging to the same net in two adjacent cell rows. The problem of choosing two terminals from each same-row net of a particular net in two adjacent cell rows is called the multirow net connection assignment problem (MCAP), which is formally defined in Section 9.3. We show that the MCAP is NP-complete and propose an efficient heuristic algorithm called ALGO-MCAP to solve it.

2. **Interval Generation for Critical and Same-Row Nets (IGC):** In this step, all the decomposed segments of the nets are classified into five categories, which intuitively indicates the difficulty involved in routing this net. After the net categorization, intervals are generated between the left-most and right-most terminals of a net in each cell row for all the critical and same-row nets.

3. **Interval Selection (IS):** In this step, we select the intervals to be routed in contiguous tracks in each cell row. However, some of the nets can be routed in two different ways, and these nets have to be routed in one way or the other. The main objective of this step is to select a set of intervals from each cell row, such that, on the whole, the number of intervals selected from all the cell rows is maximum. The complexity of the interval selection problem (ISP) is unknown, and we develop an approximate algorithm ALGO-IS to solve this problem.

4. **Track Assignment:** In order to simplify the routing, all the terminals in each row are brought to a vacant track in that row so that all the terminal points in the net intervals are connected to their respective terminals on this vacant track by a vertical strip in M2. Now the problem is to determine the appropriate track in a row to be left vacant for these terminals, so that the total length of the vertical strips in M2 is minimized. We call this

the track assignment problem, and in Section 9.6, we present an optimal algorithm to solve this problem.

5. **Interval Assignment/Same-Row Routing (IAS):** The main objective of this step is to compute the total number of contiguous tracks required in each cell row to assign all the intervals (selected in the previous step). A contiguous set of tracks in a cell row used for interval assignment of same-row nets is called a *routing block*. The tracks in a routing block are not permutable. However, the actual position of the routing block is not fixed and may be located anywhere in the cell row.

6. **Routing Block Assignment (RBA):** The routing blocks generated in the previous step are assigned to tracks in each cell row. First the densities between routing blocks belonging to two adjacent cell rows are computed. Based on the densities, the routing blocks are assigned to tracks in each cell row. We present an optimal algorithm ALGO-RBA, which runs in $O(K)$ time to solve the routing block assignment problem (RBAP).

7. **Composite VHV/HVHV OTC/Channel Router:** The previous step specifies the routing areas between two adjacent routing blocks. A composite OTC/channel router with VHV routing in OTC areas and HVHV routing in the channels is used to complete the routing between the routing blocks.

9.3 Multirow Net Connection Assignment

In this phase, the nets spanning adjacent cell rows are decomposed as same-row nets and adjacent-row nets. A multiterminal net is decomposed into three segments. The first segment is used to connect all the terminals belonging to the same net in the upper cell row. The second segment is used to connect all the terminals belonging to the same net in the lower cell row. The third segment is used to establish a connection between the first two segments.

The selection of one terminal each from the upper and lower cell rows for establishing the connectivity between the two horizontal segments that belong to the same net, but located in adjacent rows is called *multirow net connection assignment problem*. Each time two terminals (of a net) from two adjacent cell rows are selected, they form a net in that channel. The objective of the MCAP is to select terminals in all the rows for all the nets, such that sum of densities of all the channels is minimum. We have shown that the MCAP is computationally hard [BMPS93]. As a result, we propose the following heuristic for this problem. Each net is assigned a weight $w(N_i)$, $1 \le i \le n$, which refers to the criticality of a net. If a net is highly critical, it is assigned a very high weight. The nets are considered in a nonincreasing order of weights. For each net, two adjacent rows are considered sequentially, and the connections are selected as follows. If the horizontal segments of a net in the adjacent rows overlap, then a single vertical segment in the overlap region is used to connect the horizontal segments in two rows if the net has terminals in the corresponding column in both the cell rows. Otherwise, we select one terminal from each row such that they are the closest pair of terminals in the overlap region. If the horizontal segments of the nets do not overlap, then the nearest end points of the segments are connected. See Figure 9.3 for an example of MCAP.

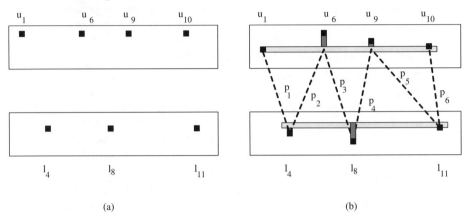

(a)   (b)

Fig. 9.3 An example for multirow net connection assignment.

9.4 Net Classification and Interval Generation

In this step, we initially classify each net as one of the five categories explained below, which intuitively indicates the difficulty involved in routing this net. After the net categorization, intervals are generated between the left-most and right-most terminals of a net in each cell row for all the critical and same row nets. Let t_1 and t_2 be the terminals of a net. The nets are considered in the nondecreasing order of their weights. The two terminal net connections can be classified as shown in Figure 9.4. In the following discussion, ROW(t) and COL(t) refer to row and column, respectively, to which the terminal t belongs. Type 0 nets are those nets in which ROW(t_1) = ROW(t_2), or COL(t_1) = COL(t_2). All the other types of nets do not satisfy the above two conditions. A net (t_1, t_2) is called a Type 1 net if either the terminal position (ROW(t_2), COL(t_1)) or (ROW(t_1), COL(t_2)), or both, are vacant. To distinguish between the three types of Type 1 nets, these nets are further classified as Type 1(a), Type 1(b), and Type 1(c) nets, depending on the condition they satisfy. They are routed in Metal 2 and Metal 3 layers. Type 1(c) nets can be routed in two different ways as shown in Figure 9.4.

If the terminal positions (ROW(t_2), COL(t_1)) and (ROW(t_1), COL(t_2)) are not vacant, then two vacant terminal positions t_3 and t_4 in ROW(t_1) and ROW(t_2), respectively, are used to route the net as a Type 2(a) net, as shown in Figure 9.4, where the vertical strip is in Metal 2. Even if t_3 and t_4 have already been used for

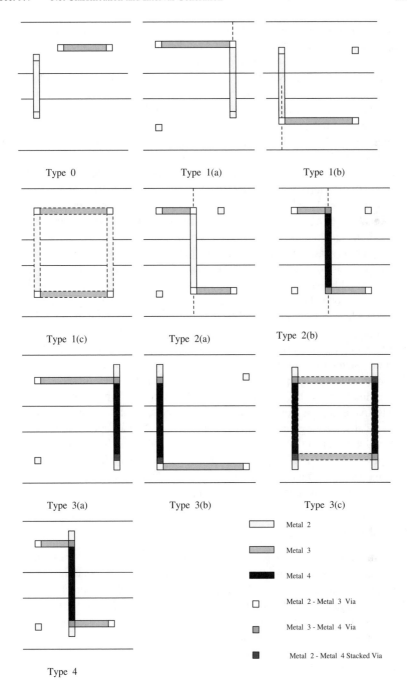

Fig. 9.4   Classification of nets.

a Type 2(a) net connecting two terminals in $ROW(t_1)$ and $ROW(t_2)$, the net under consideration can still be routed using these vacant terminal positions, so that the vertical strip is in Metal 4. This type of net is called a Type 2(b) net.

If a net cannot be routed as a Type 1 or Type 2 net, then an attempt is made to route it as a Type 3 net, as shown in Figure 9.4. For a Type 3 net, the terminal positions $(ROW(t_2), COL(t_1))$ and $(ROW(t_1), COL(t_2))$ are not vacant. It can use any one of these terminal positions if they are not already used for a similar net between $ROW(t_1)$ and $ROW(t_2)$. Like Type 1 nets, Type 3 nets can also be classified into three types. However, Figure 9.4 shows only a Type 3(a) net. If Type 3 is also not possible, then it can be routed as a Type 4 net, which is similar to Type 2, except that the terminal positions $(ROW(t_1), COL(t_3))$ and $(ROW(t_2), COL(t_4))$ are not vacant and are not already used for a similar type of net between $ROW(t_1)$ and $ROW(t_2)$. If a net cannot be routed as any of the above types, then it has to be routed in the channel between $ROW(t_1)$ and $ROW(t_2)$.

After net classification, intervals are generated for the nets in each cell row. These intervals correspond to the horizontal wire segments (Metal 3) in each row. Each interval has a via at both end points. Depending on the type of via, the end points are classified into the following types. An end point corresponding to a via t between a horizontal segment in Metal 3 and a vertical segment in Metal 2, where the Metal 2 segment spans from $ROW(t)$ to $ROW(t) - 1$, is called a Type a_1 end point. An end point corresponding to a similar type of via, but with the Metal 2 strip that spans from $ROW(t)$ to $ROW(t) + 1$, is called a Type b_1 end point. In a similar way, the end points corresponding to vias between Metal 3 and Metal 4 and Metal 2 to Metal 4 (stacked via) are classified as Type a_2 and Type b_2 end points. An end point corresponding to a via between Metal 2 and Metal 3, with no vertical segment going to another cell row, is called a Type c end point. From Figure 9.4 it is clear that a vacant column in a cell row can have only one via each of types a_1, b_1, a_2, and b_2, and no Type c vias. A column in a cell row having a terminal can have only one via each of types a_2 and b_2, and any number of Type c vias limited by the density at that column.

9.5 Interval Selection

The previous step generates two types of intervals: intervals corresponding to nets that have only one routing choice, and intervals corresponding to nets that have two routing choices. Let $R_1, R_2, \ldots, R_K$ be the cell rows in a layout, where K is the number of cell rows in the layout. Let $V_i, 1 \leq i \leq K$ be the set of intervals in R_i, corresponding to the nets that have only one routing choice. Let $V_{i\,i+1}, 1 \leq i < K$ be the set of intervals corresponding to the nets with two routing choices that can be routed in either R_i or R_{i+1}. The main objective of this step is to select the maximum number of intervals that can be assigned to contiguous tracks in each cell row, such that the intervals in each set $V_{i,i+1}, 1 \leq i < K$ are assigned to only one of the cell rows R_i and R_{i+1}. We call this the *interval selection problem*. It is quite difficult to solve the ISP optimally. At present, the complexity of ISP is unknown. We present a 0.5 approximation algorithm to solve this problem.

Let us consider two adjacent cell rows. Let V_i be the set of intervals of the nets that can be routed in cell row R_i and let V_{i+1} be the set of intervals of the nets that can be routed in the cell row R_{i+1}. Let $V_{i,i+1}$ be the set of nets that can be routed either in R_i or R_{i+1}.

The approximation algorithm for interval selection is based on MKIS algorithms in interval graphs. Although the algorithm is greedy in nature, we prove that it has a performance bound of 0.5. The details of our algorithm ALGO-IS are shown in Figure 9.5. In the algorithm, MIS refers to the MKIS in the interval graph.

Algorithm ALGO-IS( )

Input: Set of Intervals,
  $\mathcal{V} = \{V_1, V_{12}, V_2, \ldots V_{K-1\ K}, V_K\}$
Output: Interval Assignment for all cell rows

Begin
  $S_1 = \phi$;
  $S_1 = MIS(V_1 \cup V_{12})$;
  For $i = 2$ to $K - 1$
    $S_1 \cup MIS(V_i \cup V_{i\ i+1})$;
  $S_1 = S_1 \cup MIS(V_K)$;
  $S_2 = \phi$;
  $S_2 = MIS(V_1)$;
  For $i = 2$ to K
    $S_2 = S_2 \cup MIS(V_i \cup V_{i-1\ i})$;
  $S = MAX(S_1, S_2)$;
End;

Fig. 9.5   Algorithm ALGO-IS.

Theorem 40   *Let ρ be the approximation ratio of the above algorithm. Then* $\rho \geq 0.50$.

Proof: Let W_i^* be the subset of V_i, which is in the optimal solution. Similarly, assume $W_{i,i+1}^*$ is a subset of $V_{i,i+1}$, which is in the optimal solution. Each $W_{i,i+1}^*$ can be partitioned into $U_{i,i+1}^*$, $D_{i,i+1}^*$, where $U_{i,i+1}^*$ is a subset of $W_{i,i+1}^*$ assigned to cell row i, and $D_{i,i+1}$ is a subset of $W_{i,i+1}^*$ assigned to cell row $i + 1$.

The algorithm is based on two strategies as shown in Figure 9.5. The first strategy guarantees that

$$| S_1 | \geq | W_1^* | + | W_2^* | + \cdots | W_K^* |$$
$$+ | U_{12}^* | + | U_{23}^* | + \cdots + | U_{K-1,K}^* |$$

Similarly, the second strategy guarantees that

$$| S_2 | \geq | W_1^* | + | W_2 * \cdots + | W_K^* |$$
$$+ | D_{12}^* | + | D_{23}^* | + \cdots + | D_{K-1,K}^* |$$

Let α be the ratio of nets in the optimal solution, which are from sets $V_{i,i+1}, 1 \leq i \leq K - 1$. Obviously, $1 - \alpha$ is the ratio of the nets belonging to sets $V_i, 1 \leq i \leq K$, which are in the optimal set.

It is clear that both strategies select $1 - \alpha$ subset of nets. To see what fraction of α nets is chosen, notice that in the worst case, $\alpha = 1$; that is, the optimal solution may consist of nets that are only from $V_{1,2}, V_{2,3}, \ldots V_{K-1,K}$. By taking the maximum between S_1 and S_2, we guarantee that we will always select at least 0.5α. Therefore, the complete solution is

$$\rho = 1 - \alpha + 0.5\alpha$$
$$= 1 - 0.5\alpha$$

This ensures that in the worst case where $\alpha = 1$, $\rho = 0.5$. Therefore, the algorithm produces a solution which is at least 50% of the optimal.

□

9.6 Track Assignment

In order to simplify the routing, all the terminals in each cell row are brought to a vacant track in that cell row so that all the terminal points in the net intervals are connected to their respective terminals on this vacant track by a vertical strip in M2. Now the problem is to determine the appropriate track in a cell row to be left vacant for these terminals so that the total length of the vertical strips in M2 is minimized. This problem can be formally stated as follows.

Instance: Given n terminals $t_1, t_2, \ldots, t_n$ in k tracks, $T_1, T_2, \ldots, T_k$, assume that any two consecutive tracks T_i and T_{i+1} have a virtual track T_i' in between. Let $d(t_j, T_i')$ be the vertical distance between t_j in track T_m and T_i', which is given by

$$d(t_j, T_i') = | i - m | + 1$$

Problem: Find a track T'_p such that

$$\sum_{q=1}^{n} d(t_q, T'_p)$$

is minimized. We call this the *track assignment problem*. (See Figure 9.6 for an example.) This problem is similar to the *single-trunk Steiner tree problem* [CLR90], and can be solved in linear time. Therefore, we have the following result.

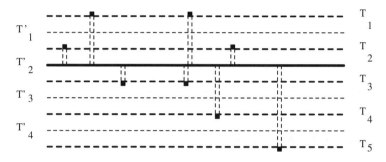

Fig. 9.6   An example of the track assignment problem.

Theorem 41   *The track assignment problem can be solved in $O(n)$ time.*

9.7 Interval Assignment/Same-Row Routing

In this phase, the intervals generated in the previous phase for each row are sorted on the left end point and assigned using a minimum number of tracks k', such that $k' \leq k$, where k is the number of tracks in a cell row. Let the tracks in a cell row be denoted by $T_1, T_2, \ldots, T_k$. The intervals should be assigned in such a way that at any column in a row, the end points of Type a_1 and Type b_1 are in tracks T_i and T_j such that $T_i < T_j$. The same rule applies to Type a_2 and Type b_2 end points in a column. Type c end points can be assigned to any track.

Figure 9.7 shows all the possible cases that can occur when assigning intervals with end points of Type a_1 and b_1. Cases I(a), II(a), III(a), and IV(a) are legal assignments and the rest are not legal. If the assignment is not legal, then let T_i and T_j be the tracks assigned to intervals I_1 and I_2 with end points of types a_1 and b_1, respectively, at a particular column, such that $T_i < T_j$. First, we try to reassign I_1 to a track T_l such that $T_l > T_j$. If this is not possible, then an attempt is made to assign I_2 to a track T_m, such that $T_m < T_i$. If this is also not possible, then the corresponding net $(\text{ROW}(i_1), \text{COL}(j_1), \text{ROW}(i_2), \text{COL}(j_2))$ is added to the list of nets, to be routed in the channel between $\text{ROW}(i_1)$ and $\text{ROW}(i_2)$. We define a routing block B_i of a cell row R_i as the set of tracks to which the intervals are assigned in a cell row.

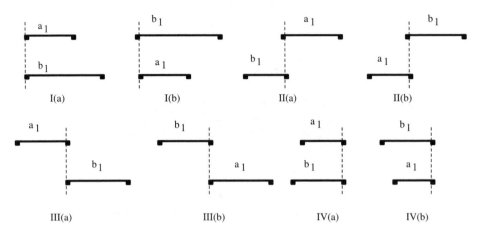

Fig. 9.7 All possible cases that occur when assigning intervals with end points of Types a_1 and b_1.

9.8 Routing Block Assignment

The routing blocks generated in the previous step are assigned to tracks in each cell row. First, the densities between routing blocks belonging to two adjacent cell rows are computed. Based on the densities, the routing blocks are assigned to tracks in each cell row. We present an optimal algorithm ALGO-RBA which runs in $O(K)$ time to solve the RBAP.

After interval assignment, routing blocks are generated for each cell row. Each routing block B_i defines a fixed set of intervals to be assigned to a contiguous set of tracks over a cell row R_i. In this step, we develop an optimal algorithm to locate the routing blocks in each cell row, such that minimum layout height is obtained. In Figure 9.8, the routing blocks are shown with dotted lines. The height of the routing block B_i is based on the intervals generated in IAS.

First, we present the formal statement of the RBAP. A greedy algorithm called ALGO-RBA is then proposed to solve the RBAP. Let us start with some terminology.

Let $\mathcal{R} = \{R_1, R_2, \ldots, R_K\}$ be the set of cell rows. Let $\mathcal{B} = \{B_1, B_2, \ldots, B_K\}$ be the set of routing blocks, such that B_i lies on cell row R_i and $|B_i|$ gives the number of tracks in B_i. Let $\mathcal{D} = \{d_1, d_2, \ldots, d_{K-1}\}$, where d_i is the density of nets between the routing blocks B_i and B_{i+1}. Let k be the total number of tracks in each cell row. The tracks in each cell row are numbered from top to bottom, in an ascending order, from 1 to k. For each routing block $B_i \in \mathcal{B}$, let TOP(B_i) and BOT(B_i) represent the top-most and bottom-most tracks, respectively, assigned to B_i. Let h_i be the height of the ith channel. h_i depends on the density not accommodated in the OTC area between the blocks B_i and B_{i+1}.

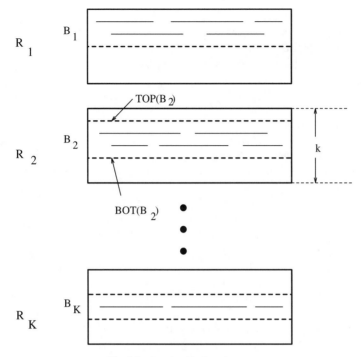

Fig. 9.8   Routing block assignment.

Instance: Given $\mathcal{R}$, $\mathcal{B}$, $\mathcal{D}$ as described above.

Problem: Assign a routing block in R_i; that is, assign TOP(B_i) to a track in R_i from $\{1, 2, \ldots, (k - \mid B_i \mid)\}$, for $i = 1$ to K, such that

$$H = \sum_{i=1}^{K-1} h_i$$

is minimized.

A top-down approach is considered for the routing block assignment. The upper-most locations are considered in each cell row for placing the routing block, depending on the net densities between the blocks. The greedy algorithm called ALGO-RBAP is formally stated in Figure 9.9.

Theorem 42   *The algorithm ALGO–RBAP generates an optimal solution for RBAP in $O(K)$ time, where K is the number of cell rows in a layout.*

Proof: Deleted for the sake of brevity.

□

Algorithm ALGO–RBA
Input: $\mathcal{R}, \mathcal{D}, \mathcal{B}$
Output: Routing-Block Assignment
on each Row.

---

begin Algorithm
    $TOP(B_1) = 1;$
    for $i = 1$ **to** $(K - 1)$ **do**
        if $d_i > (2k - TOP(B_i) - \mid B_i \mid + 1 - \mid B_{i+1} \mid)$
        then
          $h_i = \frac{d_i - (2k - TOP(B_i) - \mid B_i \mid + 1 - \mid B_{i+1} \mid)}{2};$
          $TOP(B_{i+1}) = k - \mid B_{i+1} \mid + 1;$
        else if $d_i > k - TOP(B_i) + 1$
          $h_i = 0;$
          $TOP(B_{i+1}) = d_i + TOP(B_i) + \mid B_i \mid - k$
        else
          $h_i = 0;$
          $TOP(B_{i+1}) = 1;$
end Algorithm ALGO–RBAP

Fig. 9.9 Algorithm for Routing Block Assignment.

9.9 Multilayer Routing in Full-Custom Layouts

In this section, we describe the basic methodology of our approach to multilayer routing in full-custom layouts. A full-custom layout may consist of several arbitrarily shaped rectilinear blocks. We assume that the terminals are located in M1 and poly layers. This imposes a restriction on the block design, with the result that M2 cannot be used for routing within the blocks. Therefore, it may not be possible to route all the *intrablock nets*. However, in our approach, we route the unrouted intrablock nets over the blocks, along with the *interblock nets*. By doing so, we also give the block/cell designer the flexibility of leaving the terminals of the unrouted intrablock nets at arbitrary positions in the block.

Figure 9.10 shows a full-custom layout with interblock nets and intrablock nets, which could not be routed in the blocks using M1 and poly layers. All these nets are routed over the blocks. The OTB routing reduces the area of the layout, since it decreases the area of the channels between the blocks where the interblock nets would have been routed otherwise. In fact, our router does not use the channels between the blocks in a layout for routing. The channels in this case are located on top of the active areas. The routing of such layouts using our approach involves the following steps:

1. Pseudorow generation
2. Global routing
3. Net classification
4. Interval generation
5. Interval assignment
6. Routing block assignment
7. Interrow routing

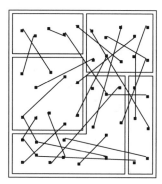

Fig. 9.10   A full-custom layout with net connections for OTB routing.

In the pseudorow generation phase, we partition the entire layout into pseudorows such that any vertical column in a pseudorow can have at most one terminal. In order to accomplish this, we use the M2 layer to position the terminals so that we have at most one terminal in each column of a pseudorow. The terminal can be located anywhere in the column. After generating the pseudorows, the rest of the steps followed are similar to those in our standard-cell routing approach. However, the M2 layer is not used any further, and the remaining routing is done in layers M3 and above.

Figure 9.11(a) shows an example of a general cell layout. The partitioning of the layout into pseudorows is shown in Figure 9.11(b). Figure 9.11(c) shows the layout after the routing block assignment phase. The interrow routing is done in the space between two adjacent routing blocks called the *channel*.

If during any of these phases it is found that the given layout is unroutable, then we have to go back to the placement phase and rearrange the blocks and route the layout again. This process is repeated until the placement makes the layout routable.

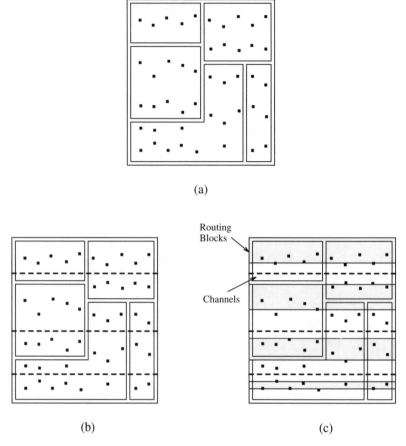

Fig. 9.11 **(a)** Partitioning a general cell layout; **(b)** pseudorows; **(c)** routing block assignment.

9.10 Topological Routing Approach

In [DDS91], Dai, Dayan, and Staepelaere developed a multilayer router based on rubber-band sketch routing. This router uses hierarchical top-down partitioning to perform global routing for all nets simultaneously. It combines this with successive refinement to help correct mistakes made before more detailed information is discovered. Layer assignment is performed during the partitioning process to generate routing that has fewer vias and is not restricted to one layer and one direction. The detailed router uses a region connectivity graph to generate shortest path rubber-band routing.

The router has been designed primarily for routing MCM substrates, which consist of multiple layers of free (channel-less) wiring space. Since MCM substrate designs have potentially large numbers of terminals and nets, a router of this nature must be able to handle large designs efficiently in both time and space. In addition, the router should be flexible and permit an incremental design process. That is, when small changes are made to the design, it should be able to be updated incrementally and not be recreated from scratch. This allows faster convergence to a final design. In order to produce designs with fewer vias, the router should be able to relax the one-layer one-direction restriction. This is an important consideration in high-speed designs, since the discontinuities in the wiring caused by bends and vias are a limiting factor for system clock speed.

In order to support the flexibility described above, the router must have an underlying data representation that models planar wiring in a way that can be updated locally and incrementally. For this reason, SURF models wiring as rubber bands [CS84, LM85]. Rubber bands provide canonical representation for planar topological wiring. Because rubber bands can be stretched or bent around objects, this representation permits incremental changes to be made that only affect a local portion of the design. A discussion of this representation is in [DKJ90].

Once the topology of the wiring is known, the rubber-band sketch can be augmented with spokes to express spatial design constraints, such as wire width, wire spacing, and via size. [DKS91]. Since successful creation of the spoke sketch guarantees the existence of geometrical wiring (Manhattan or octilinear), the final transformation to fixed-geometry wiring can be delayed until later in the design process. This allows most of the manipulation to take place in the more flexible rubber-band format. Figure 9.12 shows different views of the same wiring topology. These represent various states of the rubber-band representation.

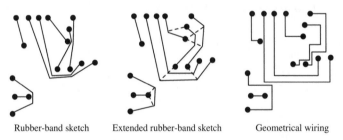

Rubber-band sketch   Extended rubber-band sketch   Geometrical wiring

Fig. 9.12   Rubber-band representations.

In this context, a topological router has been developed that produces multilayer rubber-band sketches. The input to this router is a set of terminals, a set of nets, a set of obstacles, and a set of wiring rules. These rules include geometrical design rules and constraints on the wiring topology. The topological constraints may include valid topologies (daisy chain, star, etc.), as well as absolute and relative bounds on segment lengths. The output of the router is a multilayer rubber-band sketch in which all the points of a given net are connected by wiring. However, the routability of a sketch is not guaranteed until the successful creation of spokes. At each stage, the router uses the increasingly detailed information available to generate a sketch without overflow regions. This increases the chance that the sketch can be successfully transformed into a representation (the spoke sketch) that satisfies all of the spatial constraints. In addition, the router tries to reduce overall wire length and the number of vias. A more detailed analysis of the routability of a rubber-band sketch is described in [DKS91].

9.10.1 Algorithm Overview

Topological routing is done in two steps: global routing and local routing. The global router determines rough net topology and partitions the routing area into a set of bins. The interfaces between individual bins are specified by placing crossing points on the bin boundaries for each net that crosses the boundary. These cross points specify a crossing position and layer. The local router then routes the individual bins independently.

The global routing approach employs two principles from the field of artificial intelligence: the least commitment principle and the notion of maximal use of information. The least commitment principle states that if the correct choice in a decision is not known, the decision should be delayed. This guards against making arbitrary decisions early, which, if wrong, could adversely affect the outcome. Maximal use of information states that all available relevant information should be applied to solving the problem. In early stages of partitioning, we reduce our commitment by allowing the cross points a wide range of movement along the cut line. Later, after further partitioning, we use the new and more detailed information to reassign the cross points. The only commitment at each level is the assignment of cross points to specific interfaces. When a cross point is assigned to an interface, it will remain there.

Once the global router has partitioned the problem into bins, the local routing is performed. The local routing is done one net at a time within the limits of a bin. This limits the search to a single bin and improves the time and space efficiency of the router. The local router uses a region graph representing the geometrical and topological adjacency of areas of the sketch. By using this graph, the router can efficiently find the shortest planar path through the partially routed bin. Both global and local routers rely heavily on an underlying data structure built on constrained Delaunay triangulation [Che89]. The Delaunay triangulation of a set of points is the straight-line dual of a voronoi diagram for the set. An important property of the Delaunay triangulation is that the circumcircle of each triangle contains no points in its interior. Based on the Delaunay triangulation, such problems as closest pair, all nearest neighbors, and euclidean minimum spanning tree (MST) in the plane can be solved in linear time. Constrained Delaunay triangulation includes edges that are forced to be part of the triangulation. SURF uses constrained edges to represent rubber-band segments, obstacles, etc. The global router relies on fast MST generation for calculating the cost matrices used to determine crossing point locations. The local router relies on the triangulation for performing shortest path calculations within its region graph.

9.10.2 Global Routing

The approach taken by SURF for global routing is a hierarchical top-down partitioning technique based on the approaches described in [HS85, Lau87, MS86a]. The basic approach is to repeatedly divide the routing problem into smaller subproblems (bins) until they reach an appropriate size for the local router. At every step, each of the current subproblems is partitioned into two smaller problems. The approach taken to partition a single bin is to determine a cut line that divides the bin into two pieces, and for each net in the bin that crosses the cut line, determine a place on the cut line for a net to cross. New artificial points are generated for each of the cut points. The cut line and crossing points form an interface between the two smaller bins they create. Although the cross points are tentatively assigned positions and layers along the interface, they are not fixed and may be repositioned and reordered within the limits of their interface.

The sketch is divided until all of the subproblems reach an appropriate complexity for the local router. Because the sizes and shapes of bins are flexible, the decision of when to stop cutting is made individually for each bin. One of the critical decisions in the partitioning process is selecting a proper cut line. For any given subproblem, the choice has to be made whether to cut vertically or horizontally and where along the x- or y-axis to make the cut. The decision of which cut line to select is made by examining the aspect ratio, the net flow, and the cut capacity. When dividing the bins, it is desirable to keep the bins as close to square as possible. For this reason, the orientation of the cut is chosen so that it cuts the larger axis of the bin. The flow is the number of nets that must cross the cut. The capacity is the number of net crossings that the cut can accommodate.

Once a cut line has been selected for a given bin, crossing points (tentative positions and layers) are determined for each net that crosses the cut line. The matching of nets to slots is done by modeling the problem as an integer network flow problem and determining a max-flow, min-cost solution [Lau87, MS86a]. The problem is formulated as a complete bipartite graph $V = G(V_n, V_s, E)$, where V_n is the set of nets needing to cross the cut line and V_s is the set of crossing slots. Each net n in V_n represents a unit source. Each crossing slot $s \in V_s$ is a sink with a capacity determined by the size of the slot and the obstacles in the slot. Each edge (n, s) has unit capacity and a cost associated with a desirability of assigning net n to slot s. The assignment of nets to slots is determined by finding a maximum flow assignment with minimum cost. The cost of edge (n, s) is a linear combination of three components: a wire length cost, a via cost, and a routability cost. The via cost component is used to reduce unnecessary plane switching. The routability cost is intended to increase the chances of successfully routing the sketch by encouraging wiring that is roughly horizontal to lie in one layer, and routing that is generally vertical to lie in the other layer. As the partitioning process proceeds, the interfaces between bins are repeatedly divided by the cut lines into smaller interfaces. At each step, when an interface is partitioned, the tentative positions of the crossing points in the original interface are used to assign them to one of the new smaller interfaces. Figure 9.13 shows how interface crossing point reassignment can be used to improve the positions of crossing points. Figure 9.13(a) shows a crossing point assigned before information about dense regions is discovered. Figure 9.13(b) shows the locations of crossing points on further cut lines. These are positioned to avoid cut lines. Figure 9.13(c) shows that the original crossing point has been refined using the information from (b). As the partitioning process continues, it is possible that a mistake made at a higher level will be detected at a lower level. It is possible that some locally dense regions may be overlooked at higher levels of partitioning. By the time the dense region is discovered, the problem may be too restricted to cut it without producing an overflow cut. The router will attempt to rectify the situation by backtracking and cutting the critical region at a higher level.

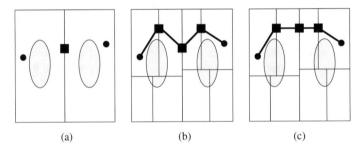

(a)   (b)   (c)

Fig. 9.13   Example of interface refinement.

9.10.3 Local Routing

Once the rough net topology is determined by the global router, the local router generates wiring that conforms to this topology. It routes the bins produced by the global router one at a time. The basic approach used by the local router is to route the nets one at a time using a shortest path algorithm. The strategy is to transform partial nets into complete nets by adding points to them one at a time. Adding a point to an existing partial net consists of two steps: (1) finding a topological path from the point to the existing partial net, and (2) transforming the topological path into a valid rubber-band path. In order to find such graph, a region graph is used to represent topological adjacency information. The region graph is derived from the rubber-band representation of the bin. The nodes in this graph are regions. Figure 9.14 shows the regions associated with two rubber-band points. The triangles denote incident regions. Incident regions are those that actually "touch" the point in question. The squares denote attached regions.

Fig. 9.14 Regions of two rubber-band
        points.

The edges in the region graph represent information about geometrical and topological adjacency. Two regions are topologically adjacent if they can be connected with planar wiring. Two regions in the region graph share an edge if they are topologically adjacent and their points share an edge in the constrained Delaunay triangulation. Once a path from a new net point to the existing partial net has been found in the region graph, a valid rubber-band path must be created. The path specified by the region search is guaranteed to be planar, but is not guaranteed to be a valid rubber band. The validation process is an iterative one that replaces each pair of invalid segments with a set of zero or more valid ones.

For multilayer routing, for each point the new region graph has regions for all the layers. Two regions are considered adjacent if they are on the same layer and adjacent in the single-layer case, or they are on different layers and belong to the same point. If the path in the region graph for the multilayer case includes vias, they are added to the sketch. The paths in each layer are then validated independently, as in the single layer-case.

9.11 Experimental Evaluation of Routers

The router generates channel-less designs for all the benchmarks it has been tested on. The routing algorithm first generates the routing blocks. The routing blocks are assigned to the tracks in each cell based on the net densities. The height of the routing blocks and their assignment for each channel of the benchmark PRIMARY1 are shown in Table 9.2. Notice that the multirow nets are split into same-row and adjacent-row nets, and the adjacent-row nets are joined using vertical lines. Hereafter, we will refer to the routing space between the routing blocks as a *channel* for simplicity, though it is over the cell. The maximum height of the routing blocks is between channels 3 through 7. In particular, channel 3 has 21 tracks. Also notice that PRIMARY1 is dense only in the top right corner of the layout and very sparse toward the left and bottom of the layout. The location of terminals for vertical tracks between two routing blocks is considered to be at the top-most and bottom-most tracks of a routing block. From these locations, M2 segments are used to connect to the actual terminals. The entire routing solution generated by routes for PRIMARY1 is shown in Figure 9.15. A total of 657 net segments have been assigned to the routing blocks for the benchmark PRIMARY1. Table 9.3 shows the number of nets assigned to each routing block. The remaining nets are routed in OTC areas between the routing blocks. The router takes 13.65 seconds to generate the channel-less solution of PRIMARY1. The running times for various phases of the routing algorithm are given in Table 9.4. Notice that 90% of the running time is used for net decomposition, categorization, and interval assignment.

Table 9.2:   Routing Block Assignments in PRIMARY1

Row No.	Routing Block		
	Height	TOP Track	BOT Track
1	12	1	12
2	14	1	14
3	16	1	16
4	21	1	21
5	15	4	18
6	15	4	18
7	15	3	17
8	11	6	16
9	9	2	10
10	10	1	10
11	8	1	8
12	8	1	8
13	7	1	7
14	11	1	11
15	7	1	7
16	12	1	12
17	7	1	7

Table 9.3:   Routing Block Net Assignment in PRIMARY1

Row No.	Routing Block Height	Number of Nets
1	12	35
2	14	40
3	16	43
4	21	44
5	15	49
6	15	38
7	15	41
8	11	42
9	9	43
10	10	37
11	8	36
12	8	37
13	7	36
14	11	35
15	7	38
16	12	37
17	7	26

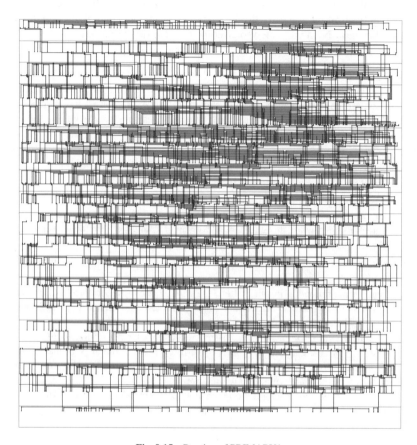

Fig. 9.15   Routing of PRIMARY1.

Table 9.4:   Time Taken for Routing PRIMARY1

Sl. No.	Routing Step	Time (sec)
1	MCA, IGC, IAS	12.23
2	RBA	0.62
3	HVHV/VHV routing	0.80
	Total Time	13.65

9.12 Summary

Chipwide and thin-film routing problems are similar in nature. The chip routing complexity is smaller, but has fixed routing layers, while the density and routing complexity is large for MCMs and has an unlimited number of layers. Essentially, these are four-, five-, or six-layer area routing problems with terminals arbitrarily located in the routing plane. In this chapter, virtual channel-based and topological approaches have been presented to effectively solve this problem.

PROBLEMS

9-1. In some high-performance designs, some OTC areas are considered critical areas ("forbidden areas") and vias are not allowed in these areas. Develop an efficient four-layer OTC router for designs with forbidden regions.

9-2. Develop a zero-skew clock routing algorithm for standard-cell designs using multilayer processes. Restrict clock lines to run in second and third metal layers only.

9-3. Extend the above problem to designs with forbidden regions.

9-4. Consider an MBC layout using a four-layer fabrication process. The blocks use all four metal layers for intrablock routing. The cells use only the first two metal layers for intracell routing, while the other metal layers in OTC areas are completely unblocked. Develop an efficient routing scheme for this problem.

9-5. Develop parallel processing-based routing algorithms for topological and channel-based routers.

9-6. Consider the following instance: Let $T_1, T_2, \ldots, T_k$ be k pseudo-terminal rows to be assigned in a layout with t tracks and arbitrarily located terminals. Let $loc(T_i)$ be the track to which T_i is assigned.

The pseudo-terminal rows are assigned such that $1 \leq loc(T_1) < loc(T_2 < \cdots < loc(T_k) \leq t$. We call this the *order constraint*. Each terminal is connected to its nearest track by a vertical metal strip. Let $w(T_i)$ be the total length of the vertical strips of wires that are connected to the terminal row T_i.

The smallest wire length virtual channel problem (SVCP) is to assign tracks to the pseudo-terminal rows without violating the order constraints, such that

$$\sum_{i=1}^{k} w(T_i)$$

is minimized.

Develop an efficient algorithm to solve SVCP.

BIBLIOGRAPHIC NOTES

Katsadas and Kinnen [KK90] present a multilayer router using OTC areas. This router uses two steps. A selected group of nets is routed in between cell areas using channel routing algorithms and the first two routing layers. Then the remaining nets are routed over the entire layout area, between cell and OTC areas using a new two-dimensional router and two metal layers.

Cong, Wong, and Liu [CWL87] present a new approach to three- and four-layer channel routing. A multilayer channel router using one-, two-, and three-layer partitions is presented in [GS88]. New models for four- and five-layer channel routing are presented in [Ho92]. A multiple-layer contour-based gridless channel router is presented in [Gro90]. Lunow [Lun88] presents a channel-less multilayer router. The rectilinear Steiner tree problem is proved to be NP-complete in [GJ77].

10

Routing Algorithms for General MCMs

MCM technology has been advancing rapidly in recent years, since it is the only feasible packaging technology that provides the performance capability that can meet the increasing demand for high-performance systems. As MCM technology develops, the number of chips mounted on an MCM substrate and the number of layers available for routing increase dramatically. For example, an MCM with more than 100 chips and 63 layers has been recently reported [RRT92]. Therefore, both the capacity and demand on the routing resources have increased rapidly and will continue to increase. Though MCM technology is developing rapidly, CAD tools for MCM lag behind. In particular, the routing tools are incapable of handling the complexity of MCM routing problems, thereby creating a bottleneck in the further development of MCM technology. Most of the existing MCM routing tools are based on either PCB or VLSI routing tools. However, PCB routing tools cannot handle the density of MCM routing problems, while VLSI routing tools are suitable for routing problems with a limited number of layers, typically two or three layers, as discussed in earlier chapters.

The MCM routing environment is distinctly different from the PCB and VLSI routing environments, and it can be characterized as a truly three-dimensional routing medium. Another aspect of the MCM routing problem is the very large number of nets and the large grid size. An MCM with over 7,000 nets and grid size of over 2,000 × 2,000 has been reported in [KC92]. It is expected that the number of layers, number of chips mounted on a substrate, number of nets, and grid size will continue to increase with fabrication technology advances. Unlike the routing problems in VLSI or PCB, where the number of layers required for routing is very critical, the number of layers required for routing is no longer the most critical factor for MCMs, as the number of layers of MCMs has increased dramatically over the years (e.g., there are MCMs with 63 layers). However, other factors, such as performance constraints, which are not dominant in VLSI or PCB routing problems, become critical in MCM problems [DF93, TTLF92]. The performance constraints include manufacturability constraints, net-length constraints, net-separation constraints, and via constraints. Yet another consideration is the growing complexity of MCM problems. As the complexity of MCM routing problems increases, the uniprocessor approach will not be suitable for very large MCM routing problems. Therefore, a parallel approach has to be adopted for the large MCM routing problems. The final consideration is early estimation. The MCM design cycle is long and expensive. Therefore, it is very critical for the routing tool to be capable of both routing the nets for an MCM as well as provide an estimation of the required routing resources as early as possible (e.g., the number of layers, the total wire length, and the length of the critical nets). By estimating the required routing resources early, the routing tool can also be used to judge the quality of the placement so that any "bad" placement can be detected at an early stage, thereby reducing the design cost. Further discussion on early estimation can be found in [LaP91].

The objective of minimizing the routing area in an IC design is no longer valid in the MCM routing environment. Instead, the objective of MCM routing is to optimize performance and minimize the number of layers, since the cost of an MCM depends on the number of layers used. Because of the long interconnect wires involved in MCM design, crosstalk and skin effect become important considerations, which are not of much concern in IC layout. In particular, in MCM-D, the skin effect of the interconnect becomes more severe. The parasitic effects also degrade the performance if not accounted for in the routing of MCMs.

Power and ground signals do not complicate global routing, because these signals are distributed on separate layers, and taps to the power supply layers are easy to make. However, overall dimensions must be tightly controlled (to fit within the package) and the packaging delay must be carefully controlled.

In summary, a good MCM routing approach should be three-dimensional, performance constraint-driven, provide early estimation, and the approach should be amiable to parallel processing if required (see Figure 10.1).

Few results have been reported for MCM routing problems. Some approaches [Dai91, HSVW90, PPC89] decompose the routing phase into a pin redistribution phase, where pins on the chip layer are first redistributed evenly, a layer assignment phase, where each net is assigned a layer or a pair of layers, and a detailed routing phase, which actually finds and assigns a signal path. Recently, a new approach has been developed by Khoo and Cong [KC92], called SLICE, which integrates pin redistribution and routing. The basic idea of SLICE is to perform planar routing on a layer at a time. The latest development in MCM routing is the V4R router developed by Khoo and Cong [KC93]. The V4R router has re-

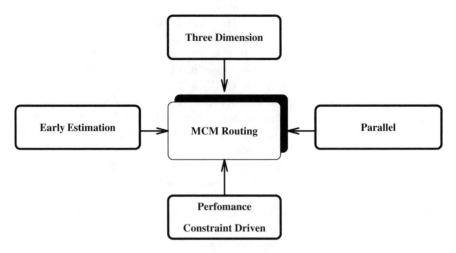

Fig. 10.1   MCM routing approach.

duced the time complexity and the number of layers used for routing as compared
to SLICE. The V4R router uses two adjacent layers and routes a column one at a
time from left to right by using four vias for most nets.

In [YBS93], a completely new routing methodology for MCM problems was
presented. Rather than converting the three-dimensional routing problems into
two-dimensional routing problems, this routing methodology decomposed three-
dimensional routing into several smaller three-dimensional problems to achieve
the best utilization of the three-dimensional routing resources.

In this chapter, we first present the different routing approaches currently
available for MCMs and then discuss them in detail.

10.1 Routing Approaches

The routing of an MCM is a three-dimensional general area routing problem where
routing can be carried out almost everywhere in the entire multilayer substrate.
However, the pitch spacing in an MCM is much smaller, and the routing is much
denser as compared to conventional PCB routing. Thus, traditional PCB routing
algorithms are often inadequate in dealing with MCM designs.

There are four distinguished approaches for general (nonprogrammable)
MCM routing problems:

1. Maze routing approach
2. Multiple stage routing approach
3. Integrated pin distribution and routing approach
4. Three-dimensional routing approach

10.2 Maze Routing Approach

The most commonly used routing method is three-dimensional maze routing. Although this method is conceptually simple to implement, it suffers from several problems. First, the quality of the maze routing solution is very much sensitive to the ordering of the nets being routed, and there is no effective algorithm for determining good net ordering in general. Moreover, since the nets are routed independently, global optimization is difficult and the final routing solution often uses a large number of vias despite the fact that there is a large number of signal layers. This is due to the fact that the maze router routes the first few nets in a planar fashion (using shorter distances), and the next few nets use a few vias each as more and more layers are utilized. The nets routed toward the end tend to use a very large number of vias, since the routing extends over many different layers. Finally, three-dimensional maze routing requires a long computational time and large memory space.

10.3 Multiple-Stage Routing Approach

In this approach, the MCM routing problem is decomposed into several subproblems. The close positioning of chips and high pin congestion around the chips require separation of pins before routing can be attempted. Pins on the chip layer are first redistributed evenly with sufficient spacing between them so that the connections between the pins of the nets can be made without violating the design rules. This redistribution of pins is done using few layers beneath the chip layer. This problem of redistributing pins to make the routing task possible is called *pin redistribution*. After the pins are distributed uniformly over the layout area using pin redistribution layers, the nets are assigned to layers on which they will be routed. This problem of assigning nets to layers is known as the *layer assignment problem*. The layer assignment problem resembles global routing of the IC design cycle. Similar to global routing, nets are assigned to layers in such a way that the routability in a layer or in a group of layers is guaranteed, and at the same time the total number of layers used is minimized. The layers on which the nets are distributed are called *signal distribution layers*. Detailed routing follows the layer assignment. Detailed routing may or may not be a reserved-layer model. Horizontal and vertical routing may be done in the same layer or in different layers. Typically, nets are distributed in such a way that each pair of layers is used for a set of nets. This pair is called an *x-y plane pair*, since one layer is used for horizontal segments, while the other one is used for vertical segments. Another approach is to decompose the net list such that each layer is assigned a planar set of nets. Thus, MCM routing problems become a set of *single-layer* problems. Yet another routing approach may combine the *x-y* plane pair and single-layer approaches. In particular, the performance-critical nets are routed in top layers using single-layer routing because x-y plane pair routing introduces vias and bends, which degrade performance.

We now discuss each of these problems in greater detail in the following subsections.

10.3.1 Pin Redistribution Problem

As mentioned before, in MCM routing, the first pins attached to each chip in the chip layer are redistributed over the uniform grid of the pin redistribution layers. The pin redistribution problem can be stated as follows. Given the placement of chips on an MCM substrate, redistribute the pins using the pin redistribution layers such that one or more of the following objectives are satisfied (depending on the design requirements):

1. Minimize the total number of pin redistribution layers.
2. Minimize the total number of signal distribution layers.
3. Minimize the crosstalks.
4. Minimize the maximum signal delay.
5. Maximize the number of nets that can routed in planar fashion.

Note that the separation between the adjacent via grid points may affect the number of layers required [CS91]. The pin redistribution problem can be illustrated by the example shown in Figure 10.2. The terminals of chips need to be connected to the vias shown in Figure 10.2(a). Usually, it is impossible to complete all the connections. In this case, we should route as many terminals as possible (shown in Figure 10.2(b)). The unrouted terminals are brought to the next layer and routed in that layer as shown in Figure 10.2(c). This procedure is repeated until each terminal is connected to some via. In [CS91], various approaches to the pin redistribution problem have been proposed.

10.3.2 Layer Assignment

The main objective of layer assignment for MCMs is to assign each net to an x-y pair of layers subject to the feasibility of routing the nets on a global routing grid on each plane pair. This step determines the number of plane pairs required for a feasible routing of nets and is therefore an important step in the design of the MCM. The cost of fabricating an MCM, as well as the cooling of the MCM when it is in operation, is directly related to the number of plane pairs in the MCM, and thus it is important to minimize the number of plane pairs. There are two approaches known to the problem of layer assignments [HSVW90, SK92]. The problem of layer assignment has been shown to be NP-complete [HSVW90].

Ho, Sarrafzadeh, Vijayan, and Wong [HSVW90] have presented several approximation and heuristic algorithms for plane pair assignment of two-terminal and multiterminal nets. In the following, we discuss the algorithm LAYER-DR for layer assignment for two-terminal nets of types U, R, and DR among six types

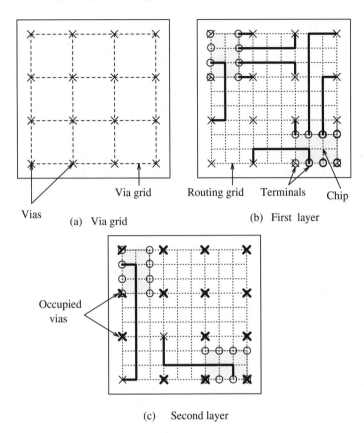

(a)   Via grid   (b)   First layer

(c)   Second layer

Fig. 10.2   Pin redistribution example.

shown in Figure 10.3. Let $\mathcal{N} = \{N_1, N_2, \ldots, N_n\}$ be a set of two-terminal nets of types U, R, and DR. The *layout environment* is represented by a two-dimensional $m \times m$ grid, with each grid of size $w \times w$. Each net has terminals in two tiles. In a layout environment, let $v(i, j)$ denote the number of nets crossing the border of tiles (i, j) and $(i, j + 1)$, $1 \le i \le m$ and $1 \le j \le m - 1$. Similarly, let $h(i, j)$ denote the number of nets crossing the border of tiles (i, j) and $(i + 1, j)$, $1 \le i \le m - 1$ and $1 \le j \le m$. $v_{\max} = \max_{i,j} v(i, j)$ and $h_{\max} = \max_{i,j} h(i, j)$ are the *vertical* and *horizontal* densities of the problem, respectively. The *layering environment* is a collection of plane pairs $P_1, P_2, \ldots, P_p$ placed on top of each other. The objective is to map each net to one of the plane pairs, with at most w nets crossing each side of a tile in any plane pair and to minimize p (the number of plane pairs used). Clearly, $p \ge \lceil d/w \rceil$. The layer assignment problem can be viewed as a net coloring problem; a net colored k means it is assigned to plane pair P_k. The algorithm LAYER-DR processes the plane grid in row-major-order, starting with tile $(1, 1)$ (see Figure 10.4(a)). Assume $p = v_{\max} + h_{\max}$ colors are available. Inductively, assume all nets going through (or having terminal in) tiles processed before tile (i, j) have been colored with at most p colors. Let (i, j) be the current tile and U_{ij}, R_{ij}, and DR_{ij} be the number of U, R, and DR types of nets in tile (i, j) that have not been colored (see Figure 10.4(b)). Let $u = U_{ij} + R_{ij} + DR_{ij}$ be the number of uncolored nets, and c the number of colored nets. Then $u + c = v(i, j) + h(i, j) \leq v_{\max} + h_{\max}$. Thus, $u + c \leq p$; therefore, there are at least $p - c$ colors available for coloring uncolored nets. Each uncolored net will be assigned a distinct color. After processing all tiles, if a net has been assigned color $qw + r, r < w$, it is assigned color $q + 1$. The algorithm LAYER-DR uses at most $\lceil v_{max} + h_{max}/w \rceil$ colors and runs in $O(nd)$ time. LAYER-DR explains the basic idea used in more general algorithms presented in [HSVW90]. Algorithms for plane pair assignment of multiterminal nets presented in [HSVW90] classify the nets into four types (see Figure 10.5).

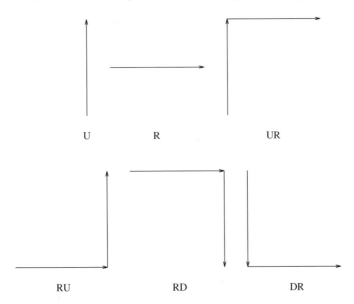

Fig. 10.3   Types of a two-terminal net.

10.3.3 Detailed Routing

After the nets have been assigned to layers, the next step is to route the nets using the signal distribution layers. Depending on the layer assignment approach, the detailed routing may differ. The routing process may be single-layer routing or x-y plane pair routing. Usually, a mixed approach is taken, in which single-layer routing is first performed for more critical nets, followed by x-y plane pair routing for less critical nets. Two models can be employed for x-y plane pair routing, namely, x-y reserved model and x-y free model. One advantage of the x-y free

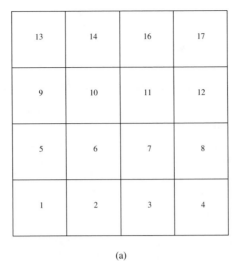

(a)

U$_{ij}$

DR$_{ij}$

R$_{ij}$

——————— colored

- - - - - - - uncolored

(b)

Fig. 10.4   (a) Row-major-order; **(b)** current tile (i, j).

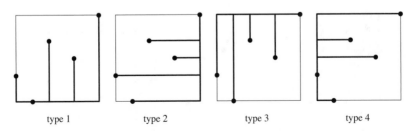

type 1   type 2   type 3   type 4

Fig. 10.5   Types of multiterminal nets.

model is that bends in nets do not necessarily introduce vias, whereas bends in nets introduce vias in the x-y reserved model. Detailed routing could be completed by using a traditional maze router.

10.4 Integrated Pin Distribution and Routing Approach

In [KC92], Khoo and Cong presented the integrated algorithm SLICE for routing in the MCM. Instead of distributing the pins before routing, the algorithm redistributes pins along with routing in each layer. The basic idea of SLICE is to perform planar routing on a layer-by-layer basis. After routing on one layer, the terminals of the unrouted nets are propagated to the next layer. The routing process is continued until all the nets are routed.

10.4.1 Algorithm Overview

The important feature of SLICE is how to compute the planar set of nets for each layer. The algorithm tries to connect as many nets as possible in each layer. For the nets that cannot be completely routed in a layer, the algorithm attempts to partially route them so that they can be completed in the next layer with shorter wires. The routing region is scanned from left to right. For each adjacent column pair, a topological planar set of nets is computed by computing a maximum weighted noncrossing matching, which consists of a set of noncrossing edges extending from the left column to the right column. Then the physical routing between the column pairs is generated based on the selected edges in the matching. This process is carried out for each column from left to right. Figure 10.6 shows an example of planer routing for a column pair. The number beside each terminal represent its net number. Routing in the first column pair is completed and the second column pair is being processed. Figure 10.6(a) shows several weighted edges extending from start points of the left column to the right column. Figure 10.6(b) shows the edges selected in the maximum weighted noncrossing matching, and Figure 10.6(c) shows the physical routing based on the selected edges.

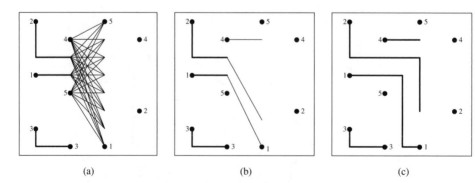

(a)   (b)   (c)

Fig. 10.6   Planer routing steps for a column pair.

After completing planar routing in a layer, the terminals of the unrouted nets are distributed so that they can be propagated to the next layer without causing local congestion. Since the left-to-right scanning operation in planar routing results in mainly horizontal wires in the solution, in order to complete the routing in the vertical direction, a restricted two-layer maze routing technique is used. The unnecessary jogs and wires are removed after each layer is routed. Figure 10.7 shows an example of the removal of simple jogs by moving the horizontal segment downward. Figure 10.8 shows an example of the removal of a complex jog, which required the displacement of another net.

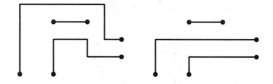

Fig. 10.7   Example of removal of simple jogs.

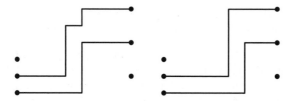

Fig. 10.8   Example of removal of a complex jog.

The terminals of the unrouted nets are propagated to the next layer. Finally, the routing region is rotated by 90° so that the scanning direction is orthogonal to the one used in the previous layer. The process is iterated until all the nets are routed. Details of the planar routing, pin redistribution, and maze routing may be found in [KC92].

10.5 Three-Dimensional Routing Approach

Yu, Badida, and Sherwani [YBS93] presented a completely new approach to MCM routing. Rather than converting the three-dimensional routing problem into a set of two-dimensional routing problems, their routing methodology decomposes three-dimensional routing into several smaller three-dimensional problems to achieve the best utilization of the three-dimensional routing space (see Figure 10.9). In addition, it incorporates several different performance constraints.

10.5.1 Performance Constraints

The primary requirements for MCM routing are to satisfy the delay, noise, and fabrication constraints. The delay constraints are introduced to ensure that the

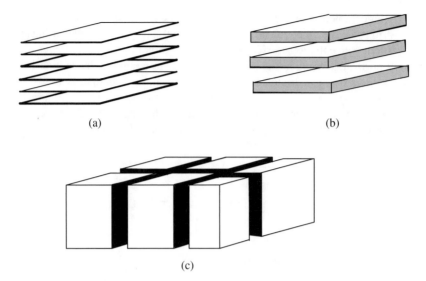

Fig. 10.9   Different routing methodologies for six layers: **(a)** layer-by-layer routing;
(b) layer-pair-by-layer-pair routing; **(c)** three-dimensional routing.

MCM operates correctly for the designed clock frequency. The noise constraints
must be met in order to avoid the inadvertent logic transitions caused by excessive noise. The fabrication constraints should be satisfied to achieve high yield.
These constraints, however, cannot be handled directly by a router. They must be
converted into geometric constraints, which are then satisfied in the routing. In
this section, we discuss the modeling of delay, noise, and fabrication constraints
as geometric constraints.

1. **Delay Constraints Converted into Net-Length Constraints:** The maximum clock frequency for which a system can operate correctly is limited
   by the delay of critical paths. For example, when two flip-flops are driven
   by the same clock source and are on a critical path, the total delay between
   them must be less than the clock period to ensure that the MCM can operate correctly. The total delay between the flip-flops is determined by the
   equation [DF93]: $t_{total} = t_s + t_{prop} + t_{settle} + t_r + t_{skew}$ where t_s is the total
   internal delay inside the devices on the source end, t_s is the corresponding
   delay on the receiving end, and t_{skew} is the clock skew. t_s, t_r, and t_{skew}
   are determined independently of the routing. t_{prop} and t_{settle} are the delays
   affected by interconnections. t_{prop} is the delay between the time when the
   output at the source end reaches the 50% point of its logic swing and the
   time when the input at the receiving end reaches the 50% point of its logic
   swing. t_{settle} is the time for the noise of the signal to settle down so that the
   signal can be safely sampled. The propagation delay t_{prop} can be further
   split into t_{flight} and $t_{rise-time\ degradation}$ where t_{flight} is the time-of-flight delay

and $t_{\text{rise}-\text{time degradation}}$ is the delay caused by the increase in rise time between the source end and receiving end of the line due to the line losses. When the line losses are controlled, (e.g., in the case of first incidence switching), the time of flight becomes the dominating factor. However, if the first incidence switching is not achieved, the propagation delay might be significantly increased, to as much as five times the time-of-flight delay. The first to signal arrival at the receiving end of the line must have sufficient voltage to switch the receiver. (The voltage should exceed V_{IH} on a $0 \rightarrow 1$ transition or V_{IL} on a $1 \leftarrow 0$ transition). The situation in which there is sufficient voltage is called *first-incidence switching* [DF93].

The first-incidence voltage at the end of a matched terminated line is calculated by the following equation [DF93]:

$$V_{\text{first}} = V_{\text{in}} \frac{Z_0}{R_{\text{out}} + Z_0} e^{-\mathcal{R}l/2Z_0}$$

where V_{in} is the open circuit output voltage swing of the driver, l is the length of the line, $\mathcal{R}$ is the unit resistance of the line, Z_0 is the impedance of the line, and R_{out} is the resistance of the output device. To achieve first-incidence switching, $V_{\text{first}} \geq V_{IH}$ or $l \leq \frac{2Z_0}{\mathcal{R}} \ln V_{\text{in}} Z_0 / V_{IH} (R_{\text{out}} + Z_0)$. When first-incidence switching is achieved, the propagation delay is mainly determined by the time of flight, which can be calculated by the following equation [Bak90]:

$$t_f = \frac{l \sqrt{\varepsilon_r}}{c_0} \tag{10.1}$$

where l is the net length, ε_r is the dielectric constant of the insulator, and c_0 is the light speed in a vacuum. Since there is a maximum time-of-flight delay (t_{max}) for each critical net to satisfy the clock frequency requirement, we have $l \sqrt{\varepsilon_r}/c_0 \leq t_{\text{max}}$ or $l \leq c_0 t_{\text{max}}/\sqrt{\varepsilon_r}$. By combining the time-of-flight delay constraint and the first-incidence switching constraint, we have *net-length constraints*: $l_i \leq \min(\frac{2Z_0}{\mathcal{R}} \ln V_{\text{in}} Z_0 / V_{IH} (R_{\text{out}} + Z_0), c_0 t_{\text{max}}/\sqrt{\varepsilon_r}$, where l_i is the length of net i.

2. **Noise Constraints Converted into Path-Separation and Parallel-Length Constraints:** Excessive electrical noise can cause inadvertent logic transition and can increase the delay of the noise settling down (t_{settle}). Therefore, the noise must be controlled to ensure that the correct output is generated and the delay is minimized.

   The noise generated by interconnections is due to crosstalk and reflection. Reflection noise occurs whenever the interconnection of a net branches, terminates, or changes direction. The level of reflection noise depends on how the line is terminated, and there is no simple model for reflection noise. However, as a general guideline, the number of vias and bends in the interconnection of a net should be minimized to reduce reflection noise. Mutual inductance and capacitance between the parallel paths of different nets create unwanted electrical coupling known as *crosstalk noise*. Whenever a signal edge travels down a signal wire, both forward and backward crosstalk noise pulses are induced in the neighboring wires.

Crosstalk noise is determined by the capacitive coupling K_C and inductive coupling K_L. The coupling between nets can be minimized by making sure no paths of two nets are in parallel for too long or are too close to each other.

Capacitive coupling K_C and inductive coupling K_L between adjacent signal paths add at the near end of the quiet line and subtract at the far end. The maximum noise voltage at the near end can be approximated by [DF93]

$$V_n \approx K_B \frac{2}{v} \frac{V_s}{T_1} l \qquad \text{if } l < \frac{v T_1}{2}$$

$$\approx K_B V_s \qquad \text{if } l > \frac{v}{2}$$

where $K_B = (K_C + K_L)/4$ is the coupling coefficient, V_s is the voltage swing on the active line, $v = c/\sqrt{\varepsilon_r}$ is the propagation velocity of electromagnetic waves in the dielectric, T_1 is 0% to 100% rise time of the signal, and l is the length for which the paths are in parallel.

The maximum noise voltage at the far end is given as [DF93]

$$V_f = K_F \frac{2}{v} \frac{V_s}{T_1} l$$

where $K_F = (K_C - K_L)/4$ is the coupling coefficient.

As can be seen from the equations for noise voltage, crosstalk noise depend on the length of the two parallel paths l and the capacitive coupling K_C and inductive coupling K_L between adjacent signal paths. K_C and K_L depend on the the length of parallel paths and the distance between the two parallel paths. As a result, the distance between paths of two nets and the length of parallel paths of different nets are the critical factors that have to be considered to minimize crosstalk noise.

We have to distinguish between the paths of nets on the same layer and the paths of nets on different layers. Therefore, the *homogeneous parallel-length constraints*, where the length of two parallel paths between nets i and j on the same layer must not exceed $\mathcal{L}_{ij}^P$, are introduced. On the other hand, if the paths of nets i and j are on different layers, the lengths of two parallel paths between nets i and j must not exceed $\mathcal{L}_{ij}^Z$, which is called the *inhomogeneous parallel-length constraint*. Both $\mathcal{L}_{ij}^P$ and $\mathcal{L}_{ij}^Z$ depend on the distance between the two parallel paths of nets i and j. The closer these two paths are, the smaller $\mathcal{L}_{ij}^P$ or $\mathcal{L}_{ij}^Z$ becomes.

In addition, if the paths of two nets are too close to each other, then the crosstalk noise increases. The *homogeneous path-separation constraints*, where the distance between the paths of nets i and j on the same layer must be greater than a certain value, and *inhomogeneous path-separation con-straints*, where the distance between the paths of nets i and j on different layers must be greater than a different value, have to be satisfied.

3. **Fabrication Constraints Converted into Layer-Number, Via-Type, and Via-Number Constraints:** Another constraint in thick-film MCMs is associated with the fabrication process. In order to achieve high yield in manufacturing MCMs, the number of routing layers, the type of vias (i.e., stacked vias, staggered vias, or spiral vias), and the maximum number of stacked vias of each net should be specified.

In summary, the delay, crosstalk noise, and fabrication constraints are converted into geometric constraints such as net-length, path-separation, parallel-length, layer-number, via-type, and via-number constraints (see Figure 10.10). The three-dimensional approach [YBS93] routes all the given nets using a minimum number of layers while satisfying all the constraints. Details about the constraints are presented in [Bak90, DF93, TTLF92].

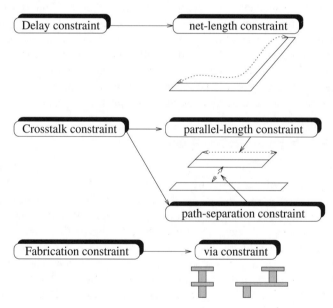

Fig. 10.10   Delay, crosstalk, and fabrication constraints are converted into geometric constraints.

10.5.2 Algorithm Overview

An overview of the approach is shown in Figure 10.11. The approach consists of several different phases. In the first phase, the substrate is partitioned into several tiles and the terminals are mapped to the tile edges. In the next phase, routing is distributed in a two-dimensional x-y plane and then in the z-dimension. This distribution (1) ensures that no specific region of the MCM routing space is overcongested or undercongested, (2) allows the net-length constraints and the manufacturing constraints to be satisfied, and (3) allows partitioning the routing problem of the entire routing space into the routing problems of several smaller regions, called *towers*. A good layer estimate is obtained after the completion of the z-dimension routing distribution. After the completion of the z-dimension routing distribution, the terminals are assigned to exact locations on the faces of each tower. During this phase, the net-separation constraints are considered to reduce crosstalk between the nets. The terminal assignment phase also maximizes the number of planar nets in each tower. After the completion of this phase, the routing problem of a complex three-dimensional problem has been converted to that of routing several small towers. Since the problem of routing in each tower is independent of the rest, all the towers can be processed in parallel. At this stage, a tower can be partitioned and the same approach is applied to the tower again if necessary (see Figure 10.12). Thus, the above approach is recursive and allocates more computing resources to the regions that require them. During the tower routing phase, a maximum set of planar nets is found for each layer, which can be directly routed on the layer using an approximation algorithm that guarantees 0.60 of the optimal solution. This reduces the running time for tower routing. Tower routing is designed to obey net-separation and via constraints. During each phase, an increasingly accurate estimate of the net lengths and number of layers is obtained. In the subsequent sections, we present the details of the three-dimensional approach.

10.5.3 Tiling and Off-Tile Routing

In this section, the tiling and off-tile routing phases are presented in detail. The tiling phase partitions the entire substrate into a set of well-defined regions called *tiles*. The purpose of the tiling phase is to simplify the complex problem of routing the entire substrate into a set of several relatively simpler and smaller routing areas (towers). In addition, tiling phase facilitates the managing of the memory requirements, which is one of the critical factors in MCM routing.

If the terminals of all the dies on an MCM are located along the perimeter of the dies, the MCM is referred to as a *periphery-terminal* MCM. The MCMs based on wire bonding and TAB technologies are periphery-terminal MCMs [DF93]. For a periphery-terminal MCM, it can be assumed that the die terminals form straight lines. As a result, the tile edges can be formed along the terminal lines.

The die terminals are located at the perimeter of the dies, whereas the I/O terminals border the MCM substrate (see Figure 10.13(a)). The inner bounding box of the I/O terminals is referred to as the *inner boundary*. The substrate boundary is simply referred to as the *boundary*. These boundaries are shown in Figure 10.13(b). The substrate is partitioned in such a way that the die terminals form the tile edges. The terminal row of each die is extended in both directions to touch the inner

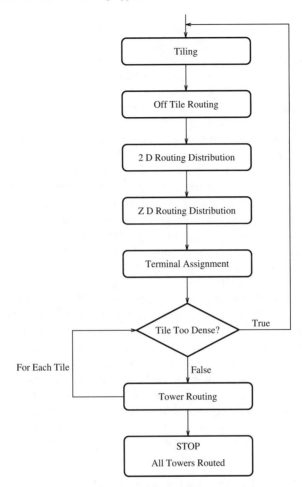

Fig. 10.11   An overview of the algorithm.

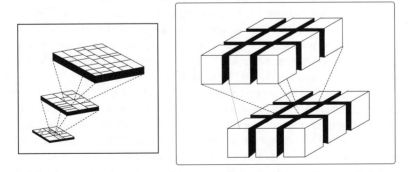

Fig. 10.12   The three-dimensional approach is recursive.

boundary, resulting in the partitioning of the substrate. This ensures that the die terminals lie on tile edges. The I/O terminals are allowed to lie within the tiles, as shown in Figure 10.13(b). In addition, if the density of nets in a tile is too large, our approach further partitions the tiles and therefore significantly reduces the complexity of routing. As a result, the MCM substrate can be defined as a set of tiles. It can be seen from Figure 10.13(b) that the sizes of the tiles differ quite significantly. This proposed approach is also capable of merging smaller tiles to reduce the number of tiles, thereby reducing the overall running time of the routing distribution phase. The following rules are used for merging tiles. The horizontal lines (or vertical lines) $1, 2, \ldots, t$ among which the distance between each pair is less than σ are combined into a single line whose coordinate is calculated as follows: $y = \sum_{i=1}^{t}(n_i y_i)/n$ (or $x = \sum_{i=1}^{t}(n_i x_i)/n$), where n_i and n are the number of terminals on line i and the total number of terminals, respectively. Figure 10.13(c) shows the effects of merging the tiles when $\sigma = 35$. Notice that as the value of σ increases, the number of tiles decreases; that is, σ is inversely proportional to the number of tiles.

In the next phase, all the terminals that are not on the tile edges are routed to the nearest available tower faces so that the characteristics of all the terminals are the same and no distinction needs to be made among them. This phase is referred to as the *off-tile* routing phase. Notice that when tiles are merged, some die terminals are located in the middle of the tiles and are routed to the tower faces during the off-tile routing phase. The length of these nets (e.g., net i in Figure 10.13(c)) may increase by as much as 2σ because of the off-tile routing, thereby increasing the delay of these nets. As a result, the value of σ must be chosen carefully to balance the tradeoff between the running time of the routing distribution and the delay of the nets. We have tested for different values of σ, and our results suggest that the running time of the routing distribution phase decreases while the length of some nets increases for larger values of σ.

10.5.4 Two-Dimensional Routing Distribution

The algorithm for distributing the nets in the x-y plane is presented in this section. During the two-dimensional routing distribution phase, the global route of each net is determined. A *global route* of a two-terminal net is a path in which each point corresponds to a tile (tower) (see Figure 10.14(a)). A *global route* of a multiterminal net is a Steiner tree in which each demand point, or Steiner point, corresponds to a tile (tower) (see Figure 10.14(b)).

More precisely, the two-dimensional routing distribution problem is defined on a routing graph. It is assumed that the global route of each net passes through the center of tiles. A routing graph $G_R = (V_R, E_R)$ is defined as follows: $v_i \in V_R$ corresponds to a tile and there is an edge $(v_i, v_j) \in E_R$ if the tiles corresponding to v_i and v_j are adjacent. It can easily be seen that the routing graph is a grid.

Let the *wiring density* of a tile be the ratio of the number of nets present in the tile to the tile area. Let the cost of each vertex $v \in V_R$ be the wiring density

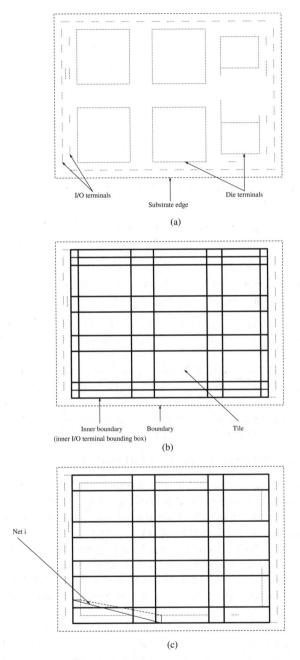

Fig. 10.13   Terminal locations of MCC1 and two different tilings of the substrate: **(a)** I/O and die terminal distribution; **(b)** substrate boundaries; **(c)** effects of merging tiles.

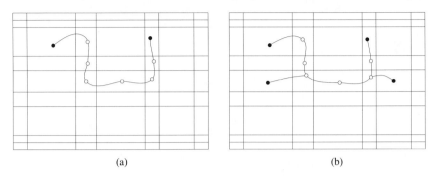

(a)   (b)

Fig. 10.14   A global route for a net: **(a)** for a two-terminal net; **(b)** for a multiterminal
net.

of the tile corresponding to v and be denoted as $c(v)$. Let $C(T_i) = \sum_{v \in T_i} c(v)$ be
the cost of Steiner tree T_i. The two-dimensional routing distribution problem is
defined as follows:

Instance: Given a net list $\mathcal{N} = \{n_1, n_2, \ldots, n_k\}$ and a routing graph $G = (V_R, E_R)$ that is a grid of size $m \times n$.

Problem: (k-minimum Steiner trees (k-MST) on an $m \times n$ grid) Find a
Steiner tree T_i for each net n_i, $1 \le i \le k$, such that the total cost ($\sum_{i=1}^{k} C(T_i)$) is
minimized.

The hierarchical routing algorithm approach of Burstein and Pelavin [AGH77,
BP83b] is extended to solve the k-MST problem. The two-dimensional routing
distribution algorithm is referred to as the *multi-net-hierarchical routing* (MNHR)
algorithm. It has been shown that the MNHR algorithm solves the k-MST problem
on an $m \times n$ grid in $O(f(k)l \log l)$ time, where $l = \max(m, n)$ and $f(k) = 4^k$.

10.5.5 *z*-Dimension Routing Distribution

In this section, we present the z-dimension routing distribution phase, which
assigns the layer on which each net passes through each tower face. The objective
of this phase is to uniformly distribute the routing density along the z-dimension
while minimizing the total number of vias for each net.

The uniform distribution of routing density is guaranteed because a capacity is
assigned to each edge e on each layer l, which is the maximum number of nets that
can pass the edge e on layer l. Let $C(e, l)$ and $c(e, l)$ be the capacity and the number
nets that have already been assigned to edge e on layer l, respectively. Clearly, for
each edge e, only those layers l where $c(e, l) < C(e, l)$ can be assigned to a net.
The available layers at each edge are modeled as shown in Figure 10.15(b). Note
that the edges that a net passes through have been determined during the x-y plane
routing distribution phase, which is as shown in Figure 10.15(a). The number of
vias required for the net in a tower is $|t_i - t_j|$, where t_i and t_j are the layers that have
been assigned to the net on the edges i and j in the tower. Assuming that the edges

that a net passes through are $1, 2, \ldots, p$, the problem of assigning layers to a net is to find t_i for $i = 1, 2, \ldots, p$, such that $c(i, t_i) < C(i, t_i)$, and $\sum_{i=1}^{p-1}(|t_i - t_{i+1}|)$ is minimized. It can be easily seen that the above problem can be solved optimally by using the multistage graph technique [HS78]. Note that the problem stated above is for a two-terminal net. The technique can be easily extended to a multiterminal net by exploring all combinations of paths at the branching point of the global route.

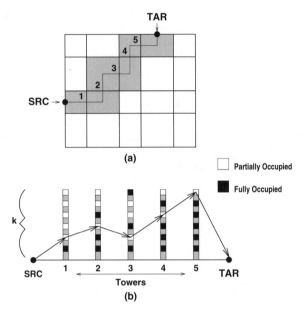

Fig. 10.15   (a) Global route of a net; (b) layer assignment for the net.

Instance: Minimum via layer assignment (MVLA) problem: Given a set of edges $i = 1, 2, \ldots, p$ and $c(i, j)$, $C(i, j)$ for $i = 1, 2, \ldots, p$, $j = 1, 2, \ldots, k$.

Question: Does there exist $1 \le t_i \le k$ for $i = 1, 2, \ldots, p$ such that $c(i, t_i) < C(i, t_i)$, and $\sum_{i=1}^{p-1}(|t_i - t_{i+1}|)$ is minimized?

It has been shown that the MVLA problem can be solved in $O(k^2 p)$ time.

10.5.6 Terminal Assignment

In this section, the details of the terminal assignment phase are presented. In this phase, the exact locations of the terminals of each net in the towers through which the net passes are assigned by using a recursive top-down approach. The objective of the terminal assignment phase is to maximize the number of planar nets in each tower while satisfying the path-separation constraint.

The terminal assignment phase is carried out by bipartitioning the substrate recursively. At each level of partitioning, a set of tiles is further partitioned into two sets of tiles that are of the same size, and the locations of terminals along the partition boundary are determined and assigned. The substrate is partitioned recursively and only the terminals along the partition boundary are assigned at each level of the hierarchy so as to maintain a global perspective of the net distribution. At the lowest level, the partitions correspond to tiles and all the terminals are completely assigned.

In the following discussion, it is assumed that the nets under consideration pass a tile edge on a layer, and the terminals of these nets on the tile edge can be permuted within the tile edge. Terminals are permuted so as to minimize the number of crossings, as well as to balance the crossings on both sides of the partition to satisfy the objective of the terminal assignment phase, which is to maximize the number of planar nets. If the line segments connecting the terminals of one net intersect with the line segment connecting the terminals of another net, then there is said to be a crossing between these two nets and the intersection between the line segments is called a *crossing*.

Figure 10.16 illustrates the permuting steps of the terminal assignment phase. An example of minimizing the number of crossings is shown in Figures 10.16 (a) and (b). Figures 10.16 (c) and (d) illustrate the idea of balancing the number of crossings on both sides of the tile edge.

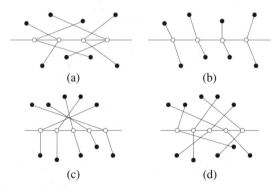

Fig. 10.16   **(a)**, **(b)** Permutation of terminals results in the minimum number of crossing; **(c)**, **(d)** permutation of terminals results in the balanced crossing on both sides.

10.5.7 Tower Routing

In this section, the details of routing within each tower is presented. Since all the terminal locations are specified and are located on the faces of the tower, all the towers can be independently routed. The objective of tower routing is to complete the required interconnections while satisfying path-separation, parallel-length, via-type, and via-number constraints.

Tower routing is completed in two steps: planar routing and three-dimensional routing.

1. **Planar Routing:** In the first step, a set of planar nets in each layer is selected and routed. Planar routing is useful in reducing the number of bends, minimizing the net lengths, and avoiding the usage of a large number of vias. As a result, planar routing is helpful in minimizing noise due to reflection and in minimizing delay.

2. **Residual Routing:** The remaining nets are routed by using a three-dimensional router based on Soukup's algorithm [Sou78] while satisfying the constraints.

10.5.7.1 Planar routing

In this section, the problem of finding a maximum planar subset for routing nets on different layers of the tower is first considered. It has been shown that the problem of finding such a maximum planar subset is NP-hard. As a result, the authors propose an approximation algorithm.

We now introduce the terminology used in our discussion. Let k be the total number of routing layers. Two types of nets are considered in the following discussion. Nets that can be routed only on layer i are denoted by $\mathcal{N}_i$, while nets that can be routed on either layer i or layer $i + 1$ are denoted by $\mathcal{N}_{i,i+1}$. For example, $\mathcal{N}_1$ consists of a set of nets that can be routed only on layer 1. Similarly, sets $\mathcal{N}_2, \mathcal{N}_3, \ldots, \mathcal{N}_k$ are defined. $\mathcal{N}_{12}$ represents a set of nets that can be routed either on layer 1 or layer 2. Similarly, sets $\mathcal{N}_{12}, \mathcal{N}_{23}, \ldots, \mathcal{N}_{k-1,k}$ are defined (see Figure 10.17). If two nets that are routable on layers i and $i + 1$ share a same pillar (see Figure 10.18), then a conflict occurs if the net with the terminal on a layer less than i (net 1 in Figure 10.18) is assigned to layer $i + 1$, where the other net is assigned to layer i. Thus, if two nets share the same pillar, the net with the terminal on a layer less than i (net 1 in Figure 10.18) can only be assigned to $\mathcal{N}_i$, while the other net can only be assigned to $\mathcal{N}_{i+1}$ to avoid the possible conflicts.

Figure 10.19 gives a formal description of the algorithm Planar-Routing. Planar-Routing finds two different solutions of planar subsets and selects the best one among the two. The first solution, represented by $\mathcal{S}_1$ (see Figure 10.19), is generated by finding two planar subsets for each layer pair by using the subroutine 2-RMPS. Finally, a maximum planar subset (MPS) is chosen for the last layer and added to $\mathcal{S}_1$ (in case the number of layers is odd). The second solution is generated by choosing an MPS for the first layer and then finding two planar subsets thereafter for each layer pair. The second solution is represented by $\mathcal{S}_2$ (see Figure 10.19). In case of an even number of layers, an MPS for the last layer is chosen and is added to $\mathcal{S}_2$. The algorithm Planar-Routing selects the maximum of $\mathcal{S}_1$ and $\mathcal{S}_2$. Once a set of two planar subsets of nets have been selected, they are routed on the respective layers while satisfying the path-separation and parallel-length constraints. In case the constraints cannot be satisfied for certain nets, then these nets are routed by the three-dimensional router in the residual routing phase.

The subroutine 2-RMPS finds two restricted MPSs, such that the first set is routable on layer i and the second set is routable on layer $i + 1$. The subroutine

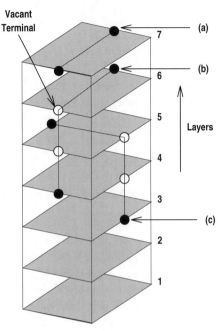

Fig. 10.17 A tower showing different types of nets: **(a)** $\mathcal{N}_7$; **(b)** $\mathcal{N}_{47}$; **(c)** $\mathcal{N}_{35}$.

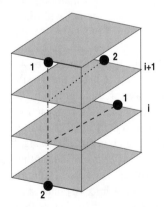

Fig. 10.18 Potential conflicts among nets.

2-RMPS selects a maximum of two planar subsets from three different strategies. The first strategy uses the bipartite subgraph of $\mathcal{N}_{12}$ generated by the algorithm MBS [CHS93] as the two planar subsets X_1 and Y_1. The performance ratio of the algorithm MBS is at least 0.75 [CHS93]. The second strategy finds the MPS of $\mathcal{N}_1 \cup \mathcal{N}_{12}$ and the MPS of $\mathcal{N}_2$ as the two planar subsets X_2 and Y_2, respectively. The subroutine MPS($\mathcal{N}_i$) finds the MPS for the given nets such that the set is routable on layer i. MPS is based on the algorithm for finding an MIS in circle graphs presented in [Sup87]. The third strategy is similar to the second strategy. The algorithm 2-RMPS then returns the best result among the three strategies.

Algorithm PLANAR-ROUTING($\mathcal{N}, k, S$)

Input: $\mathcal{N}$: Set of nets that pass through a tower
$\quad\quad k$: Number of routing layers

Output: S: Set of planar nets

---

begin
$\quad$ **if** k is odd **then** $t = 1$; **else** $t = 0$;
$\quad$ $S_1 = \phi$;
$\quad\quad$ **for** $j = 1$ **to** $\lfloor \frac{k}{2} \rfloor$
$\quad\quad\quad$ $S = $ 2-RMPS($\mathcal{N}_{2j-1} - S_1, \mathcal{N}_{2j} - S_1$);
$\quad\quad\quad$ $S_1 = S \cup S_1$;
$\quad$ **if** ($t = 1$) **then** $S_1 = S_1 \cup$ MPS($\mathcal{N}_L - S_1$);

$\quad$ $S_2 =$MPS(V_1);
$\quad\quad$ **for** $j = 1$ **to** $\lfloor \frac{k}{2} \rfloor - 1$
$\quad\quad\quad$ $S = $ 2-RMPS($\mathcal{N}_{2j} - S_2, \mathcal{N}_{2j+1} - S_2$);
$\quad\quad\quad$ $S_2 = S \cup S_2$;
$\quad$ **if** ($t = 0$) **then** $S_2 =$MPS($\mathcal{N}_L - S_2$);

$\quad$ $S = $ SELECT_MAX(S_2, S_2);
end.

Fig. 10.19   Algorithm PLANAR-ROUTING.

The time complexity of the algorithm 2-RMPS is dominated by the time complexity of MPS and the time complexity of MBS. Since the time complexity for both MPS and MBS is $O(m^2)$, the time complexity of the algorithm 2-RMPS is $O(m^2)$, where m is the number of nets. The algorithm 2-RMPS is formally described in Figure 10.20.

The overall time complexity of the algorithm Planar-Routing is dominated by the complexity of the subroutine 2-RMPS, which is $O(m^2)$. Therefore, the time complexity of the algorithm Planar-Routing is $O(km^2)$.

10.5.7.2 Provably good algorithm for 2-RMPS

In this section, we show that the restricted maximum two-planar-subset problem is NP-hard, and hence an approximation algorithm for solving such a problem is presented.

Let us start with some terminology. Given a set of nets $\mathcal{N}$, a set $\mathcal{N}' \subseteq \mathcal{N}$ is called a *planar subset* if the nets in $\mathcal{N}'$ are pairwise independent. A *maximum*

Algorithm 2-RMPS($\mathcal{N}_1, \mathcal{N}_2, \mathcal{N}_{12}$)

Input : $\mathcal{N}_1$: Set of nets that can be routed only on layer 1

$\mathcal{N}_2$: Set of nets that can be routed only on layer 2

$\mathcal{N}_{12}$: Set of nets that can be either routed on layer 1 or 2

Output : $\mathcal{R}$ Set of planar nets

---

begin

$\quad\quad M B S(\mathcal{N}_{12}, X_1, Y_1)$;

$\quad\quad\quad \mathcal{R}_1 = X_1 \cup Y_1$;

$\quad\quad M P S(\mathcal{N}_1 \cup \mathcal{N}_{12}, X_2)$;

$\quad\quad M P S(\mathcal{N}_2, Y_2)$;

$\quad\quad\quad \mathcal{R}_2 = X_2 \cup Y_2$;

$\quad\quad M P S(\mathcal{N}_1, X_3)$;

$\quad\quad M P S(\mathcal{N}_2 \cup \mathcal{N}_{12}, Y_3)$;

$\quad\quad\quad \mathcal{R}_3 = X_3 \cup Y_3$;

$\quad\quad \mathcal{R} = \text{SELECT\_MAX}(\mathcal{R}_1, \mathcal{R}_2, \mathcal{R}_3)$;

return $\mathcal{R}$;

end.

Fig. 10.20   Algorithm 2-RMPS.

planar subset (1-MPS) is one with the maximum number of nets among all planar subsets. A *k planar subset* can be defined as a set consisting of k disjoint planar subsets, and a *maximum k-planar subset* (*k*-MPS) has the maximum number of vertices among all such *k*-planar subsets.

The problem of finding the *k*-MPS in a switch box is equivalent to finding a maximum *k*-colorable subgraph in a corresponding circle graph G [YG87]. The maximum *k*-colorable-subgraph problem is known to be NP-hard for general graphs, since it includes as special cases both the 1-MPS and chromatic number problem [GJ79, YG87]. The planar subset problem in a switch box is polynomial time solvable [Sup87], but for $k = 2$, the *k*-MPS is NP-complete [SL89]. In [CHS93], a provably good algorithm was presented for the *k*-MPS problem, and the following theorem was proved:

Theorem 43   *Given a set of nets, the algorithm finds a k-planar subset such that* $\rho \geq (1 - (1 - \frac{1}{k})^k))$, *where ρ is the ratio of the solution found to optimal solution of the k-MPS problem.*

In the restricted k-MPS problem, vertices are restricted to certain layers. We call such a problem the k-RMPS problem. Since we route on a layer pair each time, we restrict ourselves to a description of 2-RMPS. Note that the problem of finding two planar subsets of nets from a given two layers of a tower is equivalent to that of finding two planar subsets in a switch box. Figure 10.21 clearly describes this transformation. Hence, algorithms for finding an MPS in switch boxes can be extended to our problem of finding two planar subsets. Let $\mathcal{N}_1$ be the set of nets that can only be routed on layer 1, $\mathcal{N}_2$ the set of nets that can only be routed on layer 2, and $\mathcal{N}_{12}$ the set of nets that can either be routed on layer 1 or layer 2.

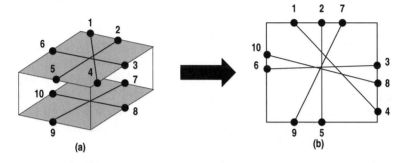

Fig. 10.21   Transformation of a tower routing problem into a switch-box problem: **(a)** a tower with two layers; **(b)** a switch-box representation of the tower.

Instance: 2-RMPS problem: Given a switch box and three sets $\mathcal{N}_1, \mathcal{N}_2, \mathcal{N}_{12}$.

Question:: Do planar subsets $N_1 \subseteq \mathcal{N}_1 \cup \mathcal{N}_{12}$ and $N_2 \subseteq \mathcal{N}_2 \cup \mathcal{N}_{12}$ exist such that $\mid N_1 \mid + \mid N_2 \mid$ is maximum among all such sets?

Theorem 44   *2-RMPS is NP-hard.*
Proof: By restricting $\mathcal{N}_1 = \mathcal{N}_2 = \phi$, the 2-RMPS problem is equivalent to 2-MPS in a circle graph, which is NP-hard [SL89]. Thus, the 2-RMPS problem is also NP-hard.

$\square$

Since the 2-RMPS problem is NP-hard, we propose an approximation algorithm that guarantees at least 60% of the optimal solution. For a 2-RMPS problem in a graph G, we define the *performance ratio* of 2-RMPS to be $\rho = S/S^*$, where S is the size of the two-restricted planar subsets obtained by the algorithm 2-RMPS, and S^* is the size of the two-restricted maximum planar subsets for a given graph G.

Theorem 45   *Let ρ be the performance ratio of the algorithm 2-RMPS. For any given instance of the problem, the algorithm 2-RMPS produces a solution such that $\rho \geq 0.6$.*

Proof: Let the contributions of $\mathcal{N}_1$, $\mathcal{N}_2$, and $\mathcal{N}_{12}$ in the optimal solution S^* be S_1^*, S_2^*, and S_{12}^*, respectively; that is, $S^* = S_1^* \cup S_2^* \cup S_{12}^*$, where $S_1^* \subset \mathcal{N}_1$, $S_2^* \subset \mathcal{N}_2$, $S_{12}^* \subset \mathcal{N}_{12}$. Let $S = S_1 \cup S_2 \cup S_{12}$ represent the planar subset obtained as a result of 2-RMPS, where $S_1 \subseteq \mathcal{N}_1$, $S_2 \subseteq \mathcal{N}_2$, and $S_{12} \subseteq \mathcal{N}_{12}$. The cardinality of any set S_{12}^* is denoted by $|S_{12}^*|$. It is shown that $|S| \geq 0.6|S^*|$.

Let α be defined as follows:

$$\alpha = \frac{|S_{12}^*|}{|S^*|} \tag{10.2}$$

and therefore

$$1 - \alpha = \frac{|S_1^* \cup S_2^*|}{|S^*|} \tag{10.3}$$

Let R1 = $X_1 \cup Y_1$ be the solution generated by the algorithm 2-RMPS (by MBS), let R2 = $X_2 \cup Y_2$ be the solution generated by the second strategy from the algorithm 2-RMPS, and let R3 = $X_3 \cup Y_3$ be the solution generated by the third strategy from the algorithm 2-RMPS (see Figure 10.20).

If the algorithm 2-RMPS selects R1, then α is greater than $1 - \alpha$. In such a case, $\rho \geq \frac{3}{4}\alpha$ [CHS93]. On the other hand, if the algorithm 2-RMPS selects $R2$ or $R3$, the performance ratio is calculated as follows. Let $S_{12}^* = S_{12}^{1^*} \cup S_{12}^{2^*}$, where $S_{12}^{1^*} \subset \mathcal{N}_1'$ and $S_{12}^{2^*} \subset \mathcal{N}_2'$. Then we have either $|S_{12}^{1^*}| \geq 0.5|S_{12}^*|$ or $|S_{12}^{2^*}| \geq 0.5|S_{12}^*|$. Without loss of generality, it is assumed that $|S_{12}^{1^*}| \geq 0.5|S_{12}^*|$ (the case when $|S_{12}^{2^*}| \geq 0.5|S_{12}^*|$ can be similarly proved). Therefore, the performance ratio

$$\rho \geq \frac{|S_1^* \cup S_2^*|}{|S^*|} + 0.5|S_{12}^*| \tag{10.4}$$

$$\rho \geq 1 - \alpha + \frac{\alpha}{2} = 1 - \frac{\alpha}{2} \tag{10.5}$$

By equating the performance ratio of MBS ($\frac{3}{4}\alpha$) and (10.5) and solving for α, we get $\alpha = 0.8$. By substituting the value of α, we obtain the performance ratio for the algorithm 2-RMPS as follows:

$$\rho = \max\{ 1 - \frac{\alpha}{2}, \frac{3}{4}\alpha \} \geq 0.6.$$

Hence, the performance of the algorithm 2-RMPS ≥ 0.6. $\qquad\square$

Figure 10.22 represents the α - ρ graph for 2-RMPS. The x-axis represents α, while the y-axis represents the performance ratio ρ. It can be clearly seen from Figure 10.22 that the performance ratio of the algorithm 2-RMPS is at least 0.6. For any given input, the solution of the 2-RMPS always lies inside the shaded region. Notice that the performance ratio at the intersection of the two lines is 0.6, which is the lower bound on the performance of 2-RMPS.

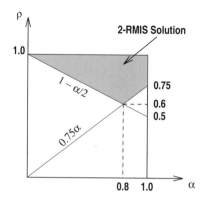

Fig. 10.22 α - ρ graph for 2-RMPS.

10.5.8 Satisfying Constraints

In this section, we describe the method used to satisfy the path-separation, via-number, and parallel-path-length constraints as proposed in [YBS93].

Noise due to crosstalk is minimized by satisfying the path-separation constraint. If the path-separation constraint is the same for all the net pairs i and j, then it is satisfied by setting the grid separation equal to the path separation value. However, the path-separation constraint is critical for nets whose signals change frequently, such as clock nets, since they are more likely to induce crosstalk to their neighboring nets. As a result, the net-separation constraint varies from net to net and hence is not uniform. The three-dimensional routing algorithm takes care of nonuniform net-separation constraints by using the *cable routing* technique. The route of a net in regular routing is a *line*. In contrast, the route of the net in cable routing is a cable consisting of a core line and several surrounding lines (see Figure 10.23(c)). Figures 10.23(a) and (b) show the cross section of a cable when the path-separation constraint is 1 and 2, respectively. Finally, the via-number and parallel-path-length constraints are satisfied by forcing the signal path of a net to change the direction in which it is routed.

10.6 Evaluation of MCM Routers

As mentioned in an earlier section, the maze router is simple to implement. However, it (1) suffers from the net ordering problem, (2) does not attempt global optimization, and (3) requires large memory space. The multiple-stage routing approach is efficient in general. However, it (1) does requires estimation of a number of layers before carrying out layer assignment, and (2) does not take into account the performance constraints. The integrated pin distribution and routing approach uses fewer vias, since it routes most of the nets in a planar fashion. It uses a two-dimensional maze router to route the nets that are not completed during the planar-routing phase. Usage of a maze router slows down the computation and

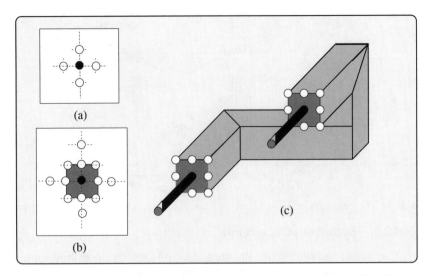

Fig. 10.23   Cable routing: **(a)** path separation = 1; **(b)** path separation = 2; **(c)** cable
routing strategy.

introduces extra vias. This router, however, runs faster than the three-dimensional
maze router. The three-dimensional routing approach differs from all the other
approaches primarily in problem decomposition. The three-dimensional approach
maintains the characteristics of the three-dimensional routing problem, while the
others convert the three-dimensional problem into a two-dimensional or two-layer
routing problem. This methodology is an inherently parallel approach, which can
be used to improve the time complexity and memory constraints.

In the following, the experimental evaluation of the routers described in this
chapter is presented. The evaluation is based on MCC1 [YBS93]. The design of
MCC1 consists of six chips and 765 I/O pins, and contains 799 signal nets, two
power nets, one ground net and a grid size of about 600×600. Figure 10.24
pictorially depicts the interconnection pattern of MCC1.

The maze router routes MCC1 in five layers. The limitation of the maze
router is the requirement of a large amount of memory. In fact, the maze router is
not completely successful in routing MCC2 [KC92].

SLICE routes MCC1 in five layers. Table 10.1 gives a comparison between
SLICE and the maze router. One of the advantages that SLICE has over the
three-dimensional maze router is its low memory requirement.

A router based on the three-dimensional approach requires 8, 10, and 12
layers to route MCC1 for separation values of 2, 3, and 4, respectively (see Table 10.3). The separation between nets is assumed to be the same for all the nets.
Figure 10.25 shows an interconnection pattern after the terminal assignment step
(three-dimensional router). As a result of the terminal assignment step, a significant number of the nets become planar. The planar nets include Type I and Type II

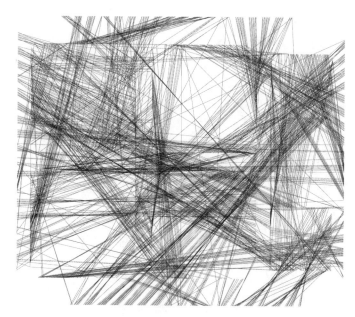

Fig. 10.24   Interconnection pattern of MCC1.

Table 10.1:   Results Reported for SLICE

Example	No. of Layers		No. of Vias		No. of Jogs		Total Wire Length		Run Time (hr:min)	
	SLICE	Maze	SLICE	Maze	SLICE	Maze	SLICE	Maze	SLICE	Maze
test1	5	4	2,013	2,975	3,453	421	109,092	107,908	0:02	0:08
test2	6	4	5,271	7,127	9,656	892	286,723	273,642	0:06	0:48
test3	6	4	6,892	9,347	13,552	1,094	459,046	441,552	0:12	1:40
mcc1	5	5	6,386	8,794	11,215	1,244	402,258	397,221	0:12	0:59
mcc2	7	-	47,864	-	108,321	-	5,902,818	-	8:15	-

nets. Type I nets are those nets that are on the same layer and do not intersect with each other, and hence can be routed on the same layer. Type II nets are those nets that are brought to a layer by using a via and are completely routed on the same layer. Table 10.2 shows the ratio of the number of Type I and Type II nets to the total number of nets. Notice that the ratio of the number of Type I nets to the total number of nets is over 42%. Also, the Type I and Type II nets account for more than 65% of the total nets, resulting in faster tower routing. The planar nets are selected from either Type I or Type II nets by using an approximation algorithm described in a previous section. It can be clearly seen from Figure 10.25 that the nets have been uniformly distributed.

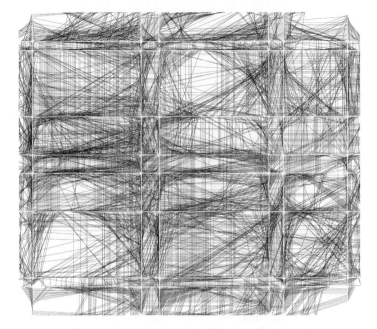

Fig. 10.25 Interconnection pattern of MCC1 after terminal assignment.

Table 10.2: Ratios of the Number of Type I Nets and the
Number of Type II Nets to the Total Number of Nets in MCC1

Type of Nets	Avg. % in a Tower	Max. % in a Tower	Min. % in a Tower
Type I	42	66	0
Type II	24	100	0

Table 10.3: Effects of Different Path-
Separation Values on the Number of
Layers in MCC1

Path Separation	Number of Layers
2	8
3	10
4	12

10.7 Summary

The general MCM routing environment is distinctly different from the VLSI routing environment, due to (1) the large number of layers available for routing, and (2) the large number of nets to be routed.

MCM routers must route all nets in a minimum number of layers while satisfying the performance constraints such as (1) manufacturability constraints, (2) net-length constraints, (3) net-separation constraints, and (4) via constraints. In addition, it should partition the routing problem into several independent routing problems to minimize the memory requirements.

In this chapter, we have discussed the MCM routing environment, problems, and performance constraints. We have also discussed four different approaches for MCM routing and presented a comparison of these approaches.

PROBLEMS

10-1. Based on the basic idea of the LAYER-DR algorithm, develop algorithms for plane pair assignment of (a) all types of two-terminal nets shown in Figure 10.3, and (b) all types of multiterminal nets shown in Figure 10.5.

10-2. Develop a good lower bound for the minimum number of layers needed to complete the routing of a given MCM.

10-3. Develop an algorithm for substrate partitioning such that the routing density is equally distributed.

10-4. Develop a simple MCM router based on a single-trunk (either horizontal or vertical) Steiner tree for each net. Trunks may be assigned to different layers, but each trunk must be in a complete layer.

BIBLIOGRAPHIC NOTES

A textbook by Tummala and Rymaszewski [TR89] covers the fundamental concepts of microelectronic packaging. A survey of electronic packaging technology appears in [Tum91]. A mathematical analysis of different system packaging parameters can be found in [Mor90]. An excellent discussion on die attachment techniques can be found in the book by Bakoglu [Bak90]. A discussion on early design analysis can be found in [CL89, LaP91]. A dscussion on testing and diagnosis of multichip modules can be found in [KT91]. A detailed analysis about the skin effect in thin-film interconnections for ULSI/VLSI packages is described in [HT91]. An electrical design methodology for multichip modules is described by Davidson [Dav84]. An excellent discussion about thermal issues in MCMs has been presented by Buschbom [Bus90]. In [KC93], Khoo and Cong have presented an MCM router that uses no more than four vias to route every net, and yet produces high-quality routing solutions.

Bibliography

[AGH77]   A. Aho, M. R. Garey, and F. K. Hwang. Rectilinear Steiner trees: Efficient special case algorithms. *Networks*, 7:37–58, 1977.

[AK89]   M. J. Atallah and S. R. Kosaraju. An efficient algorithm for maxdominance with applications. *Algorithmica*, 4:221–236, 1989.

[Ake81]   S. B. Akers. On the use of the linear assignment algorithm in module placement. *Proceedings of 18th ACM/IEEE Design Automation Conference*, pages 137–144, 1981.

[AS91]   A. Abdullah and S. Sastry. Topological via minimization and routing. *Topological Via Minimization and Routing*, 1991.

[Bak90]   H. B. Bakoglu. *Circuits, Interconnections, and Packaging for VLSI*. Addison Wesley, 1990.

[BBD86]   D. Braun, J. Burns, S. Devadas, H. K. Ma, K. Mayaram, F. Romeo, and A. Sangiovanni-Vincentelli. Chameleon: A new multi-layer channel router. *Proceedings of 23rd Design Automation Conference*, IEEE-86:495–502, 1986.

[Ber61]   C. Berge. *Farbung von graphen, deren samtilchie bzw. deren ungerade kreise starr sind, wiss.* Technical Report, Z. Martin-Luther University, Halle-Wittenberg Math. Natur, Reihe, 114–115, 1961.

[Ber83]   A. A. Bertossi. Finding Hamiltonian circuits in proper interval graphs. *Information Processing Letters*, 17:97–102, 1983.

[BK87]   A. Brandstadt and D. Kratsch. On domination problems for permutation and other graphs. *Theoretical Computer Science*. pages 181–198, 1987.

[BKPS94]   S. Bhingarde, R. Khawaja, A. Panyam, and N. Sherwani. Over-the-cell routing algorithms for industrial cell models. *Proceedings of 7th International Conference on VLSI Design*, pages 143–148, 1994.

[BL76]   K. S. Boothe and G. S. Lueker. Testing for consecutive ones property, interval graphs and graph planarity using p q-trees algorithm. *Journal of Computer and System Science*, 13:335–379, 1976.

[BMPS93]   S. Bhingarde, S. Madhwapathi, A. Panyam, and N. Sherwani. A unified approach to multilayer OTC routing. Technical Report TR/93-17, Computer Science Dept., Western Michigan University, Kalamazoo, MI 49008, 1993.

[Boo80]   K. S. Booth. Dominating sets in chordal graphs. Technical Report CS-80-34, Computer Science Dept., University of Waterloo, Waterloo, Ontario, 1980.

[BP83a]   B. S. Baker and R. Y. Pinter. An algorithm for the optimal placement and routing of a circuit within a ring of pads. *Proceedings of 24th Annual Symposium on Foundations of Computer Science*, pages 360–370, November 1983.

[BP83b]   M. Burstein and R. Pelavin. Hierarchical channel router. *Proceedings of 20th ACM/IEEE Design Automation Conference*, pages 519–597, 1983.

[BPS93a]   S. Bhingarde, A. Panyam, and N. Sherwani. Efficient over-the-cell routing algorithm for general middle terminal models. *Proceedings of ISCAS*, 1993.

[BPS93b]   S. Bhingarde, A. Panyam, and N. Sherwani. Efficient over-the-cell routing algorithm for via-less middle terminal models. *Proceedings of EDAC-EUROASIC*, pages 127–132, February 1993.

[BPS93c]   S. Bhingarde, A. Panyam, and N. Sherwani. On optimal cell models for over-the-cell routing. *Proceedings of 6th International Conference on VLSI Design*, Bombay, India, pages 94–99, January 1993.

[BS86]   P. Bruell and P. Sun. A greedy three layer channel router. *Proceedings of IEEE International Conference on CAD*, pages 495–502, 1986.

[BSC91]   U. Singh, B. S. Carlson, and C. Y. R. Chen. Optimal cell generation for dual independent layout styles. *IEEE Transactions on Computer-Aided Design*, 10, 1991.

[Buc80]   M. A. Buckingham. *Circle Graphs*. Technical Report, Courant Institute of Mathematical Sciences, New York University, New York, 1980.

[Bus90]   M. Buschbom. MCM thermal challenges. *Surface Mount Technology*, pages 30–34, 1990.

[CH90]   A. Chatterjee and R. Hartley. A new simultaneous circuit partitioning and chip placement approach based on simulated annealing. *Proceedings of Design Automation Conference*, pages 36–39, 1990.

[Che86]   H. H. Chen. Trigger: A three layer gridless channel router. *Proceedings of IEEE International Conference on Computer-Aided Design*, pages 196–199, 1986.

[Che89]   L. P. Chew. Constrained Delaunay triangulation. *Algorithmica*, pages 97–108, 1989.

[CHS93]   J. Cong, M. Hossain, and N. Sherwani. A provably good multilayer topological planar routing algorithm in IC layout design. *IEEE Transactions on Computer-Aided Design*, January 1993.

[CK86]   H. H. Chen and E. Kuh. Glitter: A gridless variable-width channel router. *IEEE Transactions on Computer-Aided Design*, CAD-5(4):459–465, 1986.

[CL84]   Y. Chen and M. Liu. Three-layer channel routing. *IEEE Transactions on Computer-Aided Design*, CAD-3(2):156–163, April 1984.

[CL88]   J. Cong and C. L. Liu. Over-the-cell channel routing. *Proceedings of International Conference on Computer-Aided Design*, pages 80–83, 1988.

[CL89]   Y. H. Chen and D. P. Lapotin. *Congestion Analysis for Wirability Improvement*. Research Report, IBM T. J. Watson Research Center, P.O. Box 218, Yorktown Heights, NY, 1989.

[CL90]   J. Cong and C. L. Liu. Over-the-cell channel routing. *IEEE Transactions on Computer-Aided Design*, pages 408–418, 1990.

[CLR90]   T. Cormen, C. E. Leiserson, and R. Rivest. *Introduction to Algorithms*. McGraw-Hill, 1990.

[CN85]   G. J. Chang and G. L. Nemhauser. Covering packing and generalized perfection. *SIAM Journal Algebraic and Discrete Methods*, 6:109–132, 1985.

[Con90]   J. Cong. *Routing Algorithms in the Physical Design of VLSI Circuits*. Ph.D. thesis, University of Illinois–Urbana-Champaign, August 1990.

[CP88]   J. Cong and B. Preas. A new algorithm for standard cell global routing. *Proceedings of IEEE International Conference on Computer-Aided Design*, pages 176–179, 1988.

[CPL90]   J. Cong, B. T. Preas, and C. L. Liu. General models and algorithms for over-the-cell routing in standard cell design. *Proceedings of the 27th ACM/IEEE Design Automation Conference*, pages 709–715, June 1990.

[CS84]   R. Cole and A. Siegel. River routing every which way, but loose. *Proceedings of 25th Annual Symposium on Foundation of Computer Science*, pages 65–73, 1984.

[CS85]   C. J. Colbourn and L. K. Stewart. Dominating cycles in series parallel graphs. *Ars Combinatorica*, 19A:107–112, 1985.

[CS90]   C. J. Colbourn and L. K. Stewart. Permutation graphs: connected
            domination and Steiner trees. *Discrete Mathematics*, 86:179–189,
            1990.

[CS91]   J. D. Cho and M. Sarrafzadeh. The pin redistribution problem in mul-
            tichip modules. *Proceedings of Fourth Annual IEEE International
            ASIC Conference and Exhibit*, pages 9-2.1–9-2.4, September 1991.

[CS93]   A. B. Cohen and M. Shechory. Pathway: a datapath layout assembler.
            IFIP Transactions A [Computer Science and Technology], pages 119–
            131, 1993.

[CSW89]   C. Chiang, M. Sarrafzadeh, and C. K. Wong. A powerful global
            router: Based on Steiner min-max trees. *Proceedings of IEEE Inter-
            national Conference on Computer-Aided Design*, pages 2–5, Novem-
            ber 7–10 1989.

[CW92]   M. S. Chang and F. H. Wang. Efficient algorithms for the maximum
            weight clique and maximum weight independent set problems on per-
            mutation graphs. *Information Processing Letters*, 43:293–295, 1992.

[CWL87]   J. Cong, D. F. Wong, and C. L. Liu. A new approach to the three-
            layer channel routing problem. *Proceedings of IEEE International
            Conference on Computer-Aided Design*, pages 378–381, 1987.

[Dai91]   W. W. Dai. Topological routing in surf: Generating a rubber-band
            sketch. *Proceedings of IEEE Design Automation Conference*, pages
            39–48, 1991.

[Dav84]   E. E. Davidson. An electrical design methodology for multichip mod-
            ules. *Proceedings of International Conference on Computer Design*,
            pages 573–578, 1984.

[DDS91]   W. W. Dai, T. Dayan, and D. Staepelaere. Topological routing in
            surf: Generating a rubber-band sketch. *Proceedings of 28th Design
            Automation Conference*, pages 39–44, 1991.

[Dew83]   A. K. Dewdney. Fast turing reductions between problems in NP,
            Chapter 4: reductions between NP-complete problems. *Report No.
            71, Dept. of Computer Science, University of Western Ontario, Lon-
            don, Ontario*, 1983.

[DF93]   D. A. Doane and P. D. Frazen. *Multichip Module Technologies and
            Alternatives*. Van Nostrand Reinhold, 1993.

[DG80]   D. N. Deutsch and P. Glick. An over-the-cell router. *Proceedings of
            Design Automation Conference*, pages 32–39, 1980.

[Dji82]   H. N. Djidjev. On the problem of partitioning planar graphs. *SIAM
            Journal of Algebraic and Discrete Methods*, 3(2):229–240, 1982.

[DKJ90]   W. W. Dai, R. Kong, and J. Jue. Rubber band routing and dynamic
            data representation. *Proceedings of 1990 International Conference
            on Computer-Aided Design*, pages 52–55, 1990.

[DKS81]   D. Dolev, K. Karplus, A. Siegel, A. Strong, and J. D. Ullman. Optimal wiring between rectangles. *Proceedings of 13th ACM Symposium on Theory of Computing*, pages 312–317, 1981.

[DKS87]   D. Dolev, K. Karplus, A. Siegel, A. Strong, and J. D. Ullman. Optimal wiring between rectangle. *Proceedings of 13th Annual ACM Symposium on Theory of Computing*, pages 312–317, May 1987.

[DKS91]   W. W. Dai, R. Kong, and M. Sato. Routability of a rubber-band sketch. *Proceedings of 28th Design Automation Conference*, pages 45–48, 1991.

[DMPS93]   S. Danda, S. Madhwapathi, A. Panyam, and N. Sherwani. *Efficient algorithms for maximum two planar subset problems*. Technical Report 93-19, Department of Computer Science, Western Michigan University, Kalamazoo, MI, 1993.

[DNA*90]   W. E. Donath, R. J. Norman, B. K. Agrawal, S. E. Bello Sang Yong Han, J. M. Kurtzberg, P. Lowy, and R. I. McMillan. Timing driven placement using complete path delays. *Proceedings of 27th ACM/IEEE Design Automation Conference*, pages 84–89, 1990.

[DNB91]   S. Das, S. C. Nandy, and B. B. Bhattacharya. An improved heuristic algorithm for over-the-cell channel routing. ISCAS, 5:3106–3109, 1991.

[DSSL93]   S. Dong, Y. Sun, S. Sato, and C. L. Liu. Two channel routing algorithms for quickly customized logic. *Proceedings of EURO-ASIC 93*, pages 122–126, 1993.

[ED86]   R. J. Enbody and H. C. Du. Near-optimal n-layer channel routing. *Proceedings of 23rd Design Automation Conference*, pages 708–714, June 1986.

[EET89]   G. H. Ehrlich, S. Even, and R. E. Tarjan. Intersection graphs of curves in the plane. *Journal of Combinatorial Theory Series*, 21:394–398, April 1989.

[EI71]   S. Even and A. Itai. Queues, stacks and graphs. *Theory of Machines and Computations*, pages 71–86, 1971.

[Ell89]   D. J. Elliott. *Integrated Circuit Fabrication Technology*. McGraw-Hill, 1989.

[EPL72]   S. Even, A. Pnnueli, and A. Lempel. Permutation graphs and transitive graphs. *Journal of the ACM*, 19:400–410, 1972.

[EY90]   M. Edahiro and T. Yoshimura. New placement and global routing algorithms for standard cell layout. *Proceedings of 27th Design Automation Conference*, pages 642–645, 1990.

[Fan93]   P. E. Riley, S. S. Peng, and L. Fang. Plasma etching of aluminum for ULSI circuits. *Solid State Technology*, pages 47–52, February 1993.

[FF62]   L. R. Ford and D. R. Fulkerson. *Flows in Networks*. Princeton University Press, 1962.

[FKar]   M. Farber and J. M. Keil. Dominations in permutation graphs. *Journal Algorithms*, to appear.

[FMMY92]   T. Fujii, Y. Mima, T. Matsuda, and T. Yoshimura. A multi-layer channel router with new style of over-the-cell routing. *Proceedings of 29th Design Automation Conference*, pages 585–588, 1992.

[Gab85]   N. H. Gabow. An almost linear time algorithm for two-processor scheduling. *Journal of the ACM*, 29(3):766–780, 1985.

[Gav72]   F. Gavril. Algorithms for a minimum coloring, maximum clique, minimum covering by cliques, and maximum independent set of a chordal graph. *SIAM Journal of Computation*, 1:180–187, 1972.

[Gav73]   F. Gavril. Algorithms for a maximum clique and a maximum independent set of circle graph. *Networks*, 3:261–273, 1973.

[Gav87]   F. Gavril. Algorithms for maximum k-coloring and k-covering of transitive graphs. *Networks*, 17:465–470, 1987.

[GCW83]   I. S. Gopal, D. Coppersmith, and C. K. Wong. Optimal wiring of movable terminals. *IEEE Transactions on Computers*, C-32:845–858, September 1983.

[GGS89]   H. N. Gabow, Z. Galil, and T. H. Spencer. Efficient implementation of graph algorithms using contraction. *Journal of the ACM*, 36(3):540–572, 1989.

[GH64]   P. C. Gilmore and A. J. Hoffman. A characterization of comparability graphs and of interval graphs. *Canadian Journal of Mathematics*, 16:539–548, 1964.

[GH89]   S. Gerez and O. Herrmann. Switchbox routing by stepwise refinement. *IEEE Transactions on Computer-Aided Design*, 8:1350–1361, December 1989.

[GHS86]   C. P. Gabor, W. L. Hsu, and K. J. Supowit. Recognizing circle graphs in polynomial time. *Proceedings of 26th IEEE Symposium on Foundation of Computer Science*, pages 106–116, 1986.

[GJ77]   M. R. Garey and D. S. Johnson. The rectilinear Steiner tree problem is NP-complete. *SIAM Journal Applied Mathematics*, 32:826–834, 1977.

[GJ79]   M. R. Garey and D. S. Johnson. *Computers and Intractability: A Guide to the Theory of NP-Completeness*. Freeman, San Francisco, 1979.

[GJMP78]   M. R. Garey, D. S. Johnson, G. L. Miller, and C. H. Papadimitriou. Unpublished results. Technical Report, 1978.

[GJMP80]   M. R. Garey, D. S. Johnson, G. L. Miller, and C. H. Papadimitriou. The complexity of coloring circular arcs and chords. *SIAM Journal of Algebraic and Discrete Methods*, 1:216–227, 1980.

[GJS76]   M. R. Garey, D. S. Johnson, and L. Stockmeyer. Some simplified NP-complete graph problems. *Theory of Computation*, pages 237–267, 1976.

[GKP90]   J. Garbers, B. Korte, H. J. Promel, E. Schwietzke, and A. Steger. VLSI-placement based on routing and timing information. *Proceedings of European Design Automation Conference*, pages 317–321, 1990.

[GLL82]   U. I. Gupta, D. T. Lee, and J. Y. T. Leung. Efficient algorithms for interval graphs. *Networks*, 12:459–467, 1982.

[GMG86]   Z. Galil, S. Micali, and H. N. Gabow. An $o(ev \log v)$ algorithm for finding a maximal weighted matching in general graphs. *SIAM Journal of Computation*, pages 120–130, 1986.

[GN87]   G. Gudmundsson and S. Ntafos. Channel routing with superterminals. *Proceedings of 25th Allerton Conference on Computing, Control and Communication*, pages 375–376, 1987.

[Gol77]   M. C. Golumbic. Complexity of comparability graph recognition and coloring. *Computing*, 18:199–208, 1977.

[Gol80]   M. C. Golumbic. *Algorithmic Graph Theory and Perfect Graphs*. Academic Press, 1980.

[Gro90]   P. Groeneveld. A multiple layer contour based gridless channel router. *IEEE Transactions on CAD*, 9, December 1990.

[GS88]   R. Greenberg and A. Sangiovanni-Vincentelli. A multilayer router using one, two and three layer partitions. *Proceedings of IEEE International Conference on Computer-Aided Design*, pages 88–91, 1988.

[GS90]   R. A. Gidwani and N. A. Sherwani. Miser: an integrated three layer gridless channel router and compacter. *Proceedings of 27th ACM/IEEE Design Automation Conference*, pages 698–703, 1990.

[Haj57]   G. Hajos. Auber eine art von graphen. *Inter. Math Nachr.*, 1957.

[Ham85]   S. E. Hambrusch. Channel routing algorithms for overlap models. *IEEE Transactions on Computer-Aided Design*, CAD-4(1):23–30, January 1985.

[HC92]   C. Y. Hou and C. Y. R. Chen. A pin permutation algorithm for improving over-the-cell channel routing. *Proceedings of 29th Design Automation Conference*, pages 594–599, 1992.

[Hei88]   D. V. Heinbuch. *CMOS3 Cell Library*. Addison-Wesley, Reading, MA, 1988.

[HIZ91]   T. T. Ho, S. S. Iyengar, and S. Q. Zheng. A general greedy channel routing algorithm. *IEEE Transactions on Computer-Aided Design*, 10(2):204–211, February 1991.

[HK73]   J. E. Hopcroft and R. M. Karp. An $n^{2.5}$ algorithm for maximum matching in bipartite graphs. *SIAM Journal of Computation*, 2:225–231, 1973.

[Hna87]   E. R. Hnatek. *Integrated Circuit Quality and Reliability*. Marcel Dekker, New York, 1987.

[HO84]   G. T. Hamachi and J. K. Ousterhout. A switchbox router with obstacle avoidance. *Proceedings of 21st ACM/IEEE Design Automation Conference*, June 1984.

[Ho92]   T. Ho. New models for four and five-layer channel routing. *Proceedings of 29th Design Automation*, pages 589–593, 1992.

[Hor91]   M. W. Horn. Antireflection layers and planarization for microlithiography. *Solid State Technology*, November 1991.

[HS71]   A. Hashimoto and J. Stevens. Wire routing by optimization channel assignment within large apertures. *Proceedings of 8th Design Automation Workshop*, pages 155–163, 1971.

[HS78]   E. Horowitz and S. Sahani. *Fundamentals of Computer Algorithms*. Computer Science Press, 1978.

[HS85]   T. C. Hu and M. T. Shing. *A Decomposition Algorithm for Circuit Routing in VLSI*. IEEE Press, 1985.

[HS91]   M. Hossain and N. A. Sherwani. On topological via minimization and routing. *Proceedings of IEEE International Conference on Computer-Aided Design*, pages 532–534, November 1991.

[HSS91a]   N. Holmes, N. A. Sherwani, and M. Sarrafzadeh. Algorithms for over-the-cell channel routing using the three metal layer process. *IEEE International Conference on Computer-Aided Design*, 1991.

[HSS91b]   N. Holmes, N. A. Sherwani, and M. Sarrafzadeh. New algorithm for over-the-cell channel routing using vacant terminals. *Proceedings of 27th ACM/IEEE Design Automation Conference*, pages 126–131, June 1991.

[Hsu83]   C. P. Hsu. Minimum-via topological routing. *IEEE Transactions on Computer-Aided Design*, CAD-2(4):235–246, 1983.

[Hsu85]   W. L. Hsu. Maximum weight clique algorithm for circular-arc graphs and circle graphs. *SIAM Journal of Computation*, 14(1):160–175, February 1985.

[HSVW90]   J. M. Ho, M. Sarrafzadeh, G. Vijayan, and C. K. Wong. Layer assignment for multichip modules. *IEEE Transactions on Computer-Aided Design*, 9(12):1272–1277, December 1990.

[HT91]   L. T. Hwang and I. Turlik. The skin effect in thin-film interconnections for ULSI/VLSI packages. *Technical Report Series TR91-13*, MCNC Research Triangle Park, NC 27709, 1991.

[HTC92]   J. Y. Hsiao, C. Y. Tang, and R. S. Chang. An efficient algorithm for finding a maximum weight 2-independent set on interval graphs. *Information Processing Letters*, pages 229–235, June 1992.

[HWF92]   S. Haruyama, D. F. Wong, and D. S. Fussell. Topological channel routing. *IEEE Transactions on Computer-Aided Design*, 10(10):1117–1197, October 1992.

[HYY90]   A. Hanafusa, Y. Yamashita, and M. Yasuda. Three dimensional routing for multilayer ceramic printed circuit boards. *Proceedings of ICCAD*, pages 386–389, November 1990.

[JG72]   E. G. Coffman, Jr., and R. L. Graham. Optimal scheduling for two processor systems. *Acta Informatica*, 1:200–213, 1972.

[JJ83]   D. W. Jepsen and C. D. Gelatt, Jr. Macro placement by Monte Carlo annealing. *Proceedings of IEEE International Conference on Computer Design*, pages 495–498, 1983.

[JK91]   S. Janson and J. Kratochvil. Thresholds for classes of intersection graphs. *Discrete Mathematics*, pages 307–326, 1991.

[Joh85]   D. S. Johnson. The NP-completeness column: an ongoing guide. *Journal of Algorithms*, 6:434–451, 1985.

[Joo86]   R. Joobbani. *An Artificial Intelligence Approach to VLSI Routing*. Kluwer Academic Publishers, 1986.

[JP89]   A. Joseph and R. Y. Pinter. Feed-through river routing. *Integration, the VLSI Journal*, 8:41–50, 1989.

[Kar84]   N. Karmakar. A new polynomial time algorithm for linear programming. *Combinatorica*, 4:373–395, 1984.

[KBPS94]   R. Khawaja, S. Bhingarde, A. Panyam, and N. Sherwani. Over-the-cell routing algorithms for industrial cell models. To appear in *7th International Conference on VLSI Design*, 1994.

[KC92]   K.-Y. Khoo and J. Cong. A fast multilayer general area router for MCM designs. *IEEE Transactions on Circuits and Systems*, 1992.

[KC93]   K.-Y. Khoo and J. Cong. An efficient multilayer MCM router based on four-via routing. *Proceedings of Design Automation Conference*, 1993.

[Kei85]   M. Keil. Finding Hamiltonian circuits in interval graphs. *Manuscipt*, 1985.

[KK79]   T. Kawamoto and Y. Kajitani. The minimum width routing of a 2-row 2-layer polycell layout. *Proceedings of 16th Design Automation Conference*, pages 290–296, 1979.

[KK90]   E. Katsadas and E. Kinnen. A multi-layer router utilizing over-cell areas. *Proceedings of 27th Design Automation Conference*, pages 704–708, 1990.

[Kro83]   H. E. Krohn. An over-the-cell gate array channel router. *Proceedings of 20th Design Automation Conference*, pages 665–670, 1983.

[KT91]   D. Karpenske and C. Talbot. Testing and diagnosis of multichip modules. *Solid State Technology*, pages 24–26, 1991.

[LaP91]   D. P. LaPotin. Early assessment of design, packaging and technology tradeoffs. *International Journal of High Speed Electronics*, 2(4):209–233, 1991.

[Lau87]   U. Lauther. Top-down hierarchical routing for channelless gate arrays based on linear assignments. *VLSI Design of Digital Systems*, 1987.

[Law76]   E. L. Lawler. *Combinatorial Optimization*. Holt, Rinehart and Winston, New York, 1976.

[Leb83]   A. Leblond. Caf: a computer-assisted floorplanning tool. *Proceeding of 20th Design Automation Conference*, pages 747–753, 1983.

[LGW92]   L. L. Larmore, D. D. Gajski, and A. C. Wu. Layout placement for sliced architecture. *IEEE Transactions on CAD of Integrated Circuits and Systems*, pages 102–114, January 1992.

[LHT89]   Y. L. Lin, Y. C. Hsu, and F. S. Tsai. Silk: a simulated evolution router. *IEEE Transactions on Computer-Aided Design*, 8:1108–1114, October 1989.

[Lin84]   C. P. Lincoln. Design of a 3-micron CMOS cell library. *Electrical Communication*, 58:384–388, 1984.

[LM85]   C. E. Leiserson and F. M. Maley. Algorithms for routing and testing routability of planar VLSI layouts. *Proceedings of 17th Annual ACM Symposium on Theory of Computing*, pages 69–78, 1985.

[LM88]   T. Lengauer and R. Muller. Linear algorithms for optimizing the layout of dynamic CMOS cells. *IEEE Transactions on Circuits and Systems*, pages 279–285, 1988.

[Lov72]   L. Lovasz. Normal hypergraphs and perfect graphs conjecture. *Discrete Math*, pages 253–267, 1972.

[LP83]   C. E. Leiserson and R. Y. Pinter. Optimal placement for river routing. *SIAM Journal of Computing*, 12, No. 3:447–462, August 1983.

[LPH92]   H. R. Lin, H. W. Perng, and H. C. Hsu. Channel height reduction by routing over the cells. *Proceedings of ISCAS*, pages 2244–2247, 1992.

[LPHL91]   M. S. Lin, H. W. Perng, C. Y. Hwang, and Y. L.-Lin. Channel density reduction by routing over the cells. *Proceedings of 28th ACM/IEEE Design Automation Conference*, pages 120–125, June 1991.

[LS81]   M. Grotschel, L. Lovasz, and A. Schrijver. The ellipsoid method and its consequences in combinatorial optimization. *Combinatorica*, pages 169–197, 1981.

[LS88]   K. W. Lee and C. Sechen. A new global router for row-based layout. *Proceedings of IEEE International Conference on Computer-Aided Design*, November 1988.

[LSL90]   R. D. Lou, M. Sarrafzadeh, and D. T. Lee. An optimal algorithm for the maximum two-chain problem. *Proceedings of First SIAM-ACM Conference on Discrete Algorithms*, 1990.

[LSL92]   R. D. Lou, M. Sarrafzadeh, and D. T. Lee. An optimal algorithm for the maximum two-chain problem. *SIAM Journal of Discrete Mathematics*, 5:285–304, May 1992.

[LT79]   R. J. Lipton and R. E. Tarjan. A separator theorem for planar graphs. *SIAM Journal of Applied Mathematics*, 36(2):177–189, 1979.

[Luk85]   W. K. Luk. A greedy switchbox router. *Integration, the VLSI Journal*, 3:129–149, 1985.

[Lun88]   R. E. Lunow. A channelless multilayer router. *Proceedings of 25th Design Automation Conference*, 1988.

[Mal90]   F. M. Maley. *Single-Layer Wire Routing and Compaction*. The MIT Press, 1990.

[Mar86]   M. Marek-Sadowska. Route planner for custom chip design. *Proceedings of IEEE International Conference on Computer-Aided Design*, pages 246–249, 1986.

[Mar92]   M. Marek-Sadowska. Switch box routing: A retrospective. *Integration, the VLSI Journal*, 13:39–65, 1992.

[MH90]   R. L. Maziasz and J. P. Hayes. Layout optimization of static CMOS functional cells. *IEEE Transactions on Computer-Aided Design*, 9:708–719, 1990.

[Mil84]   G. L. Miller. Finding small simple cycle separators for 2-connected planar graph. *Proceedings of 16th Annual ACM Symposium on Theory of Computing*, pages 376–382, 1984.

[MNKF90]   S. Masuda, K. Nakajima, T. Kashiwabra, and T. Fujisawa. Efficient algorithms for finding maximum cliques of an overlap graph. *Networks*, 20:157–171, 1990.

[Mor90]   L. L. Moresco. Electronic system packaging: The search for manufacturing the optimum in a sea of constraints. *IEEE Transactions on Computers, Hybrids, and Manufacturing Technology*, pages 494–508, 1990.

[MS77]   D. Maier and J. A. Storer. A note on the complexity of the superstring problem. Research Report No. 233, Computer Science Laboratory, Princeton University, 1977.

[MS86]   D. W. Matula and F. Shahrokhi. The maximum concurrent flow prob-
            lem and sparsest cuts. Technical Report, Southern Methodist Univ.,
            1986.

[NESY89]   K. Nakamura, Y. Enomoto, Y. Suehiro, and K. Yamashita. Advanced
            CMOS ASIC design methodologies. *Proceedings of Regional Con-
            ferences on Microelectronics and Systems*, 1989.

[NS80]   T. Akiyama, T. Nishizeki, and N. Saito. NP-completeness of the
            Hamiltonian cycle problem for bipartite graphs. *Journal Information
            Processing*, 3:73–76, 1980.

[NSHS92]   S. Natarajan, N. Sherwani, N. Holmes, and M. Sarrafzadeh. Over-the-
            cell routing for high performance circuits. *Proceedings of 29th ACM/
            IEEE Design Automation Conference*, pages 600–603, June 1992.

[Pat81]   A. M. Patel. Partitioning for VLSI placement problems. *Proceedings
            of 18th ACM/IEEE Design Automation Conference*, pages 137–144,
            1981.

[PD86]   D. P. La Potin and S. W. Director. Mason: A global floorplanning
            approach for VLSI design. *IEEE Transactions on CAD*, pages 477–
            489, October 1986.

[PL88]   B. T. Preas and M. J. Lorenzetti. Physical design automation of
            VLSI systems, *Introduction to Physical Design Automation*. Ben-
            jamin Cummings, Menlo Park, CA, 1988.

[PLE71]   A. Pnnueli, A. Lempel, and S. Even. Transitive orientation of graphs
            and identification of permutation graphs. *Canadian Journal of Math-
            ematics*, 23:160–175, 1971.

[PMK90]   M. Pedram, M. Marek-Sadowska, and E. S. Kuh. Floorplanning
            with pin assignment. *Proceedings of International Conference on
            Computer-Aided Design*, pages 98–101, 1990.

[Pol74]   S. Poljak. A note on stable sets and coloring of graphs. *Comment.
            Mathematics University Carolina*, 15:307–309, 1974.

[PP92]   R. R. Pai and S. S. S. P. Pai. An over-the-cell channel router. *IFIP
            Transactions A [Computer Science and Technology]*, A-1:327–336,
            1992.

[PPC89]   B. Preas, M. Pedram, and D. Curry. Automatic layout of silicon-on-
            silicon hybrid packages. *Proceedings of IEEE Design Automation
            Conference*, pages 394–399, 1989.

[PZ87]   V. Pitchumani and Q. Zhang. A mixed HVH-VHV algorithm for
            three-layer channel routing. *IEEE Transactions on Computer-Aided
            Design*, CAD-6(4), 1987.

[RF82]   R. Rivest and C. Fiduccia. A greedy channel router. *Proceedings
            of 19th ACM/IEEE Design Automation Conference*, pages 418–424,
            1982.

[Ric84]   D. Richards. Complexity of single-layer routing. *IEEE Transactions on Computers*, C-33(3):286–288, March 1984.

[RKN89]   C. S. Rim, T. Kashiwabara, and K. Nakajima. Exact algorithms for multilayer topological via minimization. *IEEE Transactions on Computer-Aided Design*, 8(4):1165–1184, November 1989.

[RRT92]   R. R. Tummala. High-performance glass-ceramic/copper multilayer substrate with thin film redistribution. *IBM Journal of Research Development*, pages 889–902, 1992.

[RS93]   R. Rashidi and N. Sherwani. *On planar over-the-cell routing problems*. Technical Report 93-13, Department of Computer Science, Western Michigan University, Kalamazoo, MI, 1993.

[RU81]   D. Rotem and J. Urrutia. Finding maximum cliques in circle graphs. *Networks*, 11:269–278, 1981.

[SBe59]   S. Benzer. *Graphs and Hypergraphs*. North-Holland, Amsterdam, 1959.

[SBS87]   Jeremy Spinard, Andreas Brandstadt, and Lorna Stewart. Bipartite permutation graphs. *Bipartite Permutation Graphs*, 18:279–292, 1987.

[SD81]   A. Siegel and D. Dolev. The separation for general single-layer wiring barriers. *Proceedings of Carnegie-Mellon Conference on VLSI Systems and Computations*, pages 143–152, October 1981.

[SDG92]   A. Sen, H. Deng, and S. Guha. On a graph partition problem with application to VLSI layout. *Information Processing Letters*, 43:87–94, August 1992.

[SH93]   T. Strunk and N. Holmes. Victor: An almost 'channel-less' three-layer over-the-cell router. *Proceedings of Third Great Lakes Symposium on VLSI*, 1993.

[She93]   N. A. Sherwani. *Algorithms for VLSI Design Automation*. Kluwer Academic Publishers, 1993.

[SHL90]   M. Stallmann, T. Hughes, and W. Liu. Unconstrained via minimization for topological multilayer routing. *IEEE Transactions on Computer-Aided Design*, CAD-1(1):970–980, September 1990.

[Sie83]   A. Siegel. *River routing: the theory and methodology*. Ph.D. Thesis, Stanford University, 1983.

[SK87]   E. Shragowitz and J. Keel. A global router based on multicommodity flow model. *Integration: the VLSI Journal*, 1987.

[SK92]   M. Sriram and S. M. Kang. A new layer assignment approach for MCMs. Technical Report UIUC-BI-VLSI-92-01, The Beckman Institute, University of Illinois–Urbana-Champaign, 1992.

[SL89]   M. Sarrafzadeh and D. T. Lee. A new approach to topological via min-imization. *IEEE Transactions on Computer-Aided Design*, 8:890–900, 1989.

[SL90]   M. Sarrafzadeh and R. D. Lou. Maximum k-coverings of weighted transitive graphs with applications. *Proceedings of IEEE International Symposium on Circuits and Systems*, pages 332–335, 1990.

[SL93]   M. Sarrafzadeh and R. D. Lou. Maximum k-covering of weighted transitive graphs with applications. *Algorithmica*, 9:84–100, 1993.

[Sou78]   J. Soukup. Fast maze router. *Proceedings of 15th Design Automation Conference*, pages 100–102, 1978.

[SS85]   C. Sechen and A. Sangiovanni-Vincentelli. The timber wolf place-ment and routing package. *IEEE Journal of Solid-State Circuits*, Sc-20:510–522, 1985.

[SS87]   Y. Shiraishi and Y. Sakemi. A permeation router. *IEEE Transactions on CAD*, pages 462–471, May 1987.

[Sup87]   K. J. Supowit. Finding a maximum planar subset of a set of nets in a channel. *IEEE Transactions on Computer-Aided Design*, CAD-6(1):93–94, January 1987.

[Szy85]   T. G. Szymanski. Dogleg channel routing is NP-complete. *IEEE Transactions on Computer-Aided Design*, CAD-4:31–41, January 1985.

[Tar83]   R. Tarjan. *Data Structures and Network Algorithms*. Society for In-dustrial and Applied Mathematics, 1983.

[TCLH90]   Y. Takefuji, L. Chen, K. Lee, and J. Huffman. Parallel algorithms for finding a near-maximum independent set of a circle graph. *IEEE Transactions on Neural Networks*, 1:263–267, 1990.

[TH90]   T. Tuan and S. L. Hakimi. River routing with a small number of jogs. *SIAM Journal of Discrete Mathematics*, 3(4):585–597, November 1990.

[TH93]   K. Tsai and W. Hsu. Algorithms for the dominating set problem on permutation graphs. *Algorithmica*, 9:601–614, 1993.

[Tom81]   M. Tompa. An optimal solution to a wire-routing problem. *Journal of Computer and System Sciences*, 23(2):127–150, October 1981.

[TR89]   R. R. Tummala and E. J. Rymaszewski. *Microelectronics Packaging Handbook*. Van Nostrand Reinhold, 1989.

[TTLF92]   D. Theune, R. Thiele, T. Lengaur, and A. Feldmann. Hero: Hierar-chical EMC-constrained routing. *Proceedings of IEEE International Conference on Computer-Aided Design*, 1992.

[TTNS91]   M. Terai, K. Takahashi, K. Nakajima, and K. Sato. A new model for over-the-cell channel routing with three layers. *Proceedings of*

IEEE International Conference on Computer-Aided Design, pages 432–435, 1991.

[Tuc80]   A. Tucker. An efficient test for circular-arc graphs. *SIAM Journal of Computation*, 9:1–24, 1980.

[Tum91]   R. R. Tummala. Electronic packaging in the 1990s–a perspective from America. *IEEE Transactions on Components, Hybrids, and Manufacturing Technology*, 14(2):262–271, June 1991.

[TY84]   R. E. Tarjan and M. Yannakakis. Simple linear-time algorithms to test chordability of graphs, to test acyclicity of hypergraphs and selectively reduce acyclic hypergraphs. *SIAM Journal of Computation*, 13:566–579, 1984.

[VCW89]   G. Vijayan, H. H. Chen, and C. K. Wong. On VHV-routing in channels with irregular boundaries. *IEEE Transactions on Computer-Aided Design*, CAD-8(2), 1989.

[VK83]   M. P. Vecchi and S. Kirkpatrick. Global wiring by simulated annealing. *IEEE Transanctions on Computer-Aided Design of Integrated Circuits*, CAD-2(4), 1983.

[WE85]   N. Weste and K. Eshraghian. *Principles of CMOS VLSI Design—A Systems Perspective*. Addison-Wesley, 1985.

[WHSS92]   B. Wu, N. Holmes, N. Sherwani, and M. Sarrafzadeh. Over-the-cell routers for new cell models. *Proceedings of 29th ACM/IEEE Design Automation Conference*, pages 604–607, June 1992.

[Xio88]   X. Xiong. A new algorithm for topological routing and via minimization. *Proceedings of International Conference on Computer-Aided Design*, pages 410–415, 1988.

[YBS93]   Q. Yu, S. Badida, and N. Sherwani. A performance-driven three-dimensional approach for MCM routing. Technical Report, Department of Computer Science, Western Michigan University, September 1993.

[YCHK87]   Y. Pan, Y. C. Hsu, and W. J. Kubitz. A path selection global router. *Proceedings of Design Automation Conference*, 1987.

[YG87]   M. Yannakakis and F. Gavril. The maximum k-colorable problem for chordal graphs. *Information Processing Letters*, pages 133–137, January 1987.

[YHH93a]   L. Y. Hsieh, C. Y. Hwang, and Y. Hsu. An efficient layout style for two metal CMOS leaf cells and its automatic synthesis. *IEEE Transactions on Computer-Aided Design*, 12:410–423, March 1993.

[YHH93b]   L. Y. Hsieh, C. Y. Hwang, and Y. Hsu. Lib: A CMOS cell compiler. *IEEE Transactions on Computer-Aided Design*, 10:88–91, August 1993.

[YK82]   T. Yoshimura and E. S. Kuh. Efficient algorithms for channel routing. *IEEE Transactions on Computer-Aided Design*, CAD-1(1):25–30, January 1982.

[YLL91]   S. K. Dhall, Y. Liang, C. Rhee, and S. Lakshmivaran. A new approach for the domination problem on permutation graphs. *Information Processing Letters*, 37:219–224, 1991.

[Yos84]   T. Yoshimura. An efficient channel router. *Proceedings of 21st Design Automation Conference*, pages 38–44, 1984.

[YW91]   C. Yang and D. F. Wong. Optimal channel pin assignment. *IEEE Transactions on Computer-Aided Design*, CAD 10(11):1413–1423, November 1991.

Author Index

Subject Index